TECHNIQUES IN
ORGANIC REACTION KINETICS

TECHNIQUES IN ORGANIC REACTION KINETICS

PETR ZUMAN and RAMESH C. PATEL

Department of Chemistry
Clarkson University
Potsdam, New York

A Wiley-Interscience Publication

JOHN WILEY & SONS

New York • Chichester • Brisbane • Toronto • Singapore

7227-3690

CHEMISTRY

Library of Congress Cataloging in Publication Data:

Zuman, Petr.
  Techniques in organic reaction kinetics.

  "A Wiley-Interscience publication."
  Bibliography: p.
  Includes index.
  1. Chemistry, Physical organic. 2. Chemical reaction,
Rate of. I. Patel, Ramesh, 1940–   II. Title.

QD476.Z85 1984   547.1'394   84-7450
ISBN 0-471-03556-4

Printed in the United States of America

10 9 8 7 6 5 4 3 2 1

PREFACE

The origin of this book can be traced to the experience of one of the authors during his early scientific career, when he was asked by a prominent organic chemist to collaborate on the study of the mechanism of a well-known organic reaction. He agreed, assuming that with a course in reaction kinetics and several advanced courses in organic chemistry, combined with experience in several analytical methods, he was well prepared for the task. He soon realized that very little in his background had prepared him to attack the problem of investigation of reaction mechanisms in practice. The search for an adequate textbook was unsuccessful. The only source available at that time was Volume VIII of *Technique of Organic Chemistry* (S. L. Friess and A. Weissberger, Eds.), then in its first edition, particularly the chapter entitled "Evaluation and Interpretation of Rate Data" by R. Livingston. Assuming that there were other students and young researchers in a similar predicament, he offered a course on the topic. The material was gradually developed, and the section on finding the rate equation for complex reactions was strongly influenced by *Kinetics and Mechanism* by A. A. Frost and R. G. Pearson, so much that even the present version shows its debt in this respect.

Over the period of years numerous texts were published on various aspects of reaction kinetics, but none approached it fully from the same point of view as the present text, namely making an attempt to teach the student and young researcher how to obtain the data in a kinetic study, how to treat them, and how to use variations in composition of reaction mixture in the elucidation of mechanisms of organic reactions in solutions. Basic concepts in electrochemical and fast reaction kinetic methods are explained. Emphasis is made so that the student realizes the relationships between results and conclusions and learns to plan a mechanistic study. Some aspects of this approach have been dealt with more than adequately in chapters in books (e.g., by W. P. Jencks in *Catalysis in Chemistry and Enzymology*) or in treatises (e.g., by J. F. Bunnett in the second edition of Volume VIII of the Weissberger Series). The only monograph that came close to the aim of this text is the one by K. Schwetlick called *Kinetic Methods for Investigation of Reaction Mechanisms*, which came to our attention in the closing stages of the preparation of the manuscript, is available only in German (published in East Germany), and is very hard to find in this country.

The text is geared for undergraduate students in their Junior or Senior year or for graduate students and young researchers who have some background in reaction kinetics. No attempt is made to cover the theory of reaction kinetics, and terms such as "activation energy" or "transition state" are used without explanation. Such an approach may also indicate to the student that devices such as potential energy diagrams are particularly useful for rationalization of our thoughts once the mechanism is understood. For the reason of saving space, derivation of equations for reaction rates are not given in full, and the interested reader must consult monographs on reaction kinetics. In our opinion, problems in kinetic studies should be based on original data and it is a curse of modern papers in the area that original data are not published. The editorial struggle for brevity is well understood, but the possibility to check and verify kinetic studies suffers.

An earlier version of this monograph was read by Dr. D. Forsyth (currently at Northeastern University, Boston), who suggested abbreviation of the text, proposed inclusion of important topics, and contributed to clarity in presentation. We will always be indebted to him for his time-consuming efforts. The whole manuscript was read by Lee Guterman, and we are grateful to him for his numerous suggestions. Some aspects were discussed with L. Meites and J. Fendler. Last but not least, we would like to thank Helen Tyler for the care and patience she showed throughout this period with our ever-changing text and our wives for their patience with the turmoil that accompanies the preparation of any manuscript.

PETR ZUMAN
RAMESH C. PATEL

Potsdam, New York
June 1984.

CONTENTS

TECHNIQUES IN
ORGANIC REACTION KINETICS

CHAPTER ONE

INTRODUCTION

One of the possible approaches to the teaching of organic chemistry is to categorize reactions according to similarities in their mechanisms. In such an approach students are presented with a statement that a particular reaction follows a certain mechanism. Such mechanisms are described by a series of reactions which may include arrows depicting the shifts in electron density needed for a reaction to occur. Several equations are often needed to describe a given mechanism. An example can be the iodoform reaction of ketones:

$$RCOCH_3 + OH^- \rightleftharpoons RCOCH_2^- + H_2O$$

$$RCOCH_2^- + I_2 \longrightarrow RCOCH_2I + I^-$$

$$RCOCH_2I + OH^- \rightleftharpoons RCOCHI^- + H_2O$$

$$RCOCHI^- + I_2 \longrightarrow RCOCHI_2 + I^-$$

$$RCOCHI_2 + OH^- \rightleftharpoons RCOCI_2^- + H_2O$$

$$RCOCI_2^- + I_2 \longrightarrow RCOCI_3 + I^-$$

$$RCOCI_3 + OH^- \longrightarrow RC\overset{\overset{\displaystyle O^-}{|}}{\underset{\underset{\displaystyle OH}{|}}{C}}-CI_3$$

$$RC\overset{\overset{\displaystyle O^-}{|}}{\underset{\underset{\displaystyle OH}{|}}{C}}-CI_3 \longrightarrow RC\overset{\nearrow O}{\underset{\searrow OH}{}} + CI_3^-$$

$$\underset{\text{OH}^-}{\Updownarrow} \qquad \underset{+H_2O}{\Updownarrow}$$

$$RC\overset{\nearrow O}{\underset{\searrow O^-}{}} \qquad CHI_3$$

For some simpler mechanisms a notation has been introduced (S_N2, E1, etc.) to simplify the classification procedure. Attention has often been paid to structural factors affecting individual mechanisms, whereas very little is usually said about how it is possible to deduce a given mechanism from experimental data.

Theory alone does not allow us to elucidate mechanisms, particularly of more complex reactions. Quantum chemical calculations can help to distinguish between two postulated alternatives in a reaction [1-1,1-2], but usually cannot be utilized to establish the sequence of several processes. Moreover, rigorous treatments are often restricted to simpler molecules, belonging to a limited number of classes, undergoing simple one-step processes. Studies that involve guessing individual reaction steps based upon the knowledge of the reaction mixture and identification of products are also sometimes denoted as theoretical in nature. In such studies, the existence of intermediates, for which no direct evidence is available, is often assumed. Successful postulation requires great experience, ability of logical deduction, and a seasoned chemical intuition. Reaction schemes are produced by analogy with known reactions, keeping in mind structural rules that have been previously confirmed by experiment.

In numerous instances, experienced organic chemists have deduced reaction schemes by an erudite mental process (rationalization), which have eventually been confirmed by kinetic studies. Nevertheless, to prevent confusion, it is suggested that the use of the term "mechanism" be limited to the meaning based on kinetic studies described below, alternatively utilizing the term "course of reaction" to denote the reaction schemes deduced from synthetic procedures and identification of products.

Mechanisms of organic reactions is one of the topics dealt with in the range of physical organic chemistry. This relatively new scientific discipline, existing for only three or four decades, covers the border region between physical and organic chemistry. Physical organic chemistry deals both with quantities which characterize static properties of organic molecules (i.e., dipole moments, dielectric constants, polarizabilities, wavelengths, and molar absorptivity at absorption maxima, oxidation-reduction potentials, or thermodynamic properties) and dynamic properties of organic molecules.

Studies of reaction mechanisms belong to the latter part of physical organic chemistry, namely those dealing with dynamic properties of organic molecules. Prediction and control of the course of a chemical reaction are major goals sought by organic chemists. This topic also presents problems for the physical chemist who historically has been involved with finding expressions describing the reaction rate and allowing calculation of rate constants. In the past, the chemical community was satisfied with such treatment of kinetic data. The search for rigorous mathematical treatment, particularly for complex reactions, is still an active field of research. Nevertheless, to find an appropriate rate equation has ceased to be the ultimate goal of a kinetic study and instead is considered to be the first hurdle to be passed in a modern kinetic investigation. The final goal is often the elucidation of a reaction mechanism.

The mechanism of an organic reaction can be defined as the path the molecule follows in its transformation from the initial species to the final product. Originally, the elucidation of mechanism was considered equivalent to the identification of individual collisions (or elementary processes) of molecules (or other reacting species) that take place in simultaneously or consecutively occurring steps. With the advent of advanced techniques, the meaning of the term has been expanded to include identification of particular reaction steps and the stereochemistry of the processes involved. The search is extended to special conditions in which molecules or other reactants approach each other in the time frame of the reaction. This implies consideration of the geometry of the transition state, namely the interatomic distances and valence angles in this state.

In organic synthesis the starting materials and products are species, the existence of which can be verified by our senses and their physical properties. In physical chemistry we frequently measure a quantity which has a simple relation to the matter (e.g., weight, pressure, temperature, X-ray diffraction). A similarly direct approach is, nevertheless, impossible in the study of mechanisms. There is no single, measurable quantity which can provide complete information on the mechanism of a given reaction. All our knowledge of the mechanism is deduced indirectly from quantities that are only related to some aspect of the mechanism. To help illustrate the difference between the situation faced in the usual methodology of exact sciences, where quantities are measured and conclusions are drawn, and that of mechanistic studies, let us compare a reaction mechanism to a motion picture. In this hypothetical motion picture we would see two reacting species (molecules, ions, or atoms) separated in space, then slowly moving towards each other until they reach a distance and orientation in space in which the reaction takes place. This would be followed by separation of products until they reach such a distance that their mutual interaction becomes negligible. Such a motion picture would be capable of depicting the continuous movement of each atom and of the accompanying electrons in the course of the whole reaction. Such a detailed picture cannot be obtained even for the simplest reaction. Instead of the detailed information that would correspond to a motion picture we can actually obtain only isolated information, which in our above analogy would correspond to a still photograph of the participating species, in some moments of the reaction, which we hope to be significant. A complete solution of reaction mechanism would involve determination of the stability of the reacting system, that is, its energy, in every transitory configuration. We would be able to distinquish a transition state from intermediates of limited stability and these in turn from more stable intermediates. At present this ideal goal is inaccessible by available means. We are, nevertheless, trying to get as close to this ideal as possible.

To achieve this goal and obtain as much information about the actual fate of the reactants during the reaction as possible, we must try to identify all products and determine their yields, detect all intermediates, follow reaction rates as a function of concentration of reacting species, temperature, ionic strength, acidity, and other parameters that will be considered later on. Even a well-planned selection of variation of parameters can offer experimental results that are in agreement with several mechanisms, so that it is impossible to assign one single mechanism to the reaction being studied.

In practice we are unable to investigate all possible combinations of experimental conditions because they are too numerous. Usually combinations are chosen that are expected to reveal the most useful information about the mechanism. If a new combination of experimental parameters is examined at a later date, the resulting information can confirm or contradict the proposed mechanism. If the latter is true, a new mechanism should be adopted that takes into consideration both old and new results. The history of reaction kinetics and mechanistic studies contains a number of examples where the elucidation of the mechanism and correct description of the reaction depends upon factors that are hidden, and escapes detection for years or even decades. The tactics of research as reflected by the choice of reaction conditions, model substances, and extent of variations seems to be of particular importance for successful research in this branch of physical organic chemistry.

In the process of proposing a mechanism, the fundamental criterion of evaluation rests upon its capability to deal with existing experimental evidence, and its ability to predict the results of new experiments. If, on such a basis, it is still possible to describe the reaction by more than one mechanism, the simplest form is often preferred. New experimental data can result in the necessity to exclude the simplest mechanism and replace it by a more complicated one.

Most synthetically important reactions of organic compounds take place in the liquid phase. Thus, contrary to physical chemists who frequently study reactions in the gas phase, we restrict our attention to reactions in the liquid phase, even when solvation phenomena make the interpretation of results in the liquid phase more complex. Alternatively, studies of the gas phase reaction kinetics are often experimentally more demanding. Reactions in the liquid phase can take place either as a reaction between two neat liquids or a reaction in solution. In reactions between two liquids, either one of the liquids is in excess and actually serves as a solvent, or their concentrations are comparable and the reaction takes place effectively in a mixture of two solvents. Properties of numerous liquid reactants as solvents are unknown and the interpretation of such reactions is difficult. For these reasons we restrict ourselves to discussion of organic reactions in solution.

To elucidate mechanisms of organic reactions in solution is difficult because most of the reactions are complex and consist of several elementary steps, which occur either consecutively or as parallel reactions. Even such a seemingly simple reaction as the iodoform reaction discussed above consists of eight steps. Often a stepwise approach must be used. Despite the difficulties that arise from trying to find a unique mechanism, kinetic studies will always be of pivotal importance, since they provide the means to optimize transformation of chemical compounds.

REFERENCES

[1-1] G. Klopman (Ed.), *Chemical Reactivity and Reaction Paths*, Wiley, New York, 1974.

[1-2] I. Fleming, *Frontier Orbitals and Organic Chemical Reactions*, Wiley, New York, 1976.

BIBLIOGRAPHY

L. P. Hammett, *Physical Organic Chemistry*, 2nd ed., McGraw-Hill, New York, 1970.

A. A. Frost and R. G. Pearson, *Kinetics and Mechanism*, 2nd ed., Wiley, New York, 1961.

J. W. Moore and R. G. Pearson, *Kinetics and Mechanism,* 3rd ed., Wiley, New York, 1981.

S. L. Friess, E. S. Lewis, and A. Weissberger (Eds.), *Investigation of Rates and Mechanisms of Reactions,* Part I, II, 2nd ed., *Techniques of Organic Chemistry,* Vol. VIII, Wiley-Interscience, New York, 1961.

E. S. Lewis (Ed.), *Investigation of Rates and Mechanisms,* Part I, II, 3rd ed., *Techniques of Chemistry,* Vol. VI, Wiley, New York, 1974.

C. H. Bamford and C. F. H. Tipper (Eds.), *Comprehensive Chemical Kinetics,* Vols, 1–22, Elsevier, Amsterdam, 1969–1980.

J. Hine, *Physical Organic Chemistry,* McGraw-Hill, New York, 1956.

E. S. Gould, *Mechanism and Structure in Organic Chemistry,* Holt-Dryden, New York, 1959.

R. Stewart, *The Investigation of Organic Reactions,* Prentice-Hall, Englewood Cliffs, N.J., 1966.

R. D. Gilliom, *Introduction to Physical Organic Chemistry,* Addison-Wesley, Reading, Mass., 1970.

I. Amdur and G. G. Hammes, *Chemical Kinetics,* McGraw-Hill, New York, 1966.

G. M. Fleck, *Chemical Reaction Mechanisms,* Holt, Rinehart & Winston, New York, 1971.

T. H. Lowry and K. S. Richardson, *Mechanisms and Theory in Organic Chemistry,* 2nd ed., Harper & Row, New York, 1981.

K. A. Connors, *Reaction Mechanisms in Organic Analytical Chemistry,* Wiley, New York, 1973.

W. Drenth and H. Kwart, *Kinetics Applied to Organic Reactions,* Dekker, New York, 1980.

CHAPTER TWO

TECHNIQUES OF REACTION KINETICS

When two reacting species A and B are mixed in solution, their concentrations will change with time. The change with time of the concentration of one component, $d[A]/dt$, is denoted as reaction rate (v). The reaction rate usually depends on the concentration of reacting species in the solution (2-1):

$$v = \frac{d[A]}{dt} = f([A], [B], \ldots) \tag{2-1}$$

In the case of a simple reaction such as (2-2),

$$A + B \xrightarrow{k} C \tag{2-2}$$

the relationship between the reaction rate and the concentration of A and B is simple (2-3).

$$\frac{d[A]}{dt} = -k[A][B] \tag{2-3}$$

The minus sign indicates that as the reaction proceeds the concentration of A decreases. The proportionality constant k is called the specific rate or, more frequently, the rate constant. In order to elucidate reaction mechanisms it is usually necessary to determine a number of rate constants under varying conditions.

To find the values of individual rate constants needed in mechanistic studies, it is necessary to determine changes in concentration of one or more components of the reaction mixture with time. In practice this means that measurements of concentration, or a quantity that is proportional to concentration (e.g., optical rotation, conductivity, absorbance, polarographic limiting current), have to be carried out at suitable time intervals. Both measured quantities, time and concentration, are subject to errors. The accuracy of measurements depends on the technique used and on the degree of sophistication in the application of a given technique. The choice of the technique and the attention paid to the measurement of both time and concentration depend on the mode of application of the measured constant employed and on the kind of reaction. Since fast reactions with half-lives shorter than about 2 seconds involve applications of special techniques (Chapter 4), the following discussion will be restricted to the study of slower reactions.

2.1. ACCURACY OF MEASUREMENT

To achieve the greatest economy in time spent on a given study, it is essential when planning an experiment to estimate the accuracy that is needed in the measurement of both parameters. Such estimates can play an important role in the choice of the technique and experimental conditions.

The accuracy with which a complex quantity such as the rate constant can be determined depends on the accuracy of measurement of all the concentrations involved, of measurement of time, and on the manner in which experimental data are handled to obtain the rate constant. Frequently the inaccuracy associated with the measurement of one parameter is so large that it dominates the effects due to the combined errors in all the other measured quantities. In such cases it is completely unnecessary to scrutinize these "other" measurements and to try to achieve a high accuracy of these measurements. On the other hand it is important to achieve the highest possible accuracy in the measurement of the sensitive parameter. Usually, it is possible to measure time with a greater accuracy than concentrations. Thus greatest attention is paid to concentration determination.

Similarly, it is of importance to control other factors affecting reaction rate— such as temperature, pressure, or illumination. It is usually advantageous to keep constant those factors that affect the reaction rate during the course of a reaction. It is thus only of limited advantage to introduce special, accurate analytical methods for the determination of concentration, if temperature is not sufficiently controlled and varies during the reaction.

Methods used in finding the best value of the rate constant will be discussed later (Section 3.5). It is important, nevertheless, to recall that to obtain a highly accurate value of the rate constant is frequently not the ultimate goal of a physical organic study. The required accuracy depends on the use of the found rate constants. For example, in order to compare an experimentally determined rate constant with a theoretical value obtained by the collision theory or from the absolute rate theory, it is sufficient if accuracy of the determined constant is within about half an order of magnitude, since the accuracy of the theoretical calculations is usually even smaller and such approaches offer only an order of magnitude of the value of the rate constant.

Considerably higher demands on accuracy are required when the rate constant is to be used in structural studies, for example, when the effect of substituents or steric factors on reactivity is investigated. The required accuracy depends on the susceptibility of a given reaction to the investigated structural change. Still higher accuracy is required when comparing effects of solvents, ionic strength, or of isotope effects on reaction rate with values predicted by theory. The rate constants in such cases have to be measured with an accuracy greater than ± 3%. In such instances the temperature of the reaction mixture has to be controlled in such a way that temperature differences in the series of compared kinetic runs are kept below several hundredths of a degree celsius. The accuracy can be affected by fluctuations in temperature during the course of a kinetic run. The absolute value of the temperature at which the reaction takes place need not be known to better than the nearest tenth of a degree.

It is important to realize that all published values of rate constants or other kinetic parameters become part of the chemical literature. Such data can be used later on for purposes not anticipated by the original experimenter. It is thus of primary importance to always state the exact conditions of measurement used in kinetic experiments.

Frequently the study of effects of various factors and parameters on the values of rate constants (as discussed in Sections 3.6 to 3.9) is of much greater interest and importance than an accurate measurement of the value of a single rate constant. Often more information is gained, for example, by carrying out a total of 20 measurements at 10 different pH-values with an accuracy of ± 10%, than is gained by 10 measurements at one pH for which the accuracy can be shown to be better than ± 2%. Similarly, it is better to spend the same time for measurement of rate constants at three temperatures with an accuracy of ± 10% than to obtain a value reproducible to ± 2% at a single temperature.

2.2. CONTROL OF REACTION MEDIUM

For reactions that take place in solutions, the control of reaction medium involves the control of the composition of the solution, temperature, illumination, and occasionally of some other factors such as surface, pressure etc.

Solution composition depends on the concentration of the reactants, the type and composition of the solvent (particularly its dielectric constant), the acidity of the solution (usually given as pH, H_0, or similar functions), and on the ionic strength.

In most kinetic studies well-defined substances are used, and the preparation of stock solutions of known concentration presents no difficulties. Stability of such stock solutions is checked. Usually just before the kinetic run an aliquot of the stock solution is added to the reaction mixture. An accurate determination of the concentration of the stock solution by an independent analytical method is recommended. The preparation of reaction mixtures by dilution and mixing of individual stock solutions can be a source of inaccuracies. Interchange and recalibration of glassware and pipettes is recommended.

An interesting method of preparation of the reaction mixture employs electrolysis. Monitoring the number of coulombs consumed allows very exact measurement of the amount of reactants added.

For example, when following ester hydrolysis [2-1], a platinum cathode and silver anode can be immersed into a solution slightly acidified by hydrobromic acid. When a current is applied to the system consisting of these two electrodes, hydrogen gas is evolved at the cathode and silver bromide deposited on the anode. The evolution of hydrogen disturbs the equilibrium (2.2-1)

$$H^+ + OH^- \rightleftharpoons H_2O \tag{2.2-1}$$

and hydroxide ions are generated in the solution. An excess of the ester is then added and the electrolytic current is adjusted so that the hydroxide ion concentration remains constant through the course of the reaction. This constant hydroxide ion concentration is controlled by means of a pH-meter. Under such experimental conditions, the rate of consumption of hydroxide ions due to a chemical reaction (or the rate of electrolytic generation of hydroxide ions) equals the rate of decrease in concentration of the ester molecules.

Similarly, atomic hydrogen evolved electrolytically on a platinized platinum electrode has been used for the study of the kinetics of olefin hydrogenation. The electrolysis was carried out in 0.1 M perchloric or sulfuric acid in 95% ethanol. Oxygen formed on the platinum anode was separated from the reaction mixture by a sintered glass disk. The hydrogen pressure in the solution was measured by another hydrogen electrode. The generating current was again adjusted to keep the hydrogen gas concentration in the solution, containing an excess of olefin, constant.

Alternatively, a species can be generated electrochemically which reacts rapidly with one component of the reaction mixture in a competitive reaction [2-2]. Thus the reaction of thiosulfate with methyl bromoacetate (2.2-2) can be followed by electrolytic generation of iodine at constant current. Under these conditions thiosulfate is simultaneously depleted by reaction with the ester and the iodine. The rate of removal of thiosulfate by the reaction with iodine is constant and equal to the rate at which iodine is generated. The reaction of thiosulfate with iodine is very rapid. The time at which all the thiosulfate was removed by the reaction with the bromo ester is indicated by the appearance of free iodine. This time is a measure of the rate of the reaction of thiosulfate with the ester. This method is related to "clock reactions," such as the Landolt reaction [2-3] and others [2-4, 2-5].

$$CH_2BrCOOR + S_2O_3^{2-} \rightarrow CH_2(S_2O_3)COOR^- + Br^- \tag{2.2-2}$$

If volatile substances are dealt with, the determination of initial concentration becomes particularly important. The effect of volatility is of great importance, especially in systems where it is necessary to separate a gaseous component from the reaction mixture. Frequently, it is necessary to remove atmospheric oxygen, for example, in the study of radical reactions, carbanion reactions, or reactions in which strong reducing agents are involved.

Special attention must be paid to losses due to volatility in the case of systems in which one or more reacting components are volatile at the temperature at which it is necessary to carry out the reaction. An example of this type of loss is the uncatalyzed hydrolysis of methyl halides, studied at temperatures of 100°–120°C.

Since at these temperatures both reacting species and the methanol formed are volatile, it was necessary to carry out the reaction (2.2-3) in sealed ampoules.

$$CH_3X + H_2O \rightarrow CH_3OH + H^+ + X^- \qquad (2.2\text{-}3)$$

After chosen time intervals the ampoules were taken out of the temperature controlled bath, rapidly cooled (resulting in the quenching of the reaction, Section 2.3), opened, and analyzed. Air bubbles left in sealed ampoules should be as small as possible. Evaporization of the alkyl halides from the liquid phase at higher temperatures would result in a change in concentration of these halides in the aqueous phase, which in turn would affect the reaction rate. The reaction mixture was prepared by bringing distilled water to a boil at a reduced pressure at $20°C$, the solution saturated with gaseous methyl halide, its partial pressure measured, and this solution transferred into ampoules which were then sealed.

A special case in the preparation of the reaction mixture involves the measurement of the amount of an added ion-exchange resin [2-6]. These substances have been used either in H-cycle for acid or in OH-cycle for base catalysis. First the exchange capacity of the resin is determined and a quantity of the ion-exchanger is added to the reaction mixture corresponding to the required number of milliequivalents of hydrogen or hydroxide ions.

Since the reacting species is usually present in the solution in a very low concentration compared with the molar concentration of the solvent, maintaining constant solvent composition during the reaction usually presents no problems. It is sufficient to prepare the reaction mixture in the chosen solvent, which must be carefully purified for kinetic studies since impurities can participate in the studied reaction or act as catalysts. In addition to testing the solvents for impurities associated with the commercial process, solvents should be examined also for their acidic and basic impurities. In some instances the presence of oxygen, in others the presence of water, can affect the analytical method used or the course of the studied reaction. Analytical results for all such impurities, and methods used in obtaining them, should be reported.

In numerous reactions of organic substances in solutions, components of the reaction mixture—starting materials, intermediates, or products—show acidic or basic properties. It is thus important to prove whether these substances will be present in the reaction mixture as acids or conjugate bases. To maintain the composition of the solution constant with respect to the ratio of concentrations of the conjugate acid and base pair in the course of the reaction, it is essential to keep the pH of the solution constant. This, in principle, is possible either by continuous addition of an acid or a base solution or by buffering.

Continuous addition of a relatively concentrated acid or base solution (to prevent changes in the volume of the reaction mixture) is carried out so that the acid or base consumed is exactly balanced. The pH is controlled by a pH-meter (usually using a glass electrode). In modern pH-stats the addition of the acid or base solution is carried out automatically [2-7].

Maintaining a constant pH of the reaction mixture is more frequently achieved by means of a buffering solution. The requirements for the buffers used in reaction kinetics are discussed next.

The first, seemingly trivial condition is that the buffer components must be soluble in the given reaction medium. Further, it is advantageous that the buffer components do not react with the starting materials or products (e.g., boric acid buffers are sometimes unsuitable because of complex formation and amine buffers because of condensation reactions). If a reaction between the buffer components and the species of interest takes place (e.g., in general acid–base-catalyzed reactions) it cannot be ignored and has to be dealt with quantitatively. The overall reaction in such instances is treated as a complex reaction, one pathway being the reaction involving the buffer. Furthermore, it is necessary to ensure that the presence of the buffer does not interfere with the analytical method used in the determination of the concentration of the reacting species or the product (e.g., that the buffer component does not absorb in the wavelength region in which the kinetic measurement is carried out). It is advantageous if the rate of equilibrium establishment between the acid and base component of the buffer is fast compared with the rate of the reaction being investigated. This condition is fulfilled for the majority of buffers used in the study of slow reactions (with the exception of carbonate buffers). When extremely fast reactions are investigated, complications of this type may exist; doubts have been reported concerning the suitability of citrate buffers. The influence of the reaction with the buffer components on a studied rate can be distinguished by a dependence of the rate constant on buffer capacity.

In addition to the conditions mentioned above, the buffer solutions used must show a sufficient buffer capacity. This means that the use of buffers is restricted to the pH-range where they have the best buffering properties (i.e., for pH = $pK \pm 1$). Furthermore the relative concentration of the buffer component compared with the concentration of other components has to be considered. Conditions are chosen so as to ensure that the concentration of the buffer component present in lower concentrations is at least 20 times greater than the initial concentration of any of the reacting species.

For reaction kinetics simple buffer systems (such as acetate, phosphate, or borate buffers) are preferred to composite buffers (such as Britton–Robinson or McIlvaine universal buffers). Whenever a choice has to be made between buffers composed of anions with different charges, the buffers with lower anionic charge are preferred (e.g., acetate is preferred to citrate). Anions with higher charges show a greater tendency for electrostatic interactions. Cations are invariably univalent, alkali metal cations, or ammonium ions.

For the interpretation of kinetic measurements, it is simpler not to vary the concentrations of both buffer components simultaneously (as is frequently proposed in tables of buffers designed mainly for use in analysis where the overall volume and concentration of both components are kept constant). When the pH-dependence of the reaction rate is followed, it is beneficial to keep the concentration of one buffer component constant and vary only the concentration of the second component; for example, in the preparation of acetate buffers for kinetic measurements it is recommended first to keep the acetic acid concentration constant and vary that of sodium acetate, followed by keeping the sodium acetate concentration unchanged and altering that of the acid.

In the preparation of buffer solutions, special attention has to be paid to the use of carbonate-free hydroxide. The measurement of pH before and after reaction is recommended in preliminary experiments. If there is a choice, buffers made from easily accessible chemicals are preferred (e.g., from acetate rather than phenyl-acetate).

Reactions of ions and to a lesser degree of molecules depend on ionic strength. This must be kept constant during a kinetic run. If a reaction mixture does not contain any ionizable substances, apart from the reacting components, it is sufficient to add an excess of an inert electrolyte, such as sodium perchlorate. If the reaction mixture contains considerable concentrations of other ionized substances, as for example in buffered solutions, one of two alternatives may be chosen: Either the ionic strength (μ) is calculated by equation (2.2-4):

$$\mu = \tfrac{1}{2} \sum C_i z_i^2 \qquad (2.2\text{-}4)$$

(where C_i is concentration of ionic species i and z_i is its unit charge) for all ionized components present and a sufficient amount of neutral salt is added to keep the value of μ constant, or a massive concentration of the neutral salt is added. For example, to a buffer composed of 0.01 M weak acid and its sodium salt, varying in concentration from 0.001 M to 0.1 M, a neutral salt (such as potassium chloride or sodium perchlorate) is added so that its final concentration is 1 molar. It is assumed that in the variation of the ionic strength contribution of the buffer components with pH variation can be neglected. Such an assumption is reasonably valid only in the case of buffers containing only univalent cations and univalent anions. For buffers containing multivalent anions, for example, a buffer containing HPO_4^{2-} and PO_4^{3-}, the changes in the ionic strength cannot be neglected even for the above mentioned case involving addition of excess neutral salt.

To enhance solubility of alkali metal salts in nonaqueous solvents as well as to enhance activity of anions, such as hydroxide or alkoxide ions, crown ethers are added [2-7a].

Having summarized the factors that are important in controlling the constant composition of the reaction mixture, it is possible to proceed to the techniques used in temperature control.

Most reactions are studied under isothermal conditions employing thermostats for the temperature control of baths or circulating devices. Higher quality, commonly available instruments keep the temperature constant up to $\pm$ 0.05°C to 0.02°C for temperatures ranging from 10° to 100°C, up to $\pm$ 0.1°C at lower temperatures, and up to $\pm$ 0.5°C at high temperatures up to 900°C. To control the temperature at higher temperatures is experimentally difficult; fortunately such a temperature range is rarely investigated by kinetic methods and in particular is not relevant for solution kinetics.

In a thermostat a sensor activates a heating element if the temperature decreases below the chosen value, due to heat losses by radiation or other cooling. After a predetermined time interval, the heating element is switched off, which results in a certain fluctuation of the bath temperature. Smaller fluctuations result in smaller variations in temperature, usually achieved in more expensive thermostats.

The reaction vessel is either immersed directly in the stirred bath or is surrounded by a jacket through which the thermostating liquid is pumped. In the latter case it is essential to measure the temperature and its changes directly in the reaction vessel. Connections between the bath and the reaction vessel should be thermally insulated.

For temperatures between 2° and 80°C water has been found to be the best thermoregulating liquid. Ethanol is useful at low temperatures and oil and silicones at high temperatures. The very high temperatures can be reached by using molten metals such as lead.

For thermostatic control, the use of a cooling liquid circulating in a coil through the bath is recommended. Since the tap-water temperature is usually between 10° and 20°C, it can be efficiently used for temperatures control at 25°C and higher. For lower temperatures precooled water or refrigeration units are needed.

In the past, temperature control was achieved by immersion of the reaction vessel in vapors of pure liquids at the boiling point. Such an approach, which does not involve the use of costly thermostats, is still sometimes used in undergraduate laboratories [2-8]. A dependable and experimentally less-demanding alternative is an immersion of the reaction vessel in melting ice, or benzene, or keeping the vessel at the sublimation point of carbon dioxide. Finally, if the temperatures studied are not greatly different from room temperature—provided that no high accuracy in temperature control is required—it is possible to place the thermoregulating liquid (brought to the required temperature) in a Dewar vessel. For 1–2 hr the temperature can be kept reasonably constant.

An additional parameter that must be kept constant is illumination. It is wise to check whether the rate of the studied reaction is sensitive to room illumination or sunlight. This can be easily verified by running the reaction in diffuse daylight, artificial light (using incadescent or fluorescent lamps), and in darkness (in a vessel with blackened walls). The effect of light should be excluded for reactions of compounds which are known to undergo photolysis. Even for systems for which photochemical transformations are not generally known, a preliminary test for the effect of light is recommended.

Sometimes in the presence of light, competitive unwanted reactions resulting in side products, which do not occur in nonilluminated solutions, can take place. For example, when the reaction of periodic acid with some sugars is carried out in daylight, the consumption of periodic acid is greater than when the reaction is carried out in the dark. This happens because in the presence of light, periodic acid reacts with some of the reaction products (e.g., formaldehyde, formic acid) which are formed in the reaction of sugar with one mole of periodic acid. In the dark these products do not react with periodic acid. In some instances the presence of light will cause some of the organic compounds present to react with the iodic acid formed in the reduction of periodic acid. The effect of illumination is frequently observed for reactions in which free radicals or radical ions are formed.

In some instances, in particular for radical and chain reactions, the reaction rate may depend on the surface of the reaction vessel. Not only the surface area, but the shape of the reaction vessel and the quality, particularly the roughness of the surface, can be decisive factors. In such cases it is necessary to keep the surface of the vessel, which comes into contact with the reaction mixture, constant.

This surface effect is so specific for radical chain reactions that it can aid in recognition of a radical mechanism. Covering the inner walls of the reaction vessel with paraffin, silver chloride, by a metal or graphite can lead to a decrease or increase in the reaction rate. An increase of the surface (e.g., made by placing glass beads or glass wool in the reaction mixture) can similarly result in an increase or decrease in the rate of the radical reaction depending on whether an initiation or termination of the chain reaction takes place on the surfaces.

For most common reactions in solutions it is unnecessary to consider changes in barometric pressure. Recently the effect of high pressure (several hundred or even thousands of atmospheres) on the rate of solution reactions has been investigated. Such experiments need special reaction cells and equipment. Control of reaction volume is usually not of great importance, since changes in the volume of liquids with pressure and temperature are small. Such a control is also difficult to realize due to the small coefficient of compressibility of liquids.

2.3. MEASUREMENT OF TIME INTERVALS

For reactions that take place slowly, with half-times greater than about 30 min, it is sufficient to use a precise watch which is accurate to 0.2 to 1%. For shorter intervals stopwatches are used. For half-times of the order of 30 s, the accuracy of time measurement involving the watch and operator is usually no better than $\pm$ 1%. It is important to ensure that the stopwatch is wound up and its performance checked periodically by comparison with a standard chronometer. Whenever possible, experiments should be performed using a pair of stopwatches (and two operators) to measure the time independently in two parallel kinetic runs.

The problem of time-interval measurement is the establishment of a time scale for a given reaction that often involves knowledge of the starting time of the reaction and of the time intervals elapsed between start and the time of individual measurements of concentration or quantity proportional to it. The former is of importance for any method used for following the reactant concentration. Exact knowledge of the time at which the measurement is carried out usually poses no problem when a continuous measurement is undertaken, but it can involve some uncertainty when sampling is necessary.

Reaction is usually started by rapid mixing of given volumes of two or more solutions, each of which separately contains one of the reacting components. Mixing should be done in such a way that the reaction mixture is homogeneous and the temperature is constant in the shortest interval possible after the start. Both or all solutions to be mixed are usually brought to the reaction temperature in advance, either by placing them into a water bath or in two vessels with thermostated water jackets. The mixing is achieved by pouring together, pressure of an inert gas, or by the action of gravity. Sometimes one of the components of the reaction mixture is placed in an ampoule which is then broken under the surface by hitting with a glass or metal rod or which is "shot" into the reaction vessel where it breaks against a wall. Alternatively, in the "Baffle method" a barrier (wall) separating two liquids is rapidly removed. Such mixing accompanied by vigorous stirring can result in a homogeneous reaction mixture in 10–15 ms.

If flasks calibrated for "outflow" or special rapidly flowing pipettes are used, it is possible to prepare the reaction mixture within 5 to 10 s with an accuracy of about 0.5%. To achieve complete mixing as rapidly as possible, the reaction mixture is vigorously stirred (e.g., by a stream of inert gas of the same controlled temperature as the reaction mixture) or agitated.

A somewhat complicated situation can arise if it becomes necessary to take samples from the reaction mixture. If the volume of the reaction mixture is large, samples can be taken by pipettes, or for smaller samples (usually up to 10 ml) by syringes. To quench the reaction in the sample taken it is possible to use rapid cooling, dilution (for reactions of second and higher order), change in pH, or addition of a reagent that reacts rapidly and quantitatively with one component of the reaction mixture. Efficient mixing by stirring or agitation is recommended upon addition of the sample to the quenching liquid.

If the reaction is very slow, it is possible to place small volumes of the reaction mixture in a number of sealed ampoules which are simultaneously placed in a thermostat. At selected time intervals one ampoule at a time is removed from the thermostat, rapidly cooled or broken under a quenching solution, and analyzed.

Several complications with the determination of the exact time corresponding to a given composition of the reaction mixture are eliminated, when the course of the reaction is followed continuously and one or more concentrations are monitored directly in the reaction mixture. This is discussed in connection with the use of optical and electrochemical techniques in Section 2.4.3. Continuous recording of such concentration changes is made possible by photographic, electrical, or mechanical devices. On the recorded graph of $C = f(t)$ it is possible to find the value of concentration at any time interval. For instruments with constant chart movement the time intervals can be prerecorded. The use of continuous recording with time markers decreases inaccuracies in time measurements to such a degree that the inaccuracies can be neglected. Inaccuracies in concentration determination will govern the overall accuracy of the rate constant.

For the handling and evaluation of kinetic data it is frequently necessary to know the initial or final concentration (or proportional physical quantity) of reacting species.

Initial concentration is usually known, and thus no problems are encountered when kinetics is followed by a direct determination of concentration in individual samples. When physical methods are used, the value of the measured physical quantity at time zero must be obtained either from an independent experiment in the absence of one of the reactants, at another pH or temperature, or by extrapolation to zero time. If the extrapolation is used, it is necessary to measure the physical quantity after several precisely known time intervals in the initial stage of the reaction and to ensure rapid and complete mixing of reaction components. As these time intervals should refer to the beginning of the reaction, the estimate of the time at which the mixing can be considered as complete becomes of particular importance. For systems in which the value of the quantity at time zero or infinity is too uncertain, experimental data can be handled by special treatments described in Section 3.5.1.

2.4.   TECHNIQUES OF MEASUREMENT OF CONCENTRATION

2.4.1.   General

To study concentration changes of the reaction mixture it is possible to use any chemical or physical quantity, or a combination of chemical and physical meas-urements, as long as the set characterizes the composition of the solution.

The success of a kinetic study depends to a high degree on the choice of a suitable analytical method. In some instances existing methods are not suitable and it is necessary to develop new, special procedures.

Analytical methods used in the study of reaction kinetics must be specific, sufficiently accurate, and rapid. Methods will be classified as direct when a method is used to measure the amount of the substance itself (chemical methods, meas-urement of gas evolution, precipitation, etc.) and denoted as indirect when a physical property that is a simple function of concentration (absorbance, optical rotation, polarographic limiting currents, etc.) is measured.

When direct methods are used it is important that no interference from any other reactant or reaction product affects the analytical procedure. As an example, when the oxidation of 1,2-diols by periodic acid is followed titrimetrically, the procedure used must enable a determination of periodic acid in the presence of excess iodic acid, which is one of the reduction products.

Similarly, in the determination of hydrogen sulfide formed by cleavage of ethyl mercaptan [2-9]:

$$2C_2H_5SH \xrightarrow[Al_2O_3]{} (C_2H_5)_2S + H_2S \tag{2.4-1}$$

the precipitation must be carried out by a cadmium sulfate solution in a sample acidified with sulfuric acid, since under other conditions mercaptan and even the sulfide react with heavy metals.

When the effects of the composition of the medium (pH, ionic strength, dielectric constant, and temperature) on the rate constant are followed, the change must not affect the analytical results. For example, the use of precipitation reactions is not recommended, because they can be considerably affected by the presence of neutral salts used in adjustment of ionic strength.

Similarly, it is necessary to consider the effects due to changes in the reaction medium when indirect methods are used. If it is impossible to find a technique or condition where the effect of the medium on the measured quantity is negligible, it is necessary to determine the effect of pH, viscosity, ionic strength, or solvent on the measured quantity (e.g., absorbance or limiting current) in a separate experiment.

The analytical method used should not affect the reaction rate or interfere with the studied reaction. For example, the use of conductance measurements with platinized platinum electrodes is not recommended if the studied reaction is known to be or could be catalyzed by platinum black. Similarly it is not advisable to use a photometric method in the investigation of a system which is photosensitive at the wavelength used.

Generally, in order to minimize the effects of other components in the reaction mixture on the sought quantity, a calibration of the concentration response should be carried out using synthetic mixtures of reactants and products under conditions that closely mimic those in the kinetic run.

Conditions affecting accuracy have already been discussed. It should be mentioned that good reproducibility does not necessarily mean high accuracy. For example, when the pressure of an evolved gas, which shows a tendency to condense in a glass capillary, is measured in the course of reaction by means of a McLeod manometer, the results may be reproducible—since the amount of vapors condensed in the capillary remains under otherwise identical conditions approximately constant—but not accurate. Accuracy should be tested for each newly developed analytical method. Even reliability of an old, frequently successful method used under new conditions should be evaluated. Recovery checks (i.e., comparison of concentrations found with amount taken for analysis), comparison with other methods, and the use of tested standards are most frequently used in the evaluation of analytical methods.

The time required for analysis does not affect the value of the rate constant if the reaction is quenched after samples have been taken from the reaction mixture. If the reaction is followed continuously, the recorded concentration-time curve (and hence the value of the rate constant) can be affected by the properties of the sensor and of the recording system. The sensor is that part of the analytical device that indicates concentration changes. The properties of the recording system are characterized by its hysteresis and time lag (i.e., the time needed to follow a signal). The response of the sensor (photocell, electrode) is usually sufficiently rapid (with the exception of glass and some other ion-specific electrodes). For fast reactions the properties of the recorder can be of importance. Special, rapidly responding recorders and oscilloscopes should be used in such situations. Visual readings (as in polarimetry) can be combined with dictation of the found values into a tape recorder.

If the reaction being studied is simple, it is frequently sufficient to follow the concentration change of just one component. If the investigated system consists of consecutive or side reactions and if an accumulation of one or more intermediates takes place during the course of the reaction, it is necessary to follow concentration changes of two or more components. Generally the number of parameters characterizing the solution composition to be followed should be equal to the number of reaction steps with comparable reaction rates.

When a preliminary timetable of a kinetic mechanistic study is considered, it is necessary to take into account five factors:

1. The extent of the study: It is necessary to decide whether just one reaction or a group of related reactions should be studied. In both cases it is necessary to estimate the number of compounds whose reactions are to be investigated, which effects of the solution composition should be studied for the elucidation of the mechanism, and how many repetitive measurements are required for the determination of a single rate constant.

2. Next, the average time needed for one kinetic run should be estimated, considering whether the half-times are of the order of seconds, minutes, hours, or days. Reactions with half-times of a month or longer (common in radiochemistry) are usually avoided in physical organic chemistry because of problems of logistics.

3. The speed of analyses plays an important role in the timetable for systems where it is necessary to take samples and analyze them after quenching. The difference in time requirements can be demonstrated using the case of the reaction of isothiourea with bases [2-10]. In this reaction mercaptans formed by alkaline cleavage can be determined in several ways, gravimetrically as silver, mercury, or cadmium mercaptides, by iodometric titration, by an amperometric titration using silver nitrate, or polarographically. Comparison of time requirements for the construction of one concentration–time curve with a minimum of nine points follows:

Gravimetry	1–2 days
Iodometry	1–2 hr
Amperometric titration	$\frac{1}{2}$–1 hr
Polarography	5–20 min

For the first three methods, in which samples must be taken, it is necessary to add the reaction time (5–20 min) and the time needed for the handling of samples. For very fast reactions it is furthermore impossible to take all readings during the same kinetic run, so that the time needed for a second run and for the preparation of the second reaction mixture must be added. In some cases it is necessary to prepare a new kinetic run for each reading (for each time interval).

4. An inseparable part of the planning of the timetable is consideration of the time needed for the evaluation of kinetic results. The most time-efficient approach utilizes a computer interfaced to the apparatus (on-line). In all other cases the data have to be either introduced into the computer system or dealt with by a graphical or numerical method. The time needed for handling of data depends on the analytical method used and on the type of the reaction involved. Different analytical methods require different numbers of mathematical operations. The most time-efficient procedures are those that produce data which can be directly, without further mathematical transformations, inserted into equations for reaction rate. The evaluation of more complex reactions usually involves a greater number of mathematical

operations and is thus more time-consuming. In this respect the situation has changed dramatically with the availability of computers.

The importance of the time needed for kinetic data evaluation in the overall planning becomes obvious when it is realized that for reactions with half-times of less than 20 min, followed by continuous physical methods, the time required for evaluation of the record and numerical handling of data and evaluation of the rate constant can vary between 30 min and 2 hr.

This discussion indicates the importance of the application of computers to handling of data and evaluation of rate constants. These methods will be treated in a separate chapter.

5. The last factor, which plays a special role in the above-mentioned case of repeated preparations of reaction mixtures, is the total consumption of the reacting species. It is also important for all reactions requiring less common or expensive substances. Since the preparation of several tens of milligrams of a less accessible organic compound can be considerably easier and takes substantially less time than the synthesis of several hundred milligrams or several grams, it is important to choose an analytical method which permits a high number of measurements using the smallest possible amount of the substance.

The amount of substance needed for the determination of one concentration–time curve depends on the concentration and the volume of the reaction mixture. The lowest possible concentration is usually given by the sensitivity of the analytical method used, for example, for potentiometric titration the lower limit is usually 10^{-2} to $10^{-3} M$, for optical absorption and polarography 10^{-4} to $10^{-6} M$. Sensitivity of other methods, such as dilatometry, is usually somewhat lower.

Whereas the concentration limit is frequently imposed by the analytical method used, the ingenuity of the experimenter can be demonstrated by the adaptation of the procedure to small volumes and by modification of experimental conditions (reaction vessel and mode of measurement) for micro- or even submicromethods. It is necessary to evaluate whether, in the course of a transition to microtechniques, the accuracy of analytical results is affected. Particularly the accuracy of volume measurement can be decreased by down-scaling.

There is also a large difference between consumption of the material in procedures requiring sampling as compared to those in which the concentration change is followed directly in the reaction mixture. Frequently when physical methods are used, the composition of the reaction mixture is often unaffected by analysis, and in some cases (e.g., for reversible reactions) it is possible to regenerate the starting material.

To demonstrate the great difference between the sample consumption in direct methods using sampling and indirect methods with a continuous recording, the investigation of the reaction of 1,2-diols with periodic acid is again cited. If the periodate oxidation is applied to ethylene glycol or propylene glycol the amount of the glycol used is unimportant, since both substances are cheap and easily available. The situation is somewhat different when stereoisomeric 1,2-diols are oxidized, such as the threo- and erythro epimers of butane-2,3-diol, dihydrobenzoin, and so on. The preparation of defined pure epimers is not simple, and usually it is possible to obtain several tens or hundreds of milligrams, but not gram quantities. When an iodometric titrimetric method is used, a sample is taken from the reaction mixture, the pH is adjusted by addition of sodium hydrogen carbonate, arsenic trioxide with a trace of potassium iodide [2-11] as a catalyst is added, and the excess of arsenic is titrated by iodine. From 10 mg of the diol it is possible to carry out three titrations, so that for recording one concentration–time curve it is necessary to use 30 to 50 mg of the substance.

With spectrophotometric method based on bleaching of the violet solution of 2,4,6-tri-2-pyridyl-s-triazine by periodate [2-12], 10 mg of the diol allowed recording of 1 to 5 concentration–time curves. When for the same reaction the decrease in the concentration of periodate is followed polarographically [2-13], it is possible to obtain 20 complete concentration–time curves from the measurement of the decrease in periodate limiting current using only 10 mg of the diol. If the course of reaction is followed in a smaller volume (e.g., 1 ml) it is possible to record up to 200 concentration–time curves with the same amount of diol.

To obtain 100 concentration–time curves, which should be considered somewhat less than the average number of runs needed for the study of a mechanism, more than 3 g are consumed for titrimetry as opposed to about 50 mg for polarography. This example demonstrates cost effective savings on chemicals resulting from the application of continuous physical methods.

2.4.2. Direct Measurements

2.4.2.1. Chemical Methods

In the investigation of the course of a chemical reaction using sampling and quenching, it is possible to use any common method of chemical analysis. If the stoichiometry of the reaction being studied is unknown, it is necessary to determine (at least in preliminary experiments) concentration changes of all starting materials and products. Once the stoichiometry and absence of competitive reactions has been established and it has been proven that the reaction involves a single rate determining step, it is sufficient to monitor the concentration change of one component (reactant or product) provided that the reaction follows a simple reaction rate equation. The analysis can be carried out, for example, gravimetrically, titrimetrically (with visual, potentiometric, amperometric, conductometric or spectrophotometric end-point determination), spectrophotometrically, polarographically, or coulometrically.

Titration methods are most frequently used. For example, the study of ester hydrolysis is usually followed acidimetrically by means of visual or potentiometric indication of the equivalence point [2-14]. In the study of bromination, the unreacted bromine is determined iodometrically.

An example of spectrophotometric detection is the study of the cleavage of diazonium salts by measurement of the absorbance resulting from the color reaction due to coupling of the unreacted diazonium salt in alkaline media. In a similar manner the formation of diazonium salts can be followed using another colorimetric reaction based on the determination of nitrous acid.

2.4.2.2. Gas Evolution

The rate of some reactions carried out in the liquid phase can be followed by measuring the volume of the gas formed or absorbed during the course of the reaction. In this type of study the establishment of equilibrium between the liquid and gaseous phase has to be assumed. This condition is difficult to fulfill if the solubility of the gas in the reaction mixture is too high or if the reaction rate is too fast. The use of gas volume measurement under such conditions is not recommended. Since the rate of gas evolution or absorption and that of the establishment of equilibrium between the gas and liquid phase may be limited by the rate of diffusion of the gas in the liquid, attempts are made to accelerate this process by intensive stirring of the reaction mixture and formation of as large a gas–liquid interface as is possible. The apparatus used for the study of reactions accompanied by gas evolution consists of a thermostated reaction vessel with good stirring connected to a gas burette or a manometer. If the stirring is achieved by agitation of the reaction vessel so that supersaturation of the solution by dissolved gas is prevented, the gas burette must be either fixed and connected to the reaction cell by elastic tubings or it can be firmly attached to the reaction vessel and both agitated. When a mechanical stirrer is used, it is best to use mercury seals on the shaft of the stirrer. Alternatively, an effective magnetic stirrer can be used.

To prevent errors resulting from temperature changes it is necessary to thermostat the gas burette by means of a water jacket. In some instances it is possible to use an automatic counter to monitor the number of bubbles formed.

When dealing with the measurement of gas volumes, it is necessary to recognize the sensitivity of this operation to a range of factors, including pressure, temperature, humidity, and sometimes even the nature of the gas involved. The correction due to water vapor pressure is a frequent source of uncertainty. If the reaction is carried out in a dilute aqueous solution and if the gas is collected above water at the same temperature as the reaction mixture and has therefore prractically the same water pressure, it is possible to neglect the correction due to water vapor pressure. Measurement of gas volume can be replaced by measurement of resistance [2-15]. A potentiometric wire is stretched inside an abbreviated mercury manometer. When voltage is applied to this wire, the rising mercury changes the length of the wire (not immersed in mercury) and thus its resistance which is recorded. The function of this apparatus was tested using the decomposition of o-nitrophenylazide (2.4-2):

$$ \qquad (2.4\text{-}2) $$

The evolution of nitrogen is followed gasometrically also in other reactions, for example, in the Curtius rearrangement [2-16] of benzazides [(2.4-3) and (2.4-4)]:

$$RC\overset{O}{\underset{N_3}{\big\langle}} \longrightarrow RC\overset{O^-}{\underset{N^+}{\big\langle}} + N_2 \qquad (2.4\text{-}3)$$

$$RC\overset{O^-}{\underset{N^+}{\big\langle}} \longrightarrow R\text{—}N\text{=}C\text{=}O \qquad (2.4\text{-}4)$$

or in the thermal decomposition of some diazonium salts [2-17]:

$$ArN_2^+ \xrightarrow{H_2O} Ar\text{—}OH + N_2 + H^+ \qquad (2.4\text{-}5)$$

$$ArN_2^+ + Ar\text{—}OH \longrightarrow ArN\text{=}NArOH + H^+ \qquad (2.4\text{-}6)$$

Another example involving measurement of volume of evolved carbon dioxide is the propanoic decarboxylative debromination of 3-bromo-3-phenylpropanoic acid (2.4-7) at pH 7–9 [2-18]:

$$ArCHBrCH_2COO^- \rightarrow ArCH\text{=}CH_2 + Br^- + CO_2 \qquad (2.4\text{-}7)$$

2.4.2.3. Precipitates

In some isolated instances it is possible to follow the rate in the course of a reaction in which a slightly soluble compound is formed by collecting and weighing the precipitate. Some examples of such applications were given in Section 2.4.1. of this chapter. Because of complications resulting from co-precipitation, adsorption on the solid phase, and solubility effects, this approach is not recommended if other analytical techniques are available.

2.4.3. Indirect Measurements

2.4.3.1. General

As already mentioned, rates of reactions can be followed by measurement of physical quantities that are a function of some property of the solution. Ideally, properties of the solution are chosen which are simply defined functions of the concentration of one component. Quantities that are complex or empirical functions of solution composition can be used only when special attention is paid to their calibration.

For the use of indirect methods (especially where a nonspecific quantity is measured) it is essential to have precise knowledge of the stoichiometry of the studied reaction, sequence of reaction steps, and the nature of the rate determining step. A number of studies involving accurate physical measurements have been invalidated because the investigators neglected some side reactions in the evaluation of results.

Since physical methods generally need a standard for calibration, it is important to realize that the accuracy of the measurement cannot be greater than the accuracy with which the composition of the standard is known.

Physical quantities can be measured in quenched samples, but in most cases the greatest advantage is offered when the measurement of these quantities can be carried out directly in reaction mixtures, which is possible with continuous recording. Exceptions are chromatographic methods which are used for the analysis of quenched samples.

Measurement of physical quantities can be classified into two groups: generally applicable, and specific. General methods are based on measurements of density and refractive index changes, which accompany most reactions in solution, and calorimetric measurement of temperature changes.

Applications of specific methods depend on the properties of components of the reaction mixture. Spectrophotometry, polarography, and polarimetry have found the widest acceptance. Measurement of other quantities, such as conductivity, magnetic susceptibility, light scattering, dielectric constant, and so forth, are of limited importance.

2.4.3.2. Measurement of Density and Volume

For dilute solutions the density is usually a linear function of concentration, but for more concentrated solutions it is necessary to use an empirical calibration. The measurement can be carried out after sampling and quenching, for example by using the method of falling drop or density gradient. In some instances the change in specific weight can be determined by a pycnometer or densitometer.

Usually, the change in volume is followed continuously in dilatometers (Fig. 2-1) consisting of a small, round vessel attached to a capillary. The change in volume in the flask is reflected by a change of the position of the meniscus in the capillary. Since the experimental arrangement can be adapted in such a way that the volume of the solution in the flask is far greater than the volume in the narrow capillary used, even a small change in volume of the reaction mixture is reflected by a measurable shift of the meniscus, which is read on a scale or by a cathetometer.

It is easy to achieve high sensitivity in dilatometric measurements. Therefore it is possible to employ this technique for reactions showing only a small degree of conversion. It is somewhat more difficult to avoid systematic errors. Since a dilatometer can act as a sensitive thermometer, it is necessary to keep the temperature of the reaction mixture constant to $\pm$ 0.001°C and only in some cases $\pm$ 0.005°C is sufficiently accurate. If the reaction is accompanied by evolution or absorption of considerable heat, the exchange of heat between the solution in the dilatometer and the bath can affect the meniscus in the capillary. Since the mixing of reaction components in a dilatometer is tedious, the reaction mixture must be prepared outside and the first few measurements can be uncertain and affected by temperature changes resulting from the mixing of reaction components. The capillary used must have

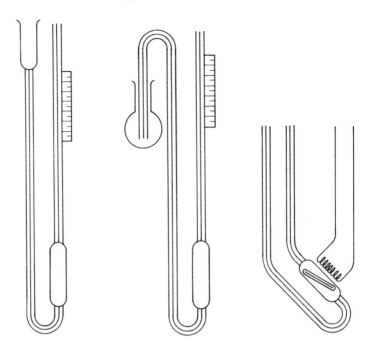

FIG. 2-1. Different types of dilatometers.

the same inner diameter over its entire length. If irregularities in the inner diameter are observed, it is necessary to apply corrections.

Dilatometric measurements can be applied to the study of a large number of solution reactions. They are particularly useful for slow and moderately fast reactions of first order and for slow reactions following a more complex kinetic behavior. The accuracy is usually sufficient if the dilatometer is well thermostated. The position of the meniscus can be read every 1–2 s without affecting the reaction using a tape to record the readings or by recording photographically. An additional consideration is that the apparatus is inexpensive and easily accessible.

2.4.3.3. *Measurement of Refractive Index*

Following concentration changes by measurement of the refractive index shows, in a manner similar to dilatometry, a sufficient accuracy and can be applied to a number of different types of reactions. The measurement can be carried out by refractometers or interferometers. Since the changes of the refractive index in the course of a reaction are usually small, the use of interferometers is preferred.

With an interferometer it is always necessary to construct a calibration curve for each system and each new pair of cuvettes. The two cells—one with sample, one with blank—must be kept at temperatures that do not differ by more than $\pm$ 0.002°C. This can be achieved by immersing both cells in the same bath. A temperature difference between cells results in ill-defined interference patterns, similar to those obtained when a temperature gradient is established in the cell containing the sample. Thus, an important operation is transfer of well-stirred reaction mixture into cells.

Compared with dilatometry, refractometry relies upon a calibration curve, needs more expensive apparatus, and is less suited for continuous recording, but the use of a less accurate thermostat is sufficient.

2.4.3.4. Thermometric Measurements

The majority of chemical reactions are accompanied by temperature changes. Measurement of the temperature of the reaction mixture enables the course of reactions to be followed.

The change of temperature is measured as a difference in temperature in two vessels placed in a precision thermostat, one containing one reactant, the other blank [2-19 to 2-24]. The same volume of reagent is added to both vessels, and the difference in temperature is measured simultaneously. Effects due to enthalpy of dilution, stirring, and dissipation of energy are thus compensated. The temperature is measured by two thermocouples or thermistors. The introduction of thermistors has resulted in an increase in the scope of use of thermometric methods. The response time of a thermistor is about 0.3 s, which makes the time-lag negligible even for the fastest reactions considered in this chapter (with half-times of about 5 s). The measurement is also sufficiently accurate (to a few micro degrees, 10^{-6} °C), which is important since the total temperature change maybe of the order of 0.001°C. From the value of the change in temperature (ΔT_∞) at sufficiently long times, it is possible to determine the reaction enthalpy. For first-order reactions the time dependence of ΔT follows first-order kinetics. Graphical method for verification, if the reaction follows first-order kinetics, has recently been described [2-25]. Thermometric methods were used for example, for the cleavage of benzenediazonium chloride [2-20], reaction of N,N-dimethylaniline with methyliodide [2-20], periodate oxidation of 1,2-diols [2-23], oxime formation [2-22], and hydrolysis of ethyl acetate under conditions of first- [2-24] and second-order kinetics [2-19]. Thermometric methods have an advantage in their universality; however, the use of these methods requires, even more so than in the case of other indirect methods, prior knowledge of the stoichiometry of the reaction, nature and sequence of reaction steps involved, and identity of the rate determining step. The presence of unknown competitive reactions can lead to false deductions.

2.4.3.5. Measurement of Absorbance

Change in absorption spectra of solutions in the course of chemical reactions is probably the most frequently used method in kinetic studies. Spectrophotometric measurements of absorption can be carried out over a wide range of wavelengths from microwave spectra, infrared spectra, to the visible and ultraviolet region.

The condition for the use of absorption spectra is that reactants, intermediates, and/or products show different absorption bands. The absorption bands followed can differ either in the molar absorptivities, in the wavelengths, or both. In the former case a decrease of an absorption band to a final finite value is observed. Such behavior can be observed for compounds with two identical groups in reactions where only one is transformed, in a partial saturation of an extended conjugated system, or in equilibrium reactions. The second alternative is more frequently encountered. In the course of the reaction the original absorption band decreases and a new absorption band is formed.

The course of the reaction can be followed either by recording whole absorption spectra in the available wavelength range, or by measuring change in absorbance at a chosen wavelength.

For a preliminary study of a reaction, identification of reacting species, intermediates, and products, the recording of whole spectra (i.e., of absorbance-wavelength plots) is strongly recommended. With recording spectrophotometers this can be carried out for moderately rapid reactions with half-lives longer than about 5 min, since the recording of one spectrum takes about 2–5 min by conventional means. For faster reactions special spectrophotometers with either rapid scanning and oscilloscopic recording of the curve or methods using photodiode arrays are needed. Such instruments enable recording of whole absorption spectra in times ranging from 5 s down to several nanoseconds (10^{-9} s).

Recording of whole spectra makes it possible to choose wavelengths at which future measurements can be carried out in subsequent systematic studies to save time. If in the course of the reaction only one component of the reaction mixture (reactant or product) shows an absorption band in the accessible wavelength range, the choice is usually a wavelength corresponding to the absorption maximum. Similarly the choice is straightforward in the case when the absorption spectrum yields two absorption bands which either both increase or both decrease and the ratio of absorbances at various wavelengths remains strictly constant. This can happen when either one species absorbing at two wavelengths is either formed or transformed, or when two absorbing substances are involved which are either formed in a stoichiometric ratio of 1:1 or which react in the same ratio. A distinction between these two possibilities can be made by comparison with the spectra of absorbing components under conditions where reaction cannot take place, such as at another pH, at a lower temperature, or in a solution containing only a single component. In both of the systems mentioned above the choice of wavelength is not essential; in general, it is possible to follow the change of the absorbance with time at any wavelength where its value changes.

Somewhat more complex is the situation in systems where several absorption maxima change with time and the ratios between absorbances change. All maxima may increase or all may decrease while ratios change, or some of the maxima may increase while some decrease. As long as the wavelengths at which the maxima are observed differ sufficiently so that the absorption bands are completely separated and do not overlap (which happens when the absorbance in the region between the individual maxima decreases to the value of blank or background), it is possible to make use of the change in the maxima in a manner similar to that of a solution in which only one absorbing substance is present [2-26]. It is, nevertheless, recommended to record spectra of a series of synthetic solutions containing varying amounts of other components to prove that the absorbance at the chosen wavelength is unaffected by the presence of other solution components.

Rather frequently, the other alternative is observed—absorption bands are not completely separated and the absorbances of individual components overlap. Absorption spectra of reactants and products are namely frequently similar, especially if the chemical transformation of a given molecule does not completely alter that part of the molecule primarily responsible for the absorption (chromophore or auxochrome).

Sometimes it is possible, even in such cases, to find a wavelength—often different from that of all the absorption maxima [2-27]—at which the measured absorbance is a function of concentration of a single component of the reaction mixture. Often such a wavelength will correspond to the long-wavelength tail of an absorption band. Sometimes the optimum wavelength corresponds to a minimum on the initial spectrum. Alternatively, it is possible to use, in some instances, measurements of the first or second derivative of the absorbance–wavelength curve. In all other cases it is necessary to measure the absorbance at two or more wavelengths. In such cases, when two absorbing substances are present in the solution, it is possible to use the equations (2.4-8) and (2.4-9):

$$\frac{A_{\lambda 1}}{d} = E_{A_{\lambda 1}}C_A + E_{B_{\lambda 1}}C_B \qquad (2.4\text{-}8)$$

$$\frac{A_{\lambda 2}}{d} = E_{A_{\lambda 2}}C_A + E_{B_{\lambda 2}}C_B \qquad (2.4\text{-}9)$$

where $A_{\lambda 1}$ and $A_{\lambda 2}$ are absorbances at wavelengths λ_1 and λ_2, E_A and E_B are molar absorptivities of substance A and B at wavelengths λ_1 and λ_2, and C_A and C_B are the unknown concentrations of these substances. It is obvious that to determine concentrations C_A and C_B it is necessary to determine the molar absorptivity of pure component A and pure component B at both wavelengths λ_1 and λ_2, at which the absorbances in samples are also measured.

In solutions containing n absorbing components it is theoretically possible to determine the concentration of all n components by measuring absorbance and molar absorptivities at n wavelengths. Nevertheless, the accuracy of the determination of concentration in solutions containing more than two components with overlapping absorptions is usually so small that such an analysis is usually meaningless.

From the change of the spectra with time it is possible to distinguish the role of the absorbing species in the reaction. If the absorbing species takes part in the reaction as a reactant, its absorbance decreases with time. If a product is the absorbing species its absorption band increases with time. The presence of an intermediate can be detected from the appearance of an absorption band which first increases and later decreases. The presence of an intermediate can also be deduced from the shape of the absorption spectra. In reactions that are simple, one-step processes that show absorption of both a reactant and a product, the absorption spectra of reactants and products may intersect in one or more points. At such a point the absorbance does not change with time and the point is called an isosbestic point (Fig. 2-2). Formation of an intermediate that does not absorb in the studied region can be demonstrated by the absence of an isosbestic point and a change in spectra as in Fig. 2-2. The existence of an isosbestic point thus suggests that the reaction in question is neither complicated by accumulation of a relatively stable intermediate nor by a side reaction, showing a rate comparable with that of the studied reaction.

In some instances it is necessary to measure the time change of the absorbance at a wavelength where the solvent, buffer, or other nonreacting components of the reaction mixture absorb. Even when such effects can be partly compensated for in double-beam instruments, it is always recommended to subtract the absorbance of blank (background) from the measured absorbance.

The cells in the spectrophotometer that are used as reaction vessels should be thermostated. This is frequently done by placing them in a thermostated metal block in the spectrophotometer. Temperature of the solution in the cell should be checked occasionally.

Since absorbance is a linear function of concentration, values obtained may be directly inserted in rate equations where the concentrations appear in ratios. The linear dependence of absorbance on concentration over the experimental concentration range should always be tested under the same conditions used in the kinetic study.

Occasionally it is possible to use spectrophotometric measurements even for cases in which neither the reacting species nor the products show absorption bands in the accessible range of wavelengths. Such an application is possible when one of the components of the reaction mixture participates in a rapidly established equilibrium with a color indicator. If the color change due to the indicator is a simple function of the concentration of the studied component (usually a linear relationship over a certain concentration range) it is possible to follow the reaction by measuring the absorbance corresponding to the indicator. Examples are provided by examining reactions in which ascorbic acid is consumed or liberated. These can be followed by changes in the color of 2,6-dichlorophenolindophenol added to the solution.

Although this discussion is not restricted to a particular kind of spectrophotometry, visible and UV spectroscopy are most frequently used for the study of solution kinetics. The reason for favoring this type of spectra lies in the fact that the solutions studied often contain water (which complicates, for example, IR measurements). Furthermore, UV spectrometers with thermostated cells are generally available, and measurements can be made for 10^{-4} to 10^{-6} M solutions. Quantitative absorbance measurements in the infrared region require high concentrations and special procedures. Microwave spectra are usually used only for the simplest molecules. Fluorescence emission is very sensitive and specific but restricted to a smaller group

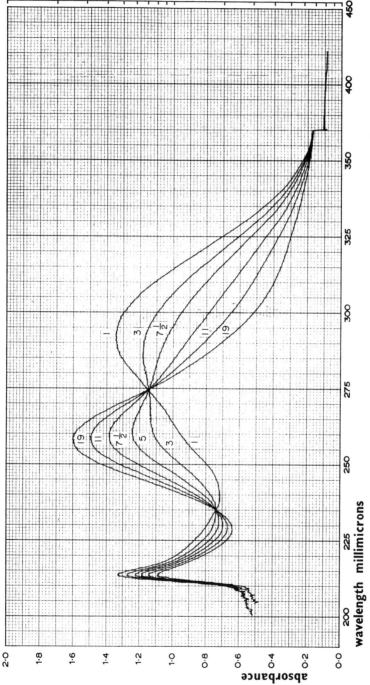

FIG. 2-2. Change of absorption spectra of diphenylacetaldehyde with time. 0.1 mM diphenylacetaldehyde in a 50% ethanolic solution containing 0.03 M LiOH with 0.47 M LiCl at 25°C. Spectra recorded after 1, 3, 5, 7½, 11, and 19 min. Two isosbestic points at 235 and 274 nm.

of compounds. Thus, for example, oxidation of thiamine to thiochrome can be followed [2-28]. Nevertheless, fluorescence can also be quenched or enhanced by some reaction products or buffer components.

Spectrophotometry makes it possible in some systems to study the transformations of several components of the reaction mixture simultaneously. It is thus particularly suitable for studies of systems of consecutive and competitive reactions that result in the formation of several intermediates or products. Modern spectrophotometers enable either repetitive recording of the spectra after chosen time intervals or continuous recording of the absorbance at a chosen wavelength as a function of time. Some instruments can be provided with attachments for several cells, so that for moderately slow reactions it is possible to follow four or five kinetic runs simultaneously.

2.4.3.6. *Polarography* [2-29], [2-30]

In polarography it is possible to follow concentration changes of substances that show waves on current-vs.-voltage curves obtained with a dropping mercury electrode (DME) (Fig. 2-3), that is, substances that are polarographically active. Such waves result from the reduction or oxidation of a solution of an organic substance at the surface of a DME. In other cases, the waves obtained can correspond to the formation of slightly soluble or complex mercury compounds or to catalytic processes at the surface of the DME. Generally speaking, the scope of application of polarography to organic systems corresponds roughly to the scope of UV and visible spectrophotometry. The main difference is found for large nonpolar molecules that are difficult to dissolve in the polar solvents used in polarography because of conductivity requirements.

On polarographic curves two quantities are measured: The half-wave potential ($E_{1/2}$), which depends on the chemical properties of the polarographically active compound, and the limiting current or wave-height (Fig. 2-3), which is usually a linear function of concentration of the electroactive substance. A formal analogy with the wavelength of an absorption maximum and with absorbance is obvious. As with electronic spectra, polarographic activity depends primarily on the electron density distribution in a given molecule rather than on the properties of a specific functional group alone. Similarly to spectrophotometry, both the characteristic quantities depend on the medium, on the solvent, pH, ionic strength, temperature, and the presence of surface active compounds.

In polarography, for most systems, a linear relationship exists between the limiting current (i_l) and the concentration

$$i_l = \kappa c \qquad (2.4\text{-}10)$$

where κ is a constant.

For diffusion currents, which are the type encountered most frequently in polarography, the value of the constant κ was derived by Ilkovič as:

$$\kappa = 0.627 \, nFD^{1/2} \, m^{2/3} t^{1/6} \qquad (2.4\text{-}11)$$

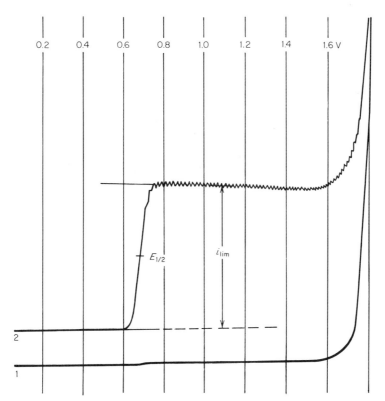

FIG. 2-3. Polarographic reduction wave of 1,4-naphthoquinone. 0.2 mM 1,4-naphthoquinone in a borate buffer pH 9.3, containing 30% ethanol. $E_{1/2}$-half-wave potential; i_{lim} limiting diffusion current (wave-height).

where n is the number of electrons transferred in the course of electrolysis; F is the Faraday charge (96,500 Coulombs); D, diffusion coefficient; m, rate of flow of mercury from the capillary; and t, the drop time (or drop-life), that is, the time between the dropping off of two consecutive drops.

A linear relationship between wave-height and concentration is also valid for the majority of the so-called kinetic currents (Section 4.3.1), the height of which is governed by the rate of a chemical reaction taking place in the vicinity of the dropping electrode. Some polarographic currents are governed by adsorption or catalytic processes. In such cases the relationship between wave-height and concentration is frequently more complicated. If such waves are used for the study of homogeneous reactions, which is rarely done, it is necessary to construct a calibration curve.

Since the diffusion coefficient D depends on temperature, it follows from the Ilkovič equation that for measurements to be compared it is necessary to keep the temperature constant at least to $\pm 0.2°C$. To keep the values of m and t constant it is necessary to keep constant also the height of the mercury reservoir connected to the capillary electrode.

The values of n and D can be affected by solvent composition and therefore it is recommended to keep the composition of the so-called supporting electrolyte constant.

The condition for the application of polarography in homogeneous kinetics is that at least one of the reactants or products gives a polarographic wave that is well separated from all the other waves of substances present in the reaction mixture. This can be achieved by a proper choice of the supporting electrolyte. Usually, when investigating reactions of organic compounds, supporting electrolytes are buffers that constitute the reaction medium at the same time. It is recommended to verify the linear relationship between the wave-height and the concentration of the studied substance in a solution of composition as similar to the reaction mixture as possible.

When the effect of a change in reaction medium (such as solvent, pH, ionic strength, or temperature) on the reaction rate is studied, it is necessary first to check the effect of the variation of the given parameter on the polarographic waves of the studied compound.

For a continuous polarographic study the reaction mixture is placed in the polarographic cell, which must be thermostated either by a water jacket or by immersing the cell into a thermostated bath.

Complete polarographic current–voltage curves should always be recorded in preliminary experiments. On polarographic curves recorded at chosen time intervals, waves for which the heights decrease with time correspond to reactants whereas the heights of waves of products increase. When the height of a wave first increases and then decreases with time, such a wave can be attributed to a reaction intermediate. This type of preliminary inspection of polarographic curves enables us to check the course of the reaction, to indicate the formation of some unexpected waves, and to choose the most suitable potential at which the change of the polarographic current with time can be studied.

For reactions with half-lives up to about 10 min the recording of the current–time curve is carried out continuously by using several polarographs (or in consecutive runs), and it is possible to follow the time changes of current simultaneously at two or more applied potentials and obtain information on concentration changes of two or more components of the reaction mixture. For slower reactions the current is usually recorded only after selected time intervals, or preferably, whole polarographic i–E curves are recorded.

Following the changes in wave-heights corresponding to individual species from recorded i–E curves is further facilitated by the additivity of polarographic limiting currents. The heights of two close but not overlapping waves can often be deconvoluted with better resolution (i.e., more accurately) than two close absorption bands. For the same reason, peak-shaped curves obtained by the modern modification of polarography, differential pulse polarography, offer advantages of higher sensitivity (allowing up to $10^{-8}\,M$ solutions to be followed) only for well-resolved peaks. Another modern variant of polarography, the normal pulse polarography, offers advantages of additivity of limiting currents, often at a higher sensitivity than classical dc polarography.

Since polarographic currents are a linear function of concentration, it is possible to use values of currents in rate equations (instead of calculating concentration first) for the evaluation of rate constants, provided that the expression used contains a ratio of concentrations, as in first-order kinetics.

Polarographic measurements have a short time lag and are suitable for reactions with half-lives over 10 s. Other variants of polarography include measurement of instantaneous current–time curves and linear sweep voltammetry. Both can be used for the study of faster reactions.

For reactions of compounds that are oxidized or reduced at positive potentials, solid electrodes such as carbon or platinum electrodes can be used. To follow concentration changes, rotating electrodes are most suitable, and changes in limiting current at a chosen potential are usually followed. Changes in the reduction current of bromine, for example, were used in a study of the kinetics of the bromination of phenols [2-31].

Polarographic measurements can be carried out at low concentrations (10^{-3}– 10^{-7} M) and in small volumes. Polarographs are usually less expensive than spectrophotometers, and temperature control, as well as changes in the volume of the reaction mixture, are simpler. Limitations are given by the condition of electroactivity, although most reactive organic compounds are electroactive. Solubility excludes only the least soluble of organic compounds.

Polarography has been shown to be complementary to spectrophotometry. For example, the hydration of the double bond in chalcone ($C_6H_5CH{=}CHCOC_6H_5$) its alkaline cleavage is preferably followed polarographically [2-32], because separate waves of the initial compound (chalcon), intermediate (ketol, $C_6H_5CHOHCH_2COC_6H_5$), and of the final products (benzaldehyde and acetophenone) are observed on polarographic curves. UV spectra show only a decrease of the initial compound (chalcone). On the other hand, when an analogous reaction of cinnamaldehyde ($C_6H_5CH{=}CHCHO$) is followed, polarographic curves of the intermediate (aldol) and products (benzaldehyde and acetaldehyde) overlap. In UV spectra, the absorption band of the aldol is observed [2-33]. Hence the kinetics of alkaline cleavage of chalcone is preferably studied by polarography, whereas that of cinnamaldehyde by UV spectra.

Application of polarography for the study of very fast reactions will be discussed in Section 4.3.1.

2.4.3.7. Optical Rotation

The measurement of optical rotation was the first reported case in which a physical measurement for a kinetic study was used. The reaction investigated was the inversion of sucrose, described by C. Wilhelmy in 1850.

Polarimetric measurements are restricted to compounds with at least one asymmetric atom. The angle α, by which the studied liquid or solution rotates the plane of polarized light, is measured. Individual compounds differ in their capacity to rotate the plane of polarized light. Using the measurement at one selected wavelength, it is impossible to distinguish several optically active substances in a mixture.

When more than one optically active compound is present in the reaction mixture, only their total effect on the plane of polarized light is measured.

The magnitude of the observed rotation α depends on the chemical properties of the optically active species and its concentration. Optical rotation can be affected by the wavelength of the source used, temperature, solvent, and pH of the solution. Because of limited specificity of the technique, traces of impurities (especially when optically active) play an important role.

Simple polarimeters work at one wavelength, either using a sodium D-line as a source (doublet 589.0 and 589.6 nm) or more recently the green line of mercury (546.1 nm). In modern polarimeters, the use of monochromators enables a choice of wavelength. Optical rotatory dispersion (ORD) curves, showing the dependence of optical rotation α on the wavelength of polarized light, are more informative for individual compounds than the value of α at a chosen wavelength. For preliminary studies the recording of ORD curves as a function of time is recommended. This enables a choice of the wavelength which is best suited for the study of a given reaction, usually a wavelength where the value of α is high and little affected by other reaction components. If the wavelength chosen is in the UV region, all the transparent parts (including those in the polarimeter) must be made of quartz or fused silica.

For substances showing no association and no change in ionization with concentration, the optical rotation is a linear function of the concentration of the optically active substance. For solutions the following relation can be used:

$$\alpha = [\alpha]_\lambda^T l/C \qquad (2.4\text{-}12)$$

where $[\alpha]_\lambda^T$ is the specific optical rotation at temperature T (usually 20° or 25°C) measured at wavelength λ, α is the measured rotation, l is the length of the polarimetric tube (in dm), C is concentration (in grams of the dissolved substance per one milliliter of solution).

Because of the linear relationship between concentration and rotation (α) it is possible to insert the values of α directly into those rate equations which involve a ratio of concentrations. This linear relationship should always be checked before the beginning of a kinetic study.

Visual polarimeters enable readings every 10 seconds or so. Modern recording polarimeters enable continuous recording of α as a function of time. Measurements at 546 nm or 589 nm are usually not very sensitive and the concentration of the optically active substance must be relatively high (0.5 to 2%), whereas measurements in the UV, in the region of the absorption maximum of the given compound, are considerably more sensitive but such measurements can be carried out only if rather expensive rotatory dispersion spectrophotometers are available. When compared with spectrophotometry and polarography, polarimetry is applicable to a smaller group of compounds, is less specific, can be affected by traces of impurities and frequently needs higher concentrations and larger volumes of the reaction mixture. On the other hand, polarimetry and measurements of the optical rotatory dispersion enable the study of some reactions which cannot be followed by any other technique.

2.4.3.8. *Measurement of Electrolytic Conductance*

When two electrodes are immersed into a solution of an electrolyte and a potential is applied, the resistance (R) of the solution to the passage of electric current depends on the nature and concentration of the electrolyte. The measured conductance ($L = 1/R$) is directly proportional to the area of the electrodes (A) and inversely to their distance (l)

$$L = \kappa \frac{A}{l} \qquad (2.4\text{-}13)$$

The proportionality constant, κ, is called specific conductance and is the conductance measured in a solution of electrolytes between two electrodes with surface area of 1 cm² which are 1 cm apart.

It can be shown that the specific conductance, κ, is directly proportional to molar concentration (C)

$$\kappa = \frac{\Lambda C}{1000} \qquad (2.4\text{-}14)$$

where Λ is the equivalent conductance.

As the measured value of conductance is thus a function of concentration of ions, the rate of reactions in which the total number of ions varies can be followed by the measurement of conductance as a function of time. The scope of reactions that can be investigated by conductance measurements is considerably larger than that of reactions followed polarimetrically. Nevertheless these two techniques have a common feature: They are neither completely general nor specific for concentration changes of a single component of the reaction mixture. If the concentrations of several ions change in the course of a reaction, conductance offers information only on the change in total amount of ionized species, provided that their mobilities are not considerably different.

Conductance depends on the reaction medium, in particular on the nature of the ionizing solvent and temperature. If high-frequency techniques are not made use of, the application of conductometric measurements is completely prevented by the presence of foreign electrolytes. This makes it extremely difficult if not impossible to use buffered solutions and to investigate the effects of ionic strength in reactions studied by conductance measurements. Since platinized platinum electrodes are most frequently used in conductometric cells, it is impossible to study conductometrically reactions that are catalyzed by platinum black.

The most common type of measurement of conductance is by means of an a.c. bridge utilizing 50 or 60 c/s frequency. Continuously recording conductometers are available, but not common. Over the past three decades the possibilities offered by the use of high-frequency conductance measurements have been explored [2-34, 2-35]. The advantage of these techniques, in addition to the limited effect of foreign electrolytes, is the possibility of using conductance cells in which the

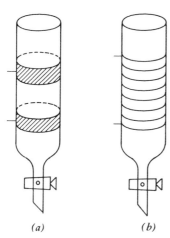

(a)   (b)

FIG. 2-4. Conductivity cells for high frequency measurements: (a) capacitor cell, (b) inductive cell.

electrodes are attached to the outer walls (Fig. 2-4) and are not in contact with the electrolyte. The high-frequency measurements are easily converted to continuous recording, for fast reactions, by means of an oscilloscope or a digital storage device. In aqueous solutions the measurement is restricted to solutions of concentration 10^{-1} to 10^{-3} M (with respect to the electrolyte).

For very dilute solutions of strong electrolytes, where the equivalent conductance (Λ) is practically constant, it is possible to consider conductance to be a linear function of concentration. Under all other conditions, in particular when weak electrolytes are present in the solution, it is necessary to find an empirical relationship between conductance and the composition of the solution. Only for an intermediate range of concentrations is it possible to use the equation for dependence of conductance on concentration, derived by Onsager, as an interpolation formula.

Conductometric measurements have been used, for example, for the study of ester or anhydride hydrolysis and similar reactions. Sometimes it is unnecessary to calculate concentrations from measured conductances, and the measured values can be directly introduced into kinetic equations. An example is alkaline ester hydrolysis in sodium hydroxide solutions [2-34, 2-35], where the conductance κ will decrease in the course of the reaction because OH^- ions with high mobility are replaced by less mobile anions of the corresponding acid in the course of the reaction. For the fraction of hydroxide ions (α) in the reaction, an expression $\alpha = (\kappa_0 - \kappa_t)/(\kappa_0 - \kappa_\infty)$ can be derived where the subscripts refer to conductance at $t = 0$, at a given time t and at $t = \infty$. This simple relation can be used for solutions in which the concentrations of the ester and hydroxide ions are comparable and under the assumption that Λ_{NaOH} and Λ_{NaA} (equivalent conductances of sodium hydroxide and the salt of the acid formed) remain practically constant over the studied concentration range.

As long as the difference between the specific conductance at a given concentration and that at infinite dilution is sufficiently large, and the measurement of conductance is carefully carried out, it is possible to achieve sufficient accuracy in the kinetic measurements.

Conductometry can be practically used either for systems where nonconducting species generate ions or in systems in which two conducting components react and form a nonconducting one. Examples of the former type are hydrolysis of acetic anhydride in water [2-36] as well as the Menschutkin reaction [2-37], in which unionized phenacyl bromide reacts with uncharged pyridine to yield ionized phenacylpyridinium bromide (2.4-15):

Conductance measurements can be carried out only for ionic substances (i.e., a relatively small number of organic compounds). They are not specific for one component of the reaction mixture and can be affected by traces of ionized impurities. Furthermore, since it is impossible to follow kinetics in buffered solutions or solutions of varying ionic strength, the importance of conductance measurements can be considered relatively limited. The evaluation of high-frequency measurements is restricted by the fact that so far only a limited number of reactions have been studied by this method and only partial theoretical treatment is available.

2.4.3.9. Measurement of the Dielectric Constant

Numerous organic reactions are accompanied by a change in the dielectric constant and the kinetics of such reactions can be followed by measuring this property. The measurement of the dielectric constant (ϵ) can be carried out either by measurement of the change in wavelength of decimeter waves in a resonance system (Lecher wires) or more frequently by measurement of a condenser capacitance, because $\epsilon = C/C_0$ where C is the measured capacitance and C_0 the capacitance in vacuum. The measurement is carried out either by means of bridge methods or oscillator circuits. For kinetic measurement cells with electrodes placed outside the walls are particularly suitable.

Because the conductivity of the investigated system should be smaller than 10^4 $\Omega^{-1}cm^{-1}$, it is possible to apply the measurement only to systems that do not contain more than trace amounts of ionized components. Such systems are encountered in reactions of two liquid organic substances or in reactions of organic substances dissolved in organic solvents with a low dielectric constant (e.g., in dioxane). For the study of reactions in aqueous or water-containing solutions such measurements are not suitable.

The relationship between the concentration of a reaction mixture component and the dielectric constant is generally not linear, and it is always necessary to construct a calibration curve.

Instruments for measurement of dielectric constants are commercially available and are not difficult to construct. Measurement is rapid and can be adapted for continuous recording. As in conductometry, the measurement of dielectric constant can be affected by traces of impurities and is nonspecific. To follow a reaction by measurement of the dielectric constant, it is necessary to know the stoichiometry of the reaction. Since the technique does not permit investigation in aqueous and other conducting solvents, but can be used for nonpolar solvents and for reactions between liquid organic components, it represents a technique that is complementary to electrochemical methods.

2.4.3.10.   Measurement of emf

Since the electrode potential is a function of concentration, respective of activity, of the corresponding ions in the solution, according to the Nernst equation (2.4-16):

$$E = E^0 - \frac{RT}{nF} \ln \frac{[\text{Red}]}{[\text{Ox}]} \qquad (2.4\text{-}16)$$

it is possible to use the measurement of emf for the study of reaction kinetics.

As with polarography and conductometry, the measurement of emf is restricted to solutions that can be made conducting.

The logarithmic relationship between the measured quantity (potential) and concentration makes it possible to obtain accurate values of concentration even when the conversion is more than 90%, where all methods giving a linear response to concentration show very limited accuracy. This makes it possible, for example, to study reactions approaching equilibrium. The potential measurement can be carried out accurately up to tenths or even hundredths of millivolts.

The indicator electrode (whose potential is measured) is coupled with another electrode (called the reference electrode), whose potential is supposed to remain constant during the course of measurement, carried out under conditions of low-current flow. In most cases, calomel, mercurous sulphate, or silver/silver chloride electrodes are used for this purpose. The potential of these electrodes is usually unaffected by the small currents flowing in the course of measurement of emf, so that these electrodes fulfill the condition for use in potentiometric measurements. The fact that it is rather difficult to prepare these electrodes so that their potential is reproducible to better than ± 1 mV is not important in kinetic measurements. Only the relative, not the absolute, values of potentials are of importance.

The most common sources of uncertainty are caused by sluggishness in the establishment of the potential of indicator electrodes. The time interval between the immersion of an electrode into a solution and attainment of constant potential of the electrode depends on the material of the electrode, on its surface, on the stirring of the solution, and on the system investigated. The nature of the electrode depends on the type of system investigated and will be discussed later. With metallic electrodes, polished surfaces prove to be best. With glass and other membrane electrodes, it is necessary to keep the surface of the membrane clean. Intensive stirring by a magnetic stirrer, vibromixer, or best by a stream of gas serve this purpose. Most frequently, the sluggishness in establishment of potentials is caused by competitive side and consecutive chemical reactions accompanying the electrode process proper. For example, the potential of a platinum electrode immersed in a solution of ascorbic acid drifts and becomes relatively constant only after 10 or 20 min. This is due to the hydration–dehydration equilibrium involving dehydroascorbic acid:

$$
\begin{array}{cc}
\underset{\substack{| \\ CO \quad CHCH_2OH \\ \diagdown\diagup \\ O}}{\overset{\substack{O \\ \| \\ C-C}}{}}{\overset{OH}{\diagup}}_{OH}
& \rightleftharpoons
\underset{\substack{| \quad | \\ CO \quad CHCH_2OH \\ \diagdown\diagup \\ O}}{\overset{\substack{O \quad O \\ \| \quad \| \\ C-C}}{}} + H_2O
\end{array}
\qquad (2.4\text{-}17)
$$

Obviously, in such cases it is impossible to use potential measurements to study faster reactions, and other methods such as spectrophotometry or polarography must be taken into consideration.

If a salt-bridge is used in the measurement, the rate of establishment of the liquid junction or membrane potentials should also be considered. The rate of establishment of the former has been found to be always faster than that of the indicator electrode (i.e., the electrode whose potential is measured). Assumptions found in the literature suggest that impurities can be introduced into the sample solution in potentiometry, polarography, and conductometry due to immersion of the electrodes and/or the salt bridge. In practice this has rarely been found to be a source of error.

Metallic electrodes of first order, defined as electrodes whose potential depends on the concentration of metal ions in solution from the same metal of which the electrode is made, are rarely used in the study of organic reactions. The most frequently used electrodes are made from precious metals (e.g., platinum, palladium, gold) and behave as indifferent electrodes. The solid phase does not undergo chemical change and the potential of such electrodes, which enable electron exchange, is governed solely by the ratio of the reduced and oxidized form in the solution.

Such electrodes have been used for instance in the study of bromination reactions where the potential of the electrode depends on the ratio $[Br^-]/[Br_2]$ [2-38] and more recently in the studies of oxidation-reduction reactions of some quinoid compounds and dyestuffs [2-39]. In the latter case the possibility of following the reaction when approaching equilibrium was fully exploited. Because of complications which may result from the possibility of adsorption of organic substances at the surface of solid electrodes, it would seem possible and desirable to use the dropping mercury electrode as an indicator electrode for potentiometric measurements, but this has been rarely employed.

It would seem simplest to use ion selective electrodes for kinetic measurements such as a glass electrode or modern membrane and semiconductor electrodes. In practice, continuous measurements of pH or activity of other cations, such as Na^+ or Ca^{2+}, are rarely used for kinetic studies. Two main reasons can be cited: First, the membrane and similar electrodes sometimes react sluggishly—the equilibrium potential may take seconds or tens of seconds to establish. Second, the potential change is sometimes a rather complex function of time as reactions involving acids and bases frequently involve more than one acid-base pair. Often in a reaction an acid will react to form another acid. The complexity of the pH-change depends not only on the initial concentration and reaction rate but also on the relative values of the two acid dissociation constants involved. Somewhat more frequent is the use of halide selective electrodes in nucleophilic substitutions and hydrolyses of organic compounds [2-40].

Nevertheless, the response of a pH-meter is sufficiently rapid in good quality instruments, and continuous recording presents no problem. Taking the above limitations into consideration, it seems that measurements of emf and, in particular, of pH present numerous opportunities which have not yet been exploited.

2.4.3.11. Measurement of Nuclear Magnetic Resonance

Nuclear magnetic resonance (NMR) has become one of the most frequently used techniques in the investigation of organic reactions. The primary importance of NMR is in the power and versatility of the technique for structural identification of organic compounds which are involved in reactions as reactants, products, or long-lived intermediates. NMR may also be used to follow concentration changes during a reaction, but for this purpose the sensitivity of NMR is a limitation. For typical NMR measurements of molecules containing ^1H nuclei, solutions must be 0.1 M or higher to obtain useful signal-to-noise ratios. However, this concentration limit has recently been lowered to about $10^{-3} M$ due to the development of Fourier Transform (FT) proton or carbon-13 NMR spectrometers which can rapidly and repeatedly apply a radio-frequency pulse to a sample, accumulate the NMR signals in a computer, and later process the data to get a normal spectrum with enhanced signal-to-noise ratios. As FT NMR instruments become more commonly available, the use of NMR for conventional kinetic measurements is likely to increase.

NMR measurements can be applied to molecules that contain atomic nuclei that possess a spin, as described by the nuclear spin number, I. A nucleus with spin behaves as a spinning, charged particle with a resulting magnetic moment. Nuclear spin is exhibited by all nuclides except those containing an even number of both protons and neutrons, for example, ^{12}C and ^{16}O. NMR spectra are most commonly measured for ^1H, but those for three other common nuclei, ^{19}F, ^{31}P, and ^{13}C, all of which have $I = \frac{1}{2}$, are finding wide acceptance. A nucleus with a spin number of $\frac{1}{2}$ will orient itself in one of two permitted directions in the presence of an externally applied magnetic field (as in a NMR spectrometer), corresponding to two spin states at discrete, quantized energy levels.

The NMR spectrum is the result of transitions from the lower energy to the upper spin state induced by the absorption of energy from electromagnetic radiation applied at the proper frequency. The types of nuclei, that is, ^1H, ^{13}C, ^{19}F, and so on, have widely differing resonance frequencies for a given magnetic field strength so that in a given NMR experiment only one type of nucleus is observed. However, for a particular type of nucleus, such as the proton, slight differences in the resonance frequency will be observed for protons in different chemical environments. NMR spectra may be generated by either keeping the magnetic field constant and varying the radiofrequency or holding the frequency constant and sweeping the applied magnetic field over a narrow range. The signals detected at the resonance frequencies are amplified and recorded as peaks rising out of the base line on a strip of calibrated chart paper.

A proton NMR spectrum of a molecule consists of a series of peaks corresponding to the resonance frequencies of chemically different protons in the molecule. These peaks may show fine structure (splitting) due to mutual interaction with nonequivalent nuclei. The peak areas are proportional to the number of protons that resonate at the particular frequencies; the number of protons is dependent on the number of chemically equivalent protons in the molecule and on the concentration of the molecule. Peak areas are usually integrated electronically and recorded with the spectra, but the areas may also be determined by other means such as cutting out the peaks from the paper and weighing them on an analytical balance.

The dependence of peak position on chemical environment, and of peak area on the number or protons, are the basis for kinetic measurements using NMR. If some of the nuclei responsible for a resonance change environments during the course of a reaction, a peak corresponding to the reactant will diminish in size and a new peak corresponding to the products will increase. For instance, a rearrangement involving the stable carbocations shown below was followed by monitoring the loss of the NMR peak due to the four methylene protons and the appearance of a split peak for the three methyl protons and a split peak for the methine proton.

$$(2.4\text{-}18)$$

NMR may also be used to monitor reactions in which a proton is replaced by a different nucleus (or vice versa), as in the acid-catalyzed isotopic exchange of hydrogen atoms on aromatic rings:

$$Ar^1H \ + \ ^2HO_2CCF_3 \ (excess) \ \longrightarrow \ Ar^2H \ + \ ^1HO_2CCF_3 \qquad (2.4\text{-}19)$$

The deuterium (^{2}H) nucleus does not have a resonance frequency within the range used for ^{1}H NMR observation, hence, the size of the peak for the aromatic protons diminishes as the protons are replaced. This type of experiment using a deuterium-labeled solvent or deuterium-labeled reactant is used in the study of many processes, for example, enolization, which involve hydrogen atom exchange with solvent.

The major limitation of the NMR technique is in the high sample concentrations typically required, as previously mentioned. Furthermore, the high expense of NMR spectrometers limits their availability for lengthy kinetic measurements. For proton spectra, it is often (although not always) necessary to use solvents that do not contain hydrogen atoms. Carbon tetrachloride, deuterium oxide, and deutero-chloroform are most frequently used, as well as other deuterated solvents such as d_6-acetone and d_6-DMSO. When kinetics of reactions in deuterium oxide are studied, deuterated acids, bases, or buffer components have to be used; presence of D^+ can affect reaction rates, and an increase in expenditure results. The temperature of the sample is controlled by a flow of heated or cooled nitrogen gas through the sample chamber. This arrangement gives access to a wide range of temperatures, typically $-100°$ to $200°C$, but control is usually no better than $\pm 1°C$. The use of NMR measurements for study of concentration changes is practical for reactions with half-times of about one minute or longer. Typically, a reaction will be followed by recording and integrating the entire spectra at various time intervals when slow reactions are involved. Faster reactions can be followed by integration of only a small region of the spectrum. NMR may also be used for kinetic measurements of some fast reactions at equilibrium, by studying line-broadening effects (see Chapter 4.3.4) using techniques of line-shape analysis.

2.4.3.12. Measurement of Electron Spin Resonance

Electron Spin Resonance (ESR) measurements can be made in a manner closely analogous to NMR measurements because the electron has a spin number of $\frac{1}{2}$ and possesses magnetic properties similar to the nuclei with spin numbers of $\frac{1}{2}$ which are commonly followed in NMR measurements. The electron behaves as a spinning charged particle, with a resulting magnetic moment, and, like the proton, will align itself in the presence of an externally applied magnetic field either with its magnetic axis parallel or antiparallel to the applied field. Excitation of the electrons from the lower to the higher energy spin state can be induced by absorption of electromagnetic radiation. The resonance frequency occurs in the microwave region in the presence of the magnetic field generated in ESR spectrometers.

ESR spectra can be obtained only for organic species that contain unpaired electrons, such as free radicals, radical ions, and molecules with triplet-state electrons. Since radicals or radical ions usually have only one unpaired electron per species, only one resonance is observed for each type of radical. In contrast ^1H NMR exhibits separate resonances for each type of proton in a molecule. However, a typical ESR spectrum shows considerable fine structure due to coupling of the electron spin with nuclear spins of atoms in the portion of the molecule which bears the unpaired electron.

ESR can be used to follow the generation or disappearances of radicals during a reaction by following the changes of intensity of the ESR signal. At a given frequency and constant applied magnetic field, the intensity of the energy absorbed in resonance is proportional to the concentration of the radical species. In order to determine the proportionality constant and obtain the exact concentration, a comparison with a separate, standard sample is made. The integrated peak area for the sample is compared with the integrated peak area for the standard, which consists of a known amount of a substance that exists only as free radicals, for example, diphenylpicrylhydrazide. The fine structure of the ESR spectrum can also be used for confirmation of structure of the radical studied.

Although ESR is limited to the study of reactions that involve radicals, for this purpose it is considered to be the most powerful technique, which is most frequently used for the study of radical reactions. Reactions where the radicals were generated chemically, photochemically, or electrochemically, have been studied by ESR. The sensitivity of this technique is much higher than NMR, so that radicals in concentrations of 10^{-5} M can be detected. Radicals with lifetimes as short as about 0.01 s have been followed by using a flow technique and mixing chamber (see discussion of flow techniques in Section 4.2) to generate the radicals continuously or intermittently before entering the cavity of the ESR spectrometer. Highly unstable radicals can also be generated photochemically or electrochemically directly in the cavity, although quantitative treatment of reactions of electrochemically generated radicals is made difficult by the uneven distribution around the electrodes inside the cell. Extremely unstable radicals can be trapped at very low temperatures, identified, and their reactions followed at somewhat higher temperatures. The particularly unstable methyl and ethyl radicals, for example, can be examined by ESR spectra when the photolysis of methyl and ethyl iodide is carried out at 4.2°K. As described for fast reactions affecting the shape of the NMR spectra, ESR spectra can be used to measure fast electron-transfer reactions by techniques of line-shape analysis.

2.4.3.13. Chromatography

All the common variants of chromatographic techniques can be used for the identification of products and intermediates and semiquantitative analysis when a sample is taken from a reaction mixture and quenched. Starting species, some stable intermediates, and products can be identified by means of paper, TLC, GLC, or HPLC techniques. An increasing spot or peak corresponds to product, decreasing to starting material, transient to an intermediate.

From a comparison of the area of spots or height and/or area of peaks of samples taken from the reaction mixture, initial information regarding the course and rate of the reaction can be obtained in a very simple and very fast manner.

For quantitive analysis of the reaction mixture for starting materials, intermediates, and/or products GLC and HPLC can be applied after quenching. As examples, determination of unreacted aldehyde and products of Cannizzaro reaction after acidification [2-41] and of ferrocene, 1-acetylferrocene and 1,1'-diacetylferrocene in acetylation of ferrocene, after rapid cooling, neutralization, and extraction [2-42], can be quoted.

It is nevertheless necessary to check that chromatographic conditions, solvent, acidity, or temperature, do not affect the composition of the reaction mixture after sampling. The user should be aware of the possibility of appearance of artifacts, resulting from chemical reactions occurring in the course of the chromatographic separation employed. Nevertheless, with proper caution, it is possible to apply chromatographic techniques successfully for investigation of kinetics of moderately fast reactions.

2.4.3.14. Miscellaneous Methods

Measurement of vapor pressure can be used for the analysis of binary mixtures, or mixtures whose behavior can be considered binary, as long as the vapor pressure of individual components differs sufficiently. Also the rate of distillation measured by the increase in volume of the liquid in the collector flask can be used [2-42, 2-44] for the study of rates of various reactions. All such reactions should be moderately slow.

Mass spectrometry has so far been used predominantly for the study of gas phase reactions. The technique is also used for the elucidation of structures of stable reaction products. Another technique used successfully for the study of reactions in the gas phase is measurement of magnetic susceptibility. This technique is very rarely used for reactions in the liquid phase, and then only in those cases where the solvent is absent—that is, for reactions between two liquids.

Typical techniques used for kinetic studies in polymer chemistry, such as measurement of viscosity, surface tension, light scattering, and nephelometry, can be occasionally used even for reactions of organic compounds of lower molecular weight.

REFERENCES

[2-1]   P. S. Farrington and D. T. Sawyer, *J. Am. Chem. Soc.* **78**, 5536 (1956).

[2-2]   G. B. Smith and G. V. Downing, Jr., *J. Phys. Chem.* **70**, 977 (1966).

[2-3]   H. Landolt, *Ber.* **19**, 1317 (1886) and **20**, 745 (1887); cf. A. Skrabal, *Homogenkinetik*, Th. von Steinkopf, Dresden, 1941.

[2-4]   G. S. Forbes, H. W. Estill, and O. J. Walker, *J. Am. Chem. Soc.* **44**, 97 (1922).

[2-5]   M. G. Burnett, *J. Chem. Ed.* **59**, 160 (1982).

[2-6]   H. Samuelson and L. P. Hammett, *J. Am. Chem. Soc.* **78**, 524 (1956).

[2-7]   M. M. Breuer and A. D. Jenkins, *Trans. Faraday Soc.* **59**, 1310 (1963).

[2-7a] A. C. Knipe, *J. Chem. Ed.* **53**, 618 (1976).

[2-8]   R. N. Dannley and L. Friedman, *J. Chem. Ed.* **53**, 265 (1976).

[2-9]   J. Limido, I. Mallah, and J. C. Jungers, *Bull. Soc. Chim. Belges,* **58**, 350 (1949).

[2-10] M. Fedoroňko and P. Zuman, *Collect. Czechoslov. Chem. Commun.* **29**, 2115 (1964).

[2-11] F. R. Duke, *J. Am. Chem. Soc.* **69**, 3054 (1947).

[2-12] P. F. Pilch and R. L. Sommerville, *J. Chem. Ed.* **54**, 449 (1977).

[2-13] P. Zuman and J. Krupička, *Collect. Czechoslov. Chem. Commun.* **23**, 598 (1958).

[2-14] F. Samhaber and H. Berbalk, *Monatsh.* **89**, 659 (1958).

[2-15] E. A. Birkhimer, B. Norup, and T. A. Bak, *Acta Chem. Scand.* **14**, 1894 (1960).

[2-16] Y. Yukawa and Y. Tsuno, *J. Am. Chem. Soc.* **79**, 5530 (1957).

[2-17] D. F. DeTar, et al., *J. Am. Chem. Soc.* **77**, 2013 (1955); **78**, 3911, 3916 (1956); **80**, 3921, 3925, 3928 (1958).

[2-18] J. E. B. McGarvey and A. C. Knipe, *J. Chem. Ed.* **57**, 155 (1980).

[2-19] T. Meites, L. Meites, and J. N. Jaitly, *J. Phys. Chem.* **73,** 3801 (1969) and references therein.

[2-20] H. J. Borchardt and F. Daniels, *J. Am. Chem. Soc.* **79,** 41 (1957).

[2-21] W. A. DeOliveira and L. Meites, *Anal. Chim. Acta* **70,** 383 (1974).

[2-22] W. A. DeOliveira, L. Meites, and T. Meites, *Anal. Chim. Acta* **100,** 245 (1978).

[2-23] D. Jeffries and J. Fresco, *J. Chem. Ed.* **51,** 545 (1974).

[2-24] E. W. Schindler, Jr., and L. Meites, *Anal. Chim. Acta* **111,** 257 (1979).

[2-25] D. F. Sargent and H. J. Moeschler, *Anal. Chem.* **52,** 365 (1980).

[2-26] A. C. Schram, *J. Chem. Ed.* **56,** 351 (1979).

[2-27] S. Ewing, *J. Chem. Ed.* **59,** 606 (1982).

[2-28] N. W. Bower, *J. Chem. Ed.* **59,** 975 (1982).

[2-29] P. Zuman, Polarography and Reaction Kinetics, in *Adv. Phys. Org. Chem.* (R. Gold, Ed.) **5,** 1 (1967).

[2-30] P. Zuman, Polarographic Methods, in *Fast Reactions in Enzymology* (K. Kustin, Ed.), Academic Press, New York, 1969, p. 121.

[2-31] R. Cohen, C. Matzek, S. Schlosser, and C. O. Huber, *J. Chem. Ed.* **58,** 823 (1981).

[2-32] P. Čársky, P. Zuman, and V. Horák, *Collect. Czechoslov. Chem. Comm.* **30,** 4316 (1965).

[2-33] L. Spritzer, Ph.D. Thesis, Clarkson College of Technology, Potsdam, New York, 1977.

[2-34] D. G. Flom and P. J. Elving, *Anal. Chem.* **25,** 541 (1953).

[2-35] P. J. Elving, *Faraday Soc. Disc.* **17,** 156 (1954).

[2-36] D. B. Greenberg, *J. Chem. Ed.* **39,** 140 (1962).

[2-37] P. W. C. Barnard and B. V. Smith, *J. Chem. Ed.* **58,** 282 (1981).

[2-38] R. H. Smith, *J. Chem. Ed.* **50,** 441 (1973).

[2-39] A. Tockstein et al., *Collect. Czechoslov. Chem. Commun.* **30,** 3621 (1965); **31,** 2466 (1966); **32,** 1309, 3089 (1967); **33,** 2715 (1968); **34,** 27, 316 (1969); **35,** 2223, 2523, 2673, 2683 (1970); **36,** 1090 (1971); **39,** 1518, 3016, 3024, 3430 (1974).

[2-40] G. J. Moody and J. D. R. Thomas, Applications of Ion-Selective Electrodes, in *Ion-Selective Electrodes in Analytical Chemistry* (H. Freiser, Ed.), Vol. 1, Plenum, New York, 1978, p. 346.

[2-41] E. Woodman, Ph.D. Thesis, Clarkson College of Technology, Potsdam, New York, 1982.

[2-42] D. T. Haworth and T. Liu, *J. Chem. Ed.* **53,** 730 (1976).

[2-43] E. F. Pratt and K. Matsuda, *J. Am. Chem. Soc.* **75,** 3739 (1953).

[2-44] E. F. Pratt and J. Lasky, *J. Am. Chem. Soc.* **78,** 4310 (1956).

BIBLIOGRAPHY

S. L. Friess, E. S. Lewis, and A. Weisberger (Eds.), *Investigation of Rates and Mechanisms of Reactions*, Part I, II, 2nd ed., *Techniques of Organic Chemistry*, Vol. VIII, Wiley-Interscience, New York, 1961.

E. S. Lewis (Ed.), *Investigation of Rates and Mechanisms*, Part I, II, 3rd ed., *Techniques of Chemistry*, Vol. VI, Wiley, New York, 1974.

C. H. Bamford and C. F. H. Tipper (Eds.), *Comprehensive Chemical Kinetics*, Vol. I, Elsevier, Amsterdam, 1969.

CHAPTER THREE

ANALYSIS
OF KINETIC DATA

3.1. INTRODUCTION

After suitable reaction conditions and the best technique are chosen, the time-dependence of the concentration of one or more components has to be measured and interpreted.

To be able to evaluate the experimental data it is first necessary to identify reacting substances, their forms, and their stoichiometric relationships. Further, the formation and chemical composition of the intermediates generated in the course of the reaction should be investigated. Finally, an attempt must be made to detect and identify all the products formed in the reactions.

After substances that take part in the reaction are identified and the ratio in which they enter the reaction established, it is possible to undertake a quantitative treatment of experimental data. The information obtained provides some facts that can aid directly in postulation of the mechanism, while others indicate which experimental parameters should be varied to obtain supporting evidence that would contribute to the elucidation of the reaction mechanism. The first group, bearing direct relationship to the mechanism of the studied reaction, is represented by the proof of the equation for reaction rate describing the concentration changes with time for the given system and determination of the reaction order with respect to individual components. The understanding of these factors is usually a condition to be fulfilled before the second step—namely the calculation of the best value of the rate constant under given conditions—is attempted. Subsequently, the effect of composition of the reaction mixture, and other factors, on the value of the rate constant is studied.

Identification of the equation for reaction rate and of the reaction order in individual components serves primarily to distinguish or verify the individual reaction steps and the ratio in which the reaction components react and are formed in each of these steps. When a complex system of reactions, consisting of several reaction steps, is dealt with, finding the proper equation for reaction rate should enable conclusions about the relative rates of individual reaction steps to be postulated and this in turn will help distinguish which of these reactions affect the overall reaction rate.

Generally speaking, more relevant questions can be directly answered by finding the correct reaction rate equation than by determining the reaction order with respect to individual components. The latter is usually sought if general rather than detailed information on the reaction is required. Sometimes the determination of the reaction order may suffice, when the system of reactions is too complex and there is little hope for a rigorous treatment—either because the differential equations are too complicated or because the scope of the study is limited by time constraints or the amount of the compound available. Determination of the order of a reaction is usually a crude process in which the concentration of a component of the reaction mixture is determined only at several time intervals; it definitely requires less time and effort than that needed for finding the rate equation. However, significant deviations in the course of reaction which can be the clue for detection of complications in the studied system can be easily missed when we restrict ourselves to the determination of reaction order.

In most cases some information about the course of the reaction is already available when a mechanistic study is planned and a working hypothesis is proposed, based, for example, on the known stoichiometry of the reaction and on analogy with similar known reactions. If such a hypothesis cannot be formulated, it is possible, in a study preliminary to performance of a detailed investigation, to find the reaction order in all the components and propose a working hypothesis based on such evidence.

When searching for a rate equation that is in the best agreement with the observed change in concentration with time, it is possible to use several approaches; usually an equation is chosen based on a trial-and-error approach. A function of concentration or measured physical quantity is plotted as a function of time and the shape of this plot is compared with the theoretical shape based on equation used. Even in this approach the choice of the equation is not completely accidental; it is based on all preliminary evidence, obtained as described above.

If for the assumed reaction scheme more than one rate equation can be fitted, depending on the rate determining step, or if, as is more frequently the case, preliminary evidence can lead to several reaction schemes of varying complexity, it proves best to first choose the scheme corresponding to the simplest rate equation. If, in the attempt to verify the applicability of this simplest equation, the experimental data do not fit the $c = f(t)$ plot corresponding to this equation, a more complex equation is chosen and tried, and this is repeated until satisfactory agreement between the shape of the theoretical plot and experimental results is found.

Comparison of the experimental results with the shape of the plot predicted theoretically for the given rate equation can be carried out numerically, graphically, or by means of analog or digital computers. In numerical calculations the chosen equation is applied to a given set of experimental data and usually the value of the rate constant is calculated for the given set. The values of the rate constant obtained for various sets for the given kinetic run are compared next. When the value of the calculated rate constant is practically the same for all sets corresponding to one kinetic run, it is deduced that the chosen rate equation can be applied to the given kinetic run. If the calculated values show, in addition to statistical fluctuations, a pronounced trend, the individual sets are to be treated by means of other equations until an equation showing sufficient agreement with experimental data is found.

Graphical methods demonstrate presence of deviations and their nature in a faster and more pronounced way. To apply these methods the rate equation is transformed so that the graphical presentation of the functional relationship of the concentration and time parameters is a simple curve, preferably a straight line. Experimental points are plotted and the best fitted curve (line) is drawn. It is then decided whether the deviations are larger than those due to statistical fluctuations and in particular whether they show any characteristic trend. Systematic deviations are easily detected and their recognition can lead to finding a more complete and exact rate equation.

The use of computers for the evaluation of kinetic data can be based either on the calculation and comparison of data which combine the two above procedures— either the value of the rate constant for each set is calculated and the variations of the rate constant observed, or a linear relationship between the two experimental parameters is checked and the slope, correlation coefficients, and other statistical characteristics evaluated—or so-called modeling and curve fitting is applied. Whereas the first two approaches based on numerical calculations can be used in conjunction with digital computers, the modeling and curve fitting can be done either by digital or analog computers. In both approaches the computer gives a plot of the time-dependence of the concentration for a programmed set of variable parameters and chosen values of constants. This is done by a program for the digital computer, and by a combination of properly chosen circuits and preset values of constants for analog computers. The curves recorded by the computer for a given type of reaction scheme are compared with experimentally obtained plots. If the shape of the calculated dependence and experimental data is identical, it can be assumed that an adequate rate equation has been found.

If the experimental data do not fit the shape of the computer generated curve, another program for another reaction scheme is tried or the circuit in the analog computer modified.

In principle, all these types of treatment can be used also for finding the best value of the rate constant. However, between these two applications there is a principal difference: when the best equation for the reaction rate is sought, the shape of the concentration–time dependence is of primary importance. What is being established is whether the shape of this dependence as computed from the rate equation used is similar to that found experimentally. Attention must be paid to deviations of the experimental data, and based on this it is decided whether these deviations are random or systematic, or whether they occur at the beginning or towards the end of the reaction. It is frequently of some importance to decide at which degree of conversion the deviations begin to occur. For example, it is not surprising to observe deviations at values corresponding to 85% conversion or greater. In this region, inaccuracies of measurement play a considerably more important role than at lower conversions. Such deviations thus do not necessarily indicate that an inappropriate equation was chosen. Alternatively, if significant and systematic deviations are observed at 50% conversion, it is probable that the considered reaction scheme is either incomplete or unacceptable and that the rate equation used does not sufficiently describe the observed concentration–time dependence.

If, on the other hand, at a later stage the best value of the rate constant is sought, it is possible to restrict the treatment to values of concentration in a given time interval, but the treatment of the experimental data in this' region is based on statistical methods. Particular stress is laid on the accuracy of the concentration and time values. When modeling of $c = f(t)$ curves is used, curves calculated for several values of rate constants are recorded and compared with the experimental plot. It is then decided which value of the rate constant is in the best agreement with the experimental data.

Recently, attempts have been made to replace such trial-and-error methods, which are not frequently applied in other fields of chemistry, by a systematic approach that would directly indicate the best equation for the reaction rate. A general method of this type has not been developed yet, and proposed methods using "dimensionless parameters" can be applied only to a restricted variation of the type of the reaction scheme and to relatively simple systems. The majority of organic reactions follow simple rate equations, and thus such a treatment is useful and will be discussed in the following sections. Some of the proposed treatments offer the advantage that it is possible to simultaneously decide which rate equation best fits the experimental results and to determine the best value of the rate constant.

This general introduction to approaches in the treatment of experimental data was considered of importance because it indicates how the problems of the choice of the equation for reaction rate, of the determination of the total reaction order, of the reaction order in individual components, and of the calculation of the best value of the rate constant are mutually interrelated, even when it will be necessary to discuss them separately.

As mentioned in the general introduction, it is usually insufficient, for the elucidation of the mechanism, to find the equation for the reaction rate and to calculate the rate constant. Additional experimental approaches have to be used and can be classified as follows:

1. Kinetic data:
   a. Stoichiometry of the reaction and intermediates.
   b. Empirical reaction rate equation.
   c. Reaction order.
   d. Rate constant.
2. Effect of reaction mixture composition:
   a. Effects of pH.
   b. Effects of temperature.

   c.   Effects of solvent and ionic strength.

   d.   Effect of pressure and illumination.

3. Structural effects:

   a.   Changes in reagent.

   b.   Changes in reactant.

       i.   Scope of reaction.

      ii.   Effect of molecular frame.

    iii.   Substituent effects.

3.1.1. Terminology of Reaction Kinetics

Kinetic data can be the source of a great deal of detailed insight into the mechanism of a reaction. While other types of experimental evidence are usually sought, the study of the kinetics of a reaction usually forms the backbone of a thorough mechanistic investigation. One important test for any postulated mechanism is that it must account for the observed kinetic behavior of the reaction.

As mentioned in Chapter 2, reaction rate is usually defined as the rate of change in concentration of a substance participating in the reaction as a function of time (t). Depending on whether the concentration change of the product or starting material is measured, the change with time t is denoted as positive or negative, so that the rate will always be a positive quantity. For example, in the simple, one-step reaction (3.1-1)

$$a\text{A} + b\text{B} \longrightarrow c\text{C} + d\text{D} \qquad (3.1\text{-}1)$$

the rates (v_i) may be expressed as

$$v_\text{A} = -\frac{d[\text{A}]}{dt}, \quad v_\text{B} = -\frac{d[\text{B}]}{dt}, \quad v_\text{C} = \frac{d[\text{C}]}{dt}, \quad v_\text{D} = \frac{d[\text{D}]}{dt} \qquad (3.1\text{-}2)$$

where the concentrations [X] are usually expressed in molarities. The values of individual rates v_i will be equal only if the stoichiometric coefficients (a, b, c, d) are the same, that is if $a = b = c = d$. In describing the reaction rate, it is therefore necessary to specify the reaction component ($v_\text{A}, v_\text{B}, \ldots$) to which it applies. Alternatively, a general rate of reaction can be defined by dividing the rate with respect to an individual component by its corresponding stoichiometric coefficient:

$$\text{Rate} = -\frac{1}{a}\frac{d[\text{A}]}{dt} = -\frac{1}{b}\frac{d[\text{B}]}{dt} = \frac{1}{c}\frac{d[\text{C}]}{dt} = \frac{1}{d}\frac{d[\text{D}]}{dt} \qquad (3.1\text{-}3)$$

In the definition of reaction rate, concentration rather than amount is used; reaction rate is an intensive quantity, independent of the size of the reacting system. Reaction rate can also be expressed in general terms as dx/dt, if reaction variable

x is introduced in place of concentration of an individual component. This reaction variable is equal to the number of equivalents per liter which reacted between the initiation of the reaction and time t. At a given temperature, the reaction rate depends on concentrations of various components of the reaction mixture, and usually only on concentrations of the starting materials. In some systems the reaction rate is affected by the concentration of products. These are classified as cases involving autocatalysis or inhibition.

The experimentally found relationship between the rate and concentration of participating species is called the *rate law* or the *reaction rate equation*. For the above reaction the equation has the form (3.1-4)

$$\text{Rate} = k[A]^m[B]^n \qquad (3.1\text{-}4)$$

where k, the proportionality constant between the rate and product of concentrations, is called the *rate constant* for the reaction. Exponents m and n define the *order* of the dependence of the rate on concentration of each reactant. A reaction is said to be mth order in A and nth order in B. When the expression for the rate can be described as a simple product of concentrations to the power of n, m, . . . , as in the reaction rate equation (3.1-4), then the reaction is said to have an overall reaction order equal to the sum of exponents, $n + m$. For more complex reactions the use of the term "overall reaction order" is not recommended.

For special cases the order in a component can be numerically equal to the stoichiometric coefficient (i.e., $m = a$, $n = b$ in the above reaction), but often, particularly for more complex reactions, there is no simple relationship between these two values.

One of the essential aspects of kinetic investigations is that they can offer information about the *molecularity of a reaction*. Molecularity can be defined as the number of molecules that approach each other and form the activated complex in the collision which leads to reaction. When one molecule of reactant is involved, the reaction is termed unimolecular; when two molecules of reactants (identical or different) participate in the collision, the reaction is called bimolecular. For certain cases, empirically obtained reaction order may be related to the molecularity of the reaction. In the simplest case of a one-step reaction in which no intermediate is formed (often referred to as an elementary reaction), the molecularity of a reaction can be the same as the reaction order. The problem faced in analysis of kinetic data is to detect whether the reaction takes place as a single-step or multistep process. Neither the algebraic form of the experimentally obtained reaction rate equation, nor the value of the rate constant indicate whether the given reaction takes place in one step or several. Additional criteria, often involving effects of composition of reaction mixture, reaction conditions, and variations in structure of reactants, are necessary to demonstrate that reaction takes place in a single step.

For a multistep reaction, there is often no simple relationship between the experimentally found reaction order and molecularity of any particular step in the mechanism. But if it is possible to distinguish individual steps in a multistep reaction, it is possible to prove whether the molecularity of each step is the same as the reaction order for that step (or elementary reaction). For a multistep reaction it is therefore necessary—if more intimate knowledge of the mechanisms is the goal—to identify individual steps, their sequence, and the reaction order in each reactant for those steps. It should be strongly emphasized that the notion of molecularity is meaningful only in reference to a single-step, elementary process.

Thus the observation that a given reaction is first order in both reactants A and B may be considered as a support for a single-step, bimolecular mechanism. But as indicated above, the fitting of experimental data to a certain equation for reaction rate and the observation of an overall second-order reaction is not on its own a sufficient proof for a bimolecular mechanism. The reaction can follow another, multistep mechanism which may exhibit the same kinetics. Only where such alternative mechanisms are excluded (based on experimental evidence other than the form of rate equation and reaction order), is it possible to deduce molecularity from reaction order.

Even when the equations for reaction rate and reaction order do not allow a direct conclusion on the molecularity of the process studied, the knowledge of these two factors restricts the number of possible mechanisms. Namely, the proposed mechanism must account for the reaction order, whether the reaction takes place in one or several steps.

Care must be taken not to confuse the stoichiometry of a reaction with the order of a reaction and its molecularity. Only proper use of these terms can avoid confusion. The *stoichiometry* is the empirical, molar ratio of reactants and products as indicated in the equation describing the overall (or net) reaction. The *reaction order* is also experimentally derived, based on kinetic measurements. As mentioned before regarding molecularity and reaction order, there is no simple relationship between reaction order and stoichiometry. A reaction may be zeroth order with respect to a reactant that appears in the net stoichiometric equation, or the reaction rate may be dependent on the concentration of a catalyst, which does not appear at all in the overall stoichiometric equation.

Contrary to the two terms discussed above, the molecularity is not an empirical factor based on experimental facts, but rather a theoretical concept. Experiments cannot be performed in an attempt to deduce the molecularity, and our conclusions about the molecularity of a reaction must be regarded as hypotheses. The knowledge of molecularities of individual steps represents a fundamental contribution to a mechanistic hypothesis.

Since it is impossible to obtain direct information about the molecularity, and since it is useful for the study of reaction rate, the first stage of a kinetic investigation should be an unequivocal establishment of the stoichiometry of the reaction to be studied. In essence, this is the procedure of finding out what is to be studied before the actual systematic investigation is commenced. At this stage, the nature of the substances that participate in the reaction, and the form in which they react, need to be identified. It is also necessary to identify all products featured and to determine their yields. Although this appears trivial, it is surprising how often the securing of such necessary information is ignored. While it is possible to measure the rate at which a particular reactant is consumed without knowing the nature of the products, the interpretation of the kinetic data with respect to the mechanism of the reaction studied requires the knowledge of the products and the stoichiometric ratios.

In addition to acquiring knowledge about the reactants and products, an attempt should be made to establish whether intermediate(s) are formed in the reaction. If this is the case, their identification represents a major contribution to the understanding of the reaction mechanisms.

As mentioned above, most organic reactions proceed through a mechanism consisting of several steps and follow complex reaction paths with one or more intermediates. For instance the formation of a molecule C may proceed through an intermediate B. In the sequence A $\longrightarrow$ B $\longrightarrow$ C, the intermediate B is essentially a product of one process and simultaneously a reactant for another.

The amount of information that can be obtained about the intermediate depends upon the relative rates of the formation and the disappearance of the intermediate. In some cases the stability of the intermediate is such that it can be examined by spectroscopic means or even isolated. In other cases the life-time of the intermediate is so short that its existence and nature can only be inferred from kinetic data and other indirect sources of information. The identification of intermediates is of particular importance because they can differ little in energy from the transition state and hence can resemble it closely in structure. This is called the *Hammond postulate*. Since the structure of intermediates is often accessible to elucidation, it can be useful in assigning the structure of transition states, which is one of the goals of mechanistic studies.

For any system at equilibrium, and for any elementary reaction whether at equilibrium or not, the favored reaction path in one direction must be the reverse of that path in the opposite direction. This is called the *principle of microscopic reversibility*. Based on this principle it is possible to study mechanisms for reactions, where the unfavorable position of equilibria prevents investigation of the forward reaction, by studying only the reverse reaction. The danger involved in this approach is that the reaction may involve a larger number of equilibria than assumed. Application of this principle to reactions occurring at steady state conditions (see Sections 3.3.2.3c and 3.3.2.5d) is more restricted [3-1].

Although advances have recently been made in theoretical procedures for the description of reaction paths, in their present state they have not yet evolved into a useful tool for the elucidation of mechanisms. In a typical procedure of this kind, molecular orbital calculations of energies are carried out for the reactants, products, and a number of possible intermediates which might occur along the reaction pathway. The minimum energy pathway from reactants to products is then found. Calculated energy surfaces depend not only on assumptions about geometry and solvation of the transition state, but also on the molecular orbital method used and simplifying assumptions made. Such calculated reaction pathways may contribute some insight into the possible structures of transition states, but cannot be used as evidence for a reaction mechanism. Any conclusions on which a choice of a reaction mechanism is based, must be derived from experimental data.

Such experimental data enable detection and identification of reaction products and intermediates as well as establishment (as clearly as possible) of the course of reaction in the preliminary investigation. This is followed by a quantitative treatment of kinetic data and their interpretation. In some cases, variations of measured rate constants with structure allow conclusions to be made about the composition of the transition state. These questions are dealt with in subsequent sections of this chapter.

3.2. REACTANTS, PRODUCTS, AND INTERMEDIATES

As mentioned above, it is essential in the study of reaction mechanisms to first establish which substances enter the reaction, and to identify substances formed as products or intermediates in the course of reaction. The stoichiometric ratios of reactants and products must also be found.

3.2.1. Reactants

Frequently, it is not particularly difficult to decide which substances enter the reaction since usually the components from which the reaction mixture is prepared are known. Nevertheless, sometimes components added to the reaction mixture undergo changes before they enter the studied reaction. Alternatively, solvent or products of its autoprotolysis participate in the reaction. In such cases it is necessary to obtain information about the species participating in the reaction in other ways, and often it is identification of the products which aids in the identification of the reactants. Thus, for example, if the products of ester hydrolysis were found to be alcohol and acetate ion, it can be deduced that a hydroxide ion is consumed in the process (3.2-1):

$$CH_3\overset{\|}{\underset{O}{C}}-OC_2H_5 + OH^- \longrightarrow CH_3\overset{\|}{\underset{O}{C}}-O^- + C_2H_5OH \qquad (3.2\text{-}1)$$

Complications can also occur in those cases of direct participation of the buffer components, salt added for control of the ionic strength, or a catalyst in the reaction.

When the consumed component is retained in the product, its involvement can usually be deduced from the nature of the products, as mentioned above. However, if the medium component participates directly in one step of a reaction to form an intermediate and then the component is regenerated in a subsequent step, the participation of that component may not be evident from the composition of the reaction products. In such cases, the participation of the medium component can be established only by the study of reaction kinetics; individual examples will be discussed in the following chapters. Here only one common principle will be mentioned: In case of suspicion of such effects it is recommended to replace the suspected component—for example, the solvent is replaced by another solvent of similar dielectric constant and hydrogen bonding ability, or the buffer is replaced by a buffer of different composition of the same pH, or the neutral salt by another neutral salt differing in cation or anion but keeping the ionic strength constant. By comparing the reaction rate under the original and modified conditions, it is frequently possible to decide whether the given component indeed participates in the reaction.

Another factor which can significantly affect the reaction rate is the presence of small amounts of impurities in the solvent or in other components of the reaction mixture. The presence of these substances, which frequently remains unknown to the experimenter, can affect the study of reaction kinetics mainly in two ways: either the impurity exerts a catalytic influence or it reacts in a fast step with one of the components of the reaction mixture and changes its concentration. The occurrence of such impurities in organic solvents is relatively common—thus alcohols frequently contain reactive aldehydes and ketones, ethers contain peroxides, amines are present in amides, and so on. Purification recommended for spectroscopic purposes may be useful (as mentioned in Section 2.2). Furthermore, the presence of dissolved oxygen, found in all solutions which are in contact with air, and of carbon dioxide can cause complications.

In some instances, the determination of the product composition makes it possible to decide which components of the reaction mixture participate unexpectedly in the reaction, as well as those compounds that were added to the reaction mixture knowingly.

For example, a Grignard reagent can react with oxygen to give either hydroperoxides or alcohols:

$$RMgX + O_2 \longrightarrow R{-}O{-}O{-}MgX \longrightarrow R{-}O{-}O{-}H \quad (3.2\text{-}2)$$
$$\searrow R{-}OH$$

The unexpected isolation of either of these products from a reaction mixture using a Grignard reagent for another purpose would clearly indicate that the oxygen was present as an impurity, and was involved in the consumption of the Grignard reagent.

Another difficulty can arise with reactions involving stereoisomers, particularly when the presence and/or content of stereoisomers is not recognized. Since stereoisomers may differ in the rate of reaction with the same reagent, in solutions where a reactant is present in two or more stereoisomeric forms, the resulting rate may be a combination of the rates of individual isomers.

Determination of the stoichiometric ratio of reacting substances and products differs according to whether the reactions involved are reversible or irreversible. For irreversible reactions it is usually sufficient to vary the ratio of reacting species and to determine for all participating reactants the ratio in which they disappear from the reaction mixture. In practice, for a reaction in which A, B, and C participate, the concentration of B and C may be kept constant and the concentration of A varied. The reaction is carried out at several concentrations of A, and at each concentration of the component A concentrations of components B and C are determined after the reaction is completed. From this procedure, it is found at which concentration of substance A all of substance B is transformed by reaction, and at which concentration of A all of substance C reacts. Stoichiometric coefficients can then be determined from the known molarity ratios.

For reversible reactions, the stoichiometric coefficients can be found if the concentration of a reactant, [B], and the concentration of a product, [P], are measured at equilibrium when the concentration of a second reactant [X] is raised. Let us consider two possible reactions (3.2-3) and (3.2-4) which differ in stoichiometric ratios of the reactants.

$$B + X \rightleftharpoons P \tag{3.2-3}$$

$$B + 2X \rightleftharpoons P \tag{3.2-4}$$

The general form of the equation for the equilibrium constant will be

$$K = \frac{[P]}{[B] [X]^n} \tag{3.2-5}$$

where n is the stoichiometric coefficient of X in the reaction. Taking logarithms, reaction (3.2-5) can be written as

$$\log K = \log \frac{[P]}{[B]} - n \log [X] \tag{3.2-6}$$

or

$$\log \frac{[B]}{[P]} = -n \log [X] - \log K \tag{3.2-7}$$

Plotting the logarithm of the fraction of conversion [B]/[P] against the negative logarithm of the concentration of the second reactant [X], the stoichiometric ratio of the reactants can be deduced (Fig. 3-1). A unit slope indicates a 1:1 ratio and a slope of 2 means $X:B = 2$. Stoichiometry can also be deduced—even for complex reactions—from equations for reaction rate [3-2].

Even though the nature of the substances added to a reaction mixture and subsequently converted to products may be known, the actual reactive form of the substance may be different. One or more of the compounds undergoing reaction may exist in solution in several forms which can interchange one into another, usually in equilibrium reactions. Individual forms frequently react with different speeds, and some of them may not react at all, so that only one or two of several possible forms take part in the reaction. If such equilibria are established either rapidly or very slowly when compared with the reaction rate, they can frequently be studied separately from the reaction proper. If the rate of the establishment of the equilibrium is comparable to the rate of the reaction followed, the equilibrium and the rate-determining step become one reaction step in the overall reaction which

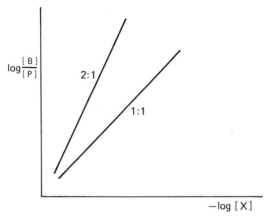

FIG. 3-1. Plot of the logarithm of the fraction of conversion [B]/[P] against the negative logarithm of the concentration of the reactant X present in excess. Plots for formation of XB and X_2B.

must be solved by kinetic methods. Sometimes the reaction proper is preceded by a rapidly established equilibrium which is shifted in such a way that the equilibrium concentration of the reacting species is very low and sometimes cannot be detected even by sensitive analytical methods. In such systems, similarly, it is necessary to elucidate the reaction scheme by kinetic analysis which enables us to detect and identify the unstable intermediates.

Among types of reactions that involve such equilibria belong tautomeric changes (in particular the keto-enol tautomeric change and similar reactions), ring formation and opening, hydration–dehydration equilibria, and, in particular, acid–base reactions. The techniques used in identification of the reactive form shall be discussed in more detail in Section 3.6.1 dealing with acid–base catalysis. The methodology depicted there can be applied to other types of equilibria.

3.2.2. Products

When products are detected and identified, an attempt should be made to establish not only the presence of those substances whose formation in the reaction was anticipated, but also to investigate whether some further compounds have been formed in the reaction which were not anticipated. Concentrated efforts should be made to detect *all* the reactants and all the products formed.

The detection and identification of products can be carried out either after isolation or directly in the reaction mixture. In the first instance, the identification is carried out by the usual methods of organic analysis, such as determinations of melting or boiling point of the analyzed substance or its derivative, molar refractivity, optical rotation, applications of chemical methods for identification of individual functional groups, and recordings of UV-visible, IR, NMR, and mass spectra. Sometimes other techniques are useful, such as Raman spectroscopy, determination of *pK*, or polarography.

Chromatographic methods of separation and identification have proven useful, especially the high pressure liquid, the gas–liquid, paper, and thin-layer variants. Chromatography often enables detection of the presence of minor reaction products.

For identification in the reaction mixture without isolation, the spectroscopic methods are of greatest value, in particular UV-visible and NMR spectroscopy, but polarography, determination of pK-values, and other methods have also been used.

The nature of products anticipated in the course of a reaction is usually deduced and proposed on the basis of analogy with the mechanisms of known reactions. This approach is successful as long as the reactions compared are sufficiently similar. Examples are nevertheless known where this approach is completely unsuccessful. Thus it would be possible to assume that reaction of furfurylchloride with cyanide ions should follow a reaction path similar to that observed for the analogous reaction of benzylchloride. The formation of a different type of reaction product in these two reactions indicates that the reactions (3.2-8) and (3.2-9) compared were not sufficiently similar:

$$\text{(furfuryl chloride)} + CN^- \longrightarrow \text{(product)} + Cl^- \qquad (3.2\text{-}8)$$

$$\text{(benzyl chloride, } CH_2Cl) + CN^- \longrightarrow \text{(product, } CH_2CN) + Cl \qquad (3.2\text{-}9)$$

The formation of different products clearly indicates differing mechanisms. Extrapolation of known facts based on analogy is a powerful research tool in many cases, but the possible limitation of this approach should always be kept in mind. There is no substitute for thorough identification of the structures of products.

Sometimes the choice of a reactant is essential for the successful application of the identification of the product. For example, evidence for the formation of aniline in the reaction of chlorobenzene with potassium amide does not make it possible to distinguish whether the reaction is a direct nucleophilic substitution of chloride by the ion NH_2^- or a reaction in which the amide ion becomes attached to another position on the benzene ring:

$$\text{(chlorobenzene, } Cl) + KNH_2 \longrightarrow \text{(aniline, } NH_2) + KCl \qquad (3.2\text{-}10)$$

When the reaction is carried out with C-14 labeled chlorobenzene, the amino group is found to be attached to the labeled carbon in only half of the aniline molecules, with the amino group attached at an adjacent carbon in the rest of the product.

This result provides strong evidence for the elimination–addition mechanism of aromatic nucleophilic substitution, involving the symmetrical benzyne molecule as an intermediate.

Following the identification of the product, its stoichiometric ratio to reactants is sought. For reversible reactions, the ratio of reactant and product can be determined in the presence of excess of the reagent in the same way as has been described for the stoichiometric ratio of reactants. For irreversible reactions, the amount of product obtained at different ratios of the substrate and reagent is determined using a suitable quantitative method. It must be considered whether the product can exist in several forms in an equilibrium.

3.2.3. Intermediates

Identification and detection of intermediates represents a more complex question than identification of reactants and products, for which no general procedure can be proposed. It is here that the general chemical knowledge of the experimenter finds greatest application.

As mentioned before, the majority of organic reactions follows complex reaction paths rather than occurring in one elementary step. The typical reaction involves one or more intermediates which correspond to valleys on the energy diagram. Formation of each intermediate must be preceded and followed in the reaction sequence by a formation of a transition state, so that in a system with five intermediates, for example, six transition states must be considered.

Several circumstances may lead to the supposition that an intermediate occurs in a reaction sequence. The existence of an intermediate may be suspected by way of analogy with reactions of a known mechanism. An intermediate may be indicated if a variation in a physical property of the reaction mixture is not accounted for by either reactants or products. The form of the reaction rate equation or the effect of composition of reaction mixture on the rate constant may imply the existence of an intermediate. The occurrence of an intermediate may be deduced from the nature of the product. Regardless of the reason for suspecting an intermediate, it is important to attempt to prove or disprove the formation of an intermediate and, if possible, to identify its structure.

The importance of detecting and identifying an intermediate is twofold. First, the knowledge or presumption of the existence of an intermediate aids in the choice of a proper rate equation. Further experiments can also then be designed to learn more about the intermediate. Second, the identification of intermediates is one of the basic goals of a mechanistic investigation, because the structure of an intermediate provides information about the reaction step leading to its formation. In many situations, an intermediate can serve as a model for the structure of the transition state for the formation or decomposition of the intermediate (cf. the Hammond postulate, Section 3.1.1), which is important because the transition state itself is experimentally inaccessible. Structurally, an intermediate may be an ordinary organic molecule which is reactive under the particular reaction conditions. On the other hand, many intermediates are reactive by virtue of their electronic structure and are rarely observed as stable species under ordinary conditions. To this group belong carbocations, carbanions, free radicals and radical ions, carbenes and nitrenes:

$$
\underset{R^2}{\overset{R^1}{\diagdown}}\overset{+}{C}\!-\!R^3 \qquad
R^3\!\!\underset{R^3}{\overset{R^1}{\diagdown}}\!\overset{-}{C}\!: \qquad
\underset{R^3}{\overset{R^1}{\underset{|}{\overset{|}{R^2\!-\!C\cdot}}}} \qquad
\underset{R^2}{\overset{R^1}{\underset{|}{\overset{|}{C\!=\!\ddot{O}}}}} \qquad
\underset{R^2}{\overset{R^1}{\underset{|}{\overset{|}{C:}}}} \quad
N:
$$

Each of these species was considered speculative at one time, proposed only to explain otherwise uninterpretable mechanisms. Presently, all of these types of species have been detected by physical methods (e.g., carbocations and carbanions by NMR, carbanions and radical ions by UV spectra, radicals by ESR and electrochemically, etc.). In some cases they can even be isolated in solid form, such as

| 1,1-Diphenylpicryl hydrazyl-radical | Tropylium tetrafluoroborate | Ethyl sodium |

This is very consoling, but great caution is still recommended when an unusual structure is attributed to an intermediate.

In addition to classification by chemical structure, intermediates may be classified according to the relation between the rate of their formation and cleavage. In particular, there is a difference in the amount of information about individual intermediates that can be derived from kinetic studies depending on the relative rates of intermediate formation and disappearance. Consider the system below in which B is an intermediate.

$$
A \underset{k_{-1}}{\overset{k_1}{\rightleftharpoons}} B \overset{k_2}{\longrightarrow} C \tag{3.2-12}
$$

Three alternative sets of conditions are possible:

1. k_{-1} *and* k_2 *are comparable.*   In this case the overall rate is affected by the rate of formation of the intermediate (k_1) and the rate of its disappearance (k_{-1} + k_2). The dependence of free energy on the reaction coordinate has the shape shown in Fig. 3-2. The intermediate in this scheme is not in equilibrium with the reactants and is known as a van't Hoff intermediate. Examples involving such intermediates are unimolecular solvolyses of R-X, hydrolysis of carboxylic acid derivatives, and some aromatic substitutions.

2. $k_2 \gg k_1$.   In this case the disappearance of the intermediate B is rapid and its formation with the rate constant k_1 becomes the slowest step. The corresponding energy diagram has the shape shown in Fig. 3-3. Here as well, the intermediate is not in equilibrium with the reactants. Situations of this type are found in the halogenation of acetone or the aldol condensation of acetaldehyde. For systems belonging to this group, very limited information is directly available about the properties of the intermediate, even when it is possible to offer sufficient evidence that an intermediate is formed. The slow step is called the rate-determining step, and the kinetics will provide information only about this step and any steps that might precede it.

3. $k_{-1} \gg k_2$.   In this last case, the cleavage or transformation of the intermediate B is slow, and this process is anteced by a more rapidly established equilibrium A $\rightleftarrows$ B. The reaction rate depends on the product of the equilibrium constant $K(K = k_1/k_{-1})$ and the rate constant k_2. The shape of the energy diagram (Fig. 3-4) shows an intermediate in equilibrium with reactants, which is called an Arrhenius intermediate. Common examples are reactions involving antecedent protonation or dissociation.

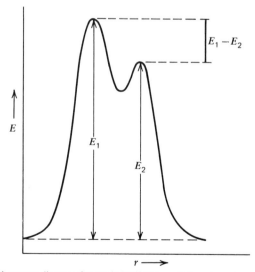

FIG. 3-2. Potential energy diagram for van't Hoff intermediate, where k_1 and k_{-1} in (3.2-12) are comparable and the intermediate is not in equilibrium with reactants.

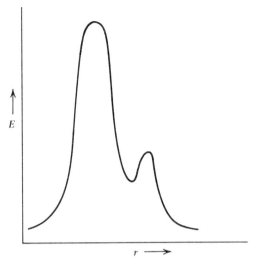

FIG. 3-3. Potential energy diagram corresponding to the formation of an intermediate which is not in equilibrium with reactants and where $k_2 \gg k_1$ [in equation (3.2-12)].

Intermediates can also be classified according to their stability. By definition, an intermediate is a reactive compound that will have a limited lifetime under the reaction conditions, but a rough classification can be made of intermediates as being stable, of limited stability, or unstable. The limits between these groups are not strictly defined and depend on the analytical method used and time-scale of measurements relative to the rate of reaction. The stability will be related to the depth

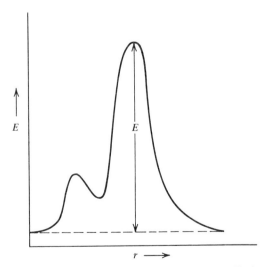

FIG. 3-4. Potential energy diagram for an Arrhenius intermediate in equilibrium with reactant, when in equation (3.2-12) $k_{-1} \gg k_2$.

of the minimum on the energy diagram (Fig. 3-2 to 3-4). If stable intermediates are formed which can be isolated from the reaction mixture, the depth of this minimum relative to the transition states corresponds to an energy difference of about 15 kcal. If the intermediate is stable enough to be detected in the reaction mixture by physical methods, but its stability is insufficient for isolation, the depth of this minimum is smaller but still corresponds to several kilocalories. Such intermediates will be denoted as intermediates of limited stability. If an attempt to directly detect the intermediate even by some rapid methods (cf. Chapter 4) is unsuccessful, the minimum is still less deep. However, such unstable intermediates may be detected in a kinetic investigation of the reaction, in particular by fitting the data to a reaction rate equation corresponding to a reaction sequence involving intermediates.

Finally, some reactions may exhibit "borderline" behavior with kinetic properties characteristic for both a reaction with an intermediate (and two transition states) and a reaction with only one transition state. This is the situation that is encountered in some nucleophilic substitution reactions which correspond neither to a unimolecular nor to a bimolecular reaction. The kinetic behavior is interpreted as either due to a varying contribution of a unimolecular and bimolecular process or due to the possibility that the reaction follows one of a number of reaction paths (indicated by a continuous series of reaction profiles, Fig. 3-5), ranging from a clearly bimolecular reaction path without an intermediate to a unimolecular reaction involving formation of a relatively stable intermediate (a carbocation in above mentioned nucleophilic substitutions).

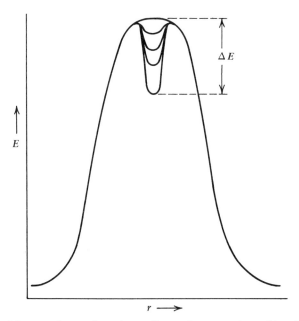

FIG. 3-5. Potential energy diagram for systems with a continuous spectrum of transition states. The value of ΔE indicates the stability of the intermediate.

The experimental approach to the identification of intermediates depends on their stability. Both chemical and physical methods can be used for a stable intermediate that can be isolated, but rapid physical methods must be used for intermediates of limited stability, and only kinetic evidence can be used for unstable intermediates. The nature of kinetic evidence for intermediates will be discussed in detail in later sections. For all types of intermediates, indirect information may be obtained from a study of the products, but it is important to gather additional, more direct information about intermediates whenever possible.

The various spectroscopic methods are the most useful methods for *identification* of intermediates in solution, although other physical measurements may aid in the *detection* of intermediates.

Electronic spectra are often most powerful for identification of intermediates of limited stability, because of the sensitivity of spectrophotometry in the UV and visible region. Intermediates are usually present in the solution only in low concentrations, but the high molar absorptivities frequently encountered in electronic spectra, especially for conjugated systems, offer sufficient sensitivity. Low molar absorptivities of IR spectra as well as low sensitivity of NMR usually limit the use of these two techniques in intermediate identification. The application of mass spectrometry necessitates separation of the studied substance from the reaction mixture and therefore is generally unsuitable for study of intermediates with limited stability. Similarly, chromatographic methods represent alternative techniques involving separations, which are only successful if the reaction can be stopped and the intermediates trapped.

Electroanalytical methods have been successfully used for detection and identification of some intermediates. Formation of free radicals and radical ions was thus first detected by potentiometry. Polarography and magnetic susceptibility measurement enable detection of radicals and radical ions of relatively short lifetimes. The most powerful technique is ESR, which makes it not only possible to follow radicals even with a very short lifetime and at a low concentration, but from the shape of recorded curves to identify the nature of the radical. A recently discovered technique demonstrating the existence of radical intermediates is CIDNP (Chemically Induced Dynamic Nuclear Polarization), in which enhanced absorptions or emissions are seen in NMR spectra of the products when the reaction involving radicals is run in the magnetic field of the NMR spectrometer.

All the standard methods of structural analysis that are used in the identification of products may be applied in the occasional situation where a stable intermediate is easily isolated. However, even with so-called stable intermediates the basic problem of their reactivity persists. Therefore, techniques are often used to increase the lifetime of intermediates so that isolation is possible or physical detection is easier. These techniques basically rely on reaction quenching—cooling, decrease of concentration by dilution (for reactions of higher order), change in pH, extraction in another solvent, or removal of those solution components which cause the instability of intermediates. For example, in the previously mentioned reaction of aliphatic Grignard reagents with oxygen, alcohols are the type of product found at room temperature. However, if the aliphatic Grignard reagent is slowly added to solvents saturated with oxygen at $-70°C$, it is possible to isolate alkyl hydroperoxide in good yield. Further, the magnesium salt of t-butylhydroperoxide reacts with an equivalent amount of t-butylmagnesium chloride to form, after hydrolysis, t-butyl alcohol. This gives reliable evidence for the formation of hydroperoxide in the course of oxidation of the Grignard reagent.

The situation is particularly advantageous in those instances where an intermediate can be prepared synthetically by an independent route. In such cases it is possible to compare chemical, physical, and kinetic properties of the intermediate isolated from the reaction mixture (or even in situ—in the reaction mixture without isolation) with those of the synthetic standard.

If an analytical method is available that enables following of the time change of the intermediate concentration, the treatment is relatively simple: the reaction rate is compared in two reaction mixtures in the same medium—one containing the original starting material, the other the synthesized intermediate. Products of both reactions must be identical. The rate of the reaction using the original starting material can be either slower than that of the intermediate (if formation of intermediate is the slowest step) or identical (when the cleavage of the intermediate is the overall rate determining step).

For example, for the alkaline degradation of chalcone, two mechanisms were considered, the first (3.2-13 to 3.2-17) involving two proton transfers:

$$C_6H_5COCH{=}CHC_6H_5 + OH^- \underset{k_{-1}}{\overset{k_1}{\rightleftharpoons}} \left[\underset{\qquad\overset{|}{OH}}{C_6H_5COCHCHC_6H_5} \right]^- \qquad (3.2\text{-}13)$$

$$\left[\underset{\qquad\overset{|}{OH}}{C_6H_5COCHCHC_6H_5} \right]^- + H_2O \overset{k_2}{\rightleftharpoons}$$

$$C_6H_5COCH_2\underset{\overset{|}{OH}}{C}HC_6H_5 + OH^- \qquad (3.2\text{-}14)$$

$$C_6H_5COCH_2\underset{\overset{|}{OH}}{C}HC_6H_5 + OH^- \overset{k_3}{\rightleftharpoons} C_6H_5COCH_2\underset{\overset{|}{O^-}}{C}HC_6H_5 + H_2O \qquad (3.2\text{-}15)$$

$$C_6H_5COCH_2\underset{\overset{|}{O^-}}{C}HC_6H_5 \underset{k_{-4}}{\overset{k_4}{\rightleftharpoons}} C_6H_5COCH_2^- + \underset{\overset{||}{O}}{C}HC_6H_5 \qquad (3.2\text{-}16)$$

$$C_6H_5COCH_2^- + H_2O \overset{k_5}{\rightleftharpoons} C_6H_5COCH_3 + OH^- \qquad (3.2\text{-}17)$$

the second involving a cyclic intermediate formed in the course of proton transfer (3.2-18) (replacing reactions (3.2-14) and (3.2-15) of the above scheme).

$$\left[\underset{\overset{:}{H\cdots O}}{C_6H_5COCHCHC_6H_5} \right]^- \rightleftharpoons C_6H_5COCH_2\underset{\overset{|}{O^-}}{C}HC_6H_5 \qquad (3.2\text{-}18)$$

where the anion formed undergoes retroaldolization fission as above. To decide between these two alternatives, kinetics of the reaction in alkaline medium was studied both for chalcone ($C_6H_5COCH{=}CHC_6H_5$) and ketol ($C_6H_5COCH_2CH(OH)C_6H_5$). Since not only the values of k_{-1} and k_4 obtained in the two kinetic runs were identical in both compared reactions, but even the pH-dependences of the two rate constants obtained from reaction of chalcone and ketol were the same, it was possible to deduce that the cyclic mechanism could be excluded [3-8].

When only the concentration change of the reactant can be measured and an assumed intermediate can be prepared synthetically, two possibilities exist: The rate of disappearance of the intermediate is the same or faster than that of the reactant. In such a case the prepared compound can be intermediate in the studied reaction. Alternatively when the rate of the conversion of the intermediate under given conditions is slower than the reaction of the reactant, the compound prepared is not likely to be an intermediate of the studied reaction.

Another general approach to the detection and identification of intermediates is *trapping* of the transient species by a chemical reaction. A reagent is added which does not react with the reactant, but reacts with the intermediate and converts it to a new product in a faster reaction than the conversion of the intermediate in the reaction in absence of the trapping agent. The structure of the new product provides clues about the structure of the intermediate. The use of a trapping reagent is most informative when the intermediate is trapped after the rate-determining step, so that the overall rate of reaction is not altered. For instance, the existence of a carbocationic intermediate in a solvolysis reaction of an alkyl methanesulfonate is suggested when the highly nucleophilic azide ion is added and no change in the rate of solvolysis is observed other than that due to a normal salt effect.

$$R{-}OSO_2CH_3 \xrightarrow{\text{slow}} R^+{-}OSO_2CH_3 \underset{N_3^-}{\overset{H_2O}{\underset{-H^+}{\rightleftarrows}}} \begin{matrix} ROH \\[4pt] RN_3 \end{matrix} \qquad (3.2\text{-}19)$$

The azide ion traps some of the cation, diverting it to alkyl azide product instead of alcohol. Since no change in the rate of loss of the sulfonate occurs in the presence of azide, the azide must be reacting with an intermediate formed after the rate-determining step, rather than directly with the sulfonate.

Another type of trapping reagent is the free radical *scavenger,* which is useful even though the kinetics of the reaction are usually altered. Scavengers are substances such as molecular oxygen or benzoquinone that react very rapidly with many free radical and radical ion intermediates. Since radical reactions are often chain reactions, the presence of scavengers (also called inhibitors) can entirely suppress a reaction involving radical intermediates, or at least decrease the rate of reaction or lengthen the induction period.

3.2.4. Products as Criteria for the Presence and Identification of Intermediates

Products often provide clues to the existence and structure of intermediates. Trapping experiments are one example, wherein the addition of a new reagent results in a new product with no change in the overall rate. In other types of observation, that may suggest the existence of an intermediate, the formation of two or more products from one reactant, the formation of an identical product or product mixture from two different reactants, and changes in stereochemistry during the reaction are found. It must be emphasized again that a product study is not sufficient to establish a mechanism or to provide evidence for an intermediate, but it may provide first indication of or supporting evidence for the existence of an intermediate as one possible explanation of experimental data.

The observation of multiple products from a single reactant is an example of evidence which may suggest formation of an intermediate but is not sufficient to prove it. Side reactions proceeding directly from a reactant without a common intermediate will also lead to multiple products. However, the formation of multiple products from one reactant may occur because of competing reactions involving a single intermediate. For instance, carbocation intermediates frequently give more than one product under certain reaction conditions.

$$R^+ \quad
\begin{array}{ll}
\xrightarrow{\text{nucleophile}} & \text{substitution or addition} \\
\xrightarrow{\text{base}} & \text{elimination} \\
\xrightarrow{} & \text{rearrangement} \\
\xrightarrow{\text{olefin}} & \text{polymerization}
\end{array}
\qquad (3.2\text{-}20)$$

The cation may combine with nucleophiles to give products of substitution or addition (depending on whether the cation was generated from a saturated or unsaturated molecule); a proton may be abstracted by a base to give an olefin; the cation may rearrange and then undergo further reaction; or, it may combine with an olefin in the first step of a polymerization reaction.

The formation of an identical product or product mixture from two or more different reactants can indicate the possibility of formation of a common intermediate. Thus, for example, the product distribution in the solvolysis of t-butyl halides, sulfonates, ammonium ions, and sulfonium ions is about the same.

$$
\begin{array}{c}
(CH_3)_3C\!-\!Br \xrightarrow[20\% \ H_2O]{80\% \ C_6H_5OH} \searrow \\[2em]
(CH_3)_3C\!-\!\overset{+}{S}\!\!\begin{array}{c}CH_3\\ \\ CH_3\end{array} \xrightarrow[20\% \ H_2O]{80\% \ C_6H_5OH} \nearrow
\end{array}
\quad [(CH_3)_3C^+]
\begin{array}{c}
\nearrow \ CH_2=C(CH_3)_2 \ \ 36\% \\[1em]
\quad\ (CH_3)_3COH \\
\quad\ + \\
\searrow \ (CH_3)_3COC_2H_5 \ \ 64\%
\end{array}
\qquad (3.2\text{-}21)
$$

The formation of the product mixture occurs from a common intermediate, where the nature of the leaving group is not important. Similarly, if two stereoisomers react to give selectively one stereoisomeric product where more than one stereoisomeric product is possible, then a common intermediate must be involved somewhere along the reaction path.

Stereochemistry is one of the most important probes when trying to relate the product to a mechanism and attempting detection of intermediates. The formation of an optically active product from an optically active reactant is evidence that a symmetrical intermediate or transition state is *not* formed in a reaction path, whereas racemization shows that a symmetrical intermediate or transition state *must* be involved. Determining product stereochemistry is essential in establishing a mechanism for any reaction of olefins or reactions forming olefins.

3.3. EMPIRICAL REACTION RATE EQUATION

In this chapter the most commonly encountered reaction rate equations (RRE) and discussion of methods of their verification are surveyed. For this purpose the equations will first be divided into those corresponding to simple and complex reactions. Simple reactions are classified as those where the reaction rate is given by $dx/dt = k[A]^a[B]^b[C]^c$. All other types of reactions will be considered among complex reactions.

3.3.1. Simple Reactions

The rate of a simple reaction which is described in a given reaction medium and temperature by equation (3.3-1):

$$aA + bB + cC + \longrightarrow pP + rR + \cdots \qquad (3.3\text{-}1)$$

and can be expressed by means of a simple differential equation

$$\frac{dx}{dt} = k([A]_0 - x)^a \left([B]_0 - \frac{b}{a}x\right)^b \left([C]_0 - \frac{c}{a}x\right)^c \cdots \qquad (3.3\text{-}2)$$

where a, b, c is the order of reaction in A, B, and C, coefficients b/a, c/a, . . . follow from the stochiometric equation, x is the decrease in concentration of A in the course of reaction, $[A]_0$, $[B]_0$, $[C]_0$, . . . are initial concentrations of A, B, C, . . . and k is the rate constant.

Integrated forms of the most common examples of such equations are summarized in Table 3-1, where expressions for half-times of individual reactions are also given. They show that the values of the half-times are a function of the initial concentrations of reactants with the exception of first-order reactions where the half-time is independent of initial concentrations.

To decide which of the RRE's best fits the experimentally found time-dependence of concentration of the reactant, and hence which reaction scheme is involved, a graphical treatment has proven to be most useful. The general approach is to plot the expression on the left-hand side of the equation for the definite integral given in Table 3-1 against time, for example, $\log[A]_0/([A]_0 - x)$ as a function of time.

Since the graphical treatment of experimental data proved in practice very useful in distinguishing whether experimental data obtained best fit a given RRE, this sort of treatment will be briefly discussed. In practice a function N of the reaction variable x is sought that is a linear function of time. The choice of the type of function N is based on our preliminary knowledge of the reaction. A good linear plot of $N = f(t)$ for experimental data is considered a proof of the chosen RRE corresponding to N. From the slope M/P (Fig. 3-6) of the linear plot of N versus t it is then usually possible to evaluate the rate constant k. The advantage of this kind of treatment is its speed and the possibility to decide whether observed deviations from the linear plot are random or systematic and indicating an unsatisfactory choice of the RRE.

For a simple first-order reaction (3.3-3):

$$A \xrightarrow{\ k\ } P \tag{3.3-3}$$

it is possible to plot, as a function of time (Fig. 3-6), the expression $N = \log[A]_0/([A]_0 - x)$ and obtain from the slope of the linear plot $M/P = k/2.3$. The treatment can be simplified, as $N = \log[A]_0 - \log([A]_0 - x)$, where $\log[A]_0$ is constant for a given composition of the reaction mixture. Hence it is possible to plot only $N' = \log([A]_0 - x)$ as a function of time (Fig. 3-7). The evaluation of the rate constant $M/P = k/2.3$ remains the same as above.

If the reaction (3.3-3) is followed under conditions where the initial concentration of A remains practically unchanged, in the course of the reaction, then the expression plotted against the time (Fig. 3-6) becomes $N = x$. The slope of the linear N–x plot, M/P, is then equal to the value of the rate constant k. Such plots are observed when the system is buffered in substance A (e.g., if the substance A can be formed from a precursor present in excess in a rapidly established equilibrium) or, for most types of reactions, in the initial stage where the conversion of substance A is small.

The plot of $N' = log([A]_0 - x)$ as a function of time can be used for reaction

$$A + B \xrightarrow{\ k\ } P \tag{3.3-4}$$

provided that a large excess of the reagent B is present in the reaction mixture. Then the slope of the linear plot $M/P = k' = k[B]_0$.

When the initial concentration $[B]_0$ is larger, but not much larger than $[A]_0$, it is possible [3-3] to obtain the value of k from k' using expression (3.3-5):

$$k = \frac{k'}{[A]_0 - m} \tag{3.3-5}$$

TABLE 3-1

Some Special Cases of Integral Forms of Equation for the Rate of Reaction $aA + bB + cC \longrightarrow \cdots$ Corresponding to Differential Equation

$$dx/dt = k\,(A_0 - x)^n \left(B_0 - \frac{b}{a}x\right)^m \left(C_0 - \frac{c}{a}x\right)^0$$

(Taking place in constant volume and at constant pressure)

Type	Order			Stechiometric Coefficient			Differential Equation	Infinite Integral	Limited Integral	$\tau_{1/2}$	Units of k
	m	n	o	a	b	c					
A	1	0	0	1	—	—	$dx/dt = k(A_0 - x)$	$\ln(A_0 - x) = C - kt$	$\ln A_0/(A_0 - x) = kt$	$1/k \ln 2$	s^{-1}
2A	2	0	0	2	—	—[a]	$dx/dt = k(A_0 - x)^2$	$1/(A_0 - x) = C + kt$	$x/A_0(A_0 - x) = kt$	$1/kA_0$	$s^{-1} l\,mol^{-1}$
A + B	1	1	0	1	1	—	$dx/dt = k(A_0 - x)(B_0 - x)$	$1/(B_0 - A_0)\ln\dfrac{B_0 - x}{A_0 - x}$ $= C + kt$	$\dfrac{1}{B_0 - A_0}\ln\dfrac{A_0\,(B_0 - x)}{B_0(A_0 - x)}$ $= kt$	$\dfrac{1}{k(B_0 - A_0)}$ $\times \ln\dfrac{2B_0 - A_0}{B_0}$	$s^{-1} l\,mol^{-1}$
A + 2B	1	1	0	1	2	—	$dx/dt = k(A_0 - x)(B_0 - 2x)$	$\dfrac{1}{B_0 - 2A_0}\ln\dfrac{B_0 - 2x}{A_0 - x}$ $= C + kt$	$\dfrac{1}{B_0 - 2A_0}\ln\dfrac{A_0(B_0 - 2x)}{B_0(A_0 - x)}$ $= kt$	$\dfrac{1}{k(B_0 - 2A_0)}$ $\times \ln\dfrac{2(B_0 - A_0)}{B_0}$	$s^{-1} l\,mol^{-1}$

	Differential equation	Integrated (C form)	Integrated ($=kt$ form)	$t_{1/2}$	Units
3A — 3 0 0 3 — [b]	$\dfrac{dx}{dt} = k(A_0 - x)^3$	$\dfrac{1}{(A_0-x)^2} = C + 2kt$	$\dfrac{2A_0 x - x^2}{A_0^2(A_0-x)^2} = 2kt$	$3/(2k\,A_0^2)$	$\text{s}^{-1}\text{l}^2\,\text{mol}^{-2}$
A+B+C — 1 1 1 1 1 1[c]	$\dfrac{dx}{dt} = k(A_0-x)(B_0-x)(C_0-x)$	$\begin{aligned}&\dfrac{C_0-B_0}{T}\ln(A_0-x)\\[2pt]&+\dfrac{A_0-C_0}{T}\ln(B_0-x)\\[2pt]&+\dfrac{B_0-A_0}{T}\ln(C_0-x)\\[2pt]&= C + kt\end{aligned}$	$\begin{aligned}&\dfrac{B_0-C_0}{T}\ln\dfrac{A_0}{A_0-x}\\[2pt]&+\dfrac{C_0-A_0}{T}\ln\dfrac{B_0}{B_0-x}\\[2pt]&+\dfrac{A_0-B_0}{T}\ln\dfrac{C_0}{C_0-x}\\[2pt]&= kt\end{aligned}$	$\begin{aligned}&\dfrac{B_0-C_0}{kT}\ln 2\\[2pt]&+\dfrac{C_0-A_0}{kT}\ln\dfrac{B_0}{B_0-A_0/2}\\[2pt]&+\dfrac{A_0-B_0}{kT}\ln\dfrac{C_0}{C_0-A_0/2}\end{aligned}$	$\text{s}^{-1}\text{l}^2\,\text{mol}^{-2}$
mA — $m>1$ 0 0 m — —	$\dfrac{dx}{dt} = k(A-x)^m$	$\dfrac{1}{(A_0-x)^{m-1}} = (m-1)kt + C$	$\dfrac{1}{(A_0-x)^{m-1}} - \dfrac{1}{A_0^{m-1}} = (m-1)kt$	$\dfrac{2^{m-1}-1}{kA_0^{m-1}}$	$\text{s}^{-1}\text{l}^{m-1}$ mol^{1-m}
0 — 0 0 0 1 — —	$\dfrac{dx}{dt} = k$	$-(A_0-x) = kt + C$	$x = kt$	$A_0/2k$	$\text{s}^{-1}\text{l}^{-1}$ mol
1/2A — 1/2 0 0 1 — —	$\dfrac{dx}{dt} = k(A_0-x)^{1/2}$	$-(A_0-x)^{1/2} = \dfrac{kt}{2} + C$	$A_0^{1/2}-(A_0-x)^{1/2} = \dfrac{kt}{2}$	$3A_0^{1/2}/2k$	$\text{s}^{-1}\,\text{l}^{1/2}$ $\text{mol}^{1/2}$

[a] Valid also for A + B $\longrightarrow$ if [A] = [B].

[b] Valid also for A + 2B $\longrightarrow$ if 2 [A] = [B] and A + B + C $\longrightarrow$ for [A] = [B] = [C].

[c] $T = A_0 B_0(A_0-B_0) + A_0 C_0(C_0-A_0) + B_0 C_0(B_0-C_0)$.

Note: Table modified from R. Livingstone, Evaluation and Interpretation of Rate Data, in *Investigation of Rates and Mechanisms of Reactions* (S. L. Friess, E. S. Lewis, and A. Weissberger, Eds.), 2nd ed., Part I, Interscience, New York, 1961, p. 114.

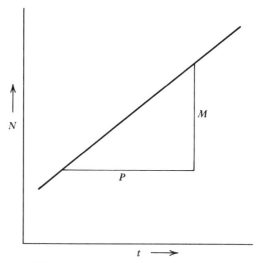

FIG. 3-6. Graphical method for the treatment of a simple reaction based on linearization. Plot of the function of the measured quantity N as a function of time. From the slope M/P it is possible to obtain the rate constant.

where m is the mean value of x over the range of conversion for which experimental data are available. For $[A]_0 : [B]_0 = 1:10$ the error in the value of the second-order rate constant k is less than 0.1% up to 70% conversion and less than 1% up to 90% conversion; for $[A]_0 : [B]_0 = 1:5$ the error is smaller than 1.1% up to 60% conversion and smaller than 2.5% up to 80% conversion. Even for $[A]_0 : [B]_0 = 1:2$ the error is less than 2% up to 60% conversion.

For reaction (3.3-6):

$$2A \xrightarrow{k} P \tag{3.3-6}$$

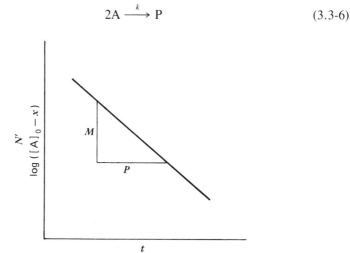

FIG. 3-7. Simplified graphical treatment of reactions following first-order kinetics. Plot of $N' = \log ([A]_0 - x)$ as a function of time. $M/P = k/2.3$.

or for reaction (3.3-5) under conditions that $[A]_0 = [B]_0$, the expression $N = X/[A]_0([A]_0 - x)$ may be plotted against the time and the slope M/P equals the rate constant k.

Since

$$\frac{x}{[A]_0([A]_0 - x)} = \frac{1}{[A]_0 - x} - \frac{1}{[A]_0} \qquad (3.3\text{-}7)$$

it is sufficient to plot $1/([A]_0 - x)$ as a function of time and the slope will still give $M/P = k$.

If the reaction (3.3-5) takes place under conditions when initial concentrations of A and B are not equal, then it is necessary to plot $N = \log([B]_0 - x)/([A]_0 - x)$ and the slope $M/P = ([B]_0 - [A]_0)k$.

Two simplifications of this treatment are possible: If the initial concentration of compound A is half of that of compound B (i.e., when $[A]_0 = [B]_0/2$) then it is possible again to plot $1/([A]_0 - x)$ against time and the slope of the linear dependence is $M/P = 2k$. Comparison with the above discussion thus indicates that under such conditions plots of both $1/([A]_0 - x)$ and $\log([A]_0 - x)$ as a function of time are practically linear up to 60% conversion.

On the other hand, if $[A]_0$ differs only a little from $[B]_0$ it is possible to derive the expression:

$$kt = \frac{1}{[A]_0 - x} - \frac{1}{[A]_0}$$
$$- \frac{[B]_0 - [A]_0}{2} \left[\frac{1}{([A]_0 - x)^2} - \frac{1}{[A]_0} \right] + \cdots \qquad (3.3\text{-}8)$$

It is then usually sufficient to neglect higher terms of the series and to plot $N = 1/([A]_0 - x)$ against time and estimate the value of the rate constant from the slope of this dependence.

Consistency of the general equation (3.3-9):

$$kt = \frac{1}{[B]_0 - [A]_0} \ln \frac{[A]_0([B]_0 - x)}{[B]_0([A]_0 - x)} \qquad (3.3\text{-}9)$$

can be verified using l'Hôpital's rule [3-4].

For reaction

$$A + 2B \xrightarrow{k} P \qquad (3.3\text{-}10)$$

the expression $N = \log([B]_0 - 2x)/([A]_0 - x)$ is plotted in Fig. 3-6 as a function of time and the slope $M/P = ([B]_0 - 2[A]_0)k/2.3$.

Finally, for reactions

$$3A \xrightarrow{k} P \tag{3.3-11}$$

or

$$A + B + C \xrightarrow{k} P \tag{3.3-12}$$

when initial concentrations are equal (i.e., $[A]_0 = [B]_0 = [C]_0$) the expression $N = (2[A]_0 x - x^2)/[A]_0^2([A]_0 - x)^2$ is plotted in Fig. 3-6 against time, and the slope M/P equals $2k/2.3$.

If, in the latter reaction, the initial concentrations are not equal (or at least two of them), the system becomes complex and unsuitable for simple graphical treatment.

A simplification for conditions, where $[A]_0 = [B]_0 = [C]_0$, is possible because

$$\frac{2[A]_0 x - x^2}{[A]_0^2([A]_0 - x)^2} = \frac{1}{([A]_0 - x)^2} - \frac{1}{[A]_0^2} \tag{3.3-13}$$

It is thus possible to plot $1/([A]_0 - x)^2$ against time rather than the more complex expression quoted above.

Generally, for a reaction of the type (3.3-14):

$$nA \xrightarrow{k} P \tag{3.3-14}$$

the value of $1/([A]_0 - x)^{n-1}$ is plotted against time and the slope is given by $M/P = (n - 1)k$, provided that $n \neq 1$.

In the majority of the expressions for N discussed above, the value of the initial concentration $[A]_0$ is involved when the decrease in concentration of one of the reagents is followed. In systems in which the concentration change of the reacting component is directly determined, the acquisition of the value of $[A]_0$ presents no difficulties.

When a physical quantity (ϕ) is measured which is a linear function of concentration of the reacting species, but reaches a finite value (ϕ_∞) at $t = \infty$, then

$$[A] = \text{const}(\phi - \phi_\infty) \tag{3.3-15}$$

and at $t = 0$

$$[A]_0 = \text{const}(\phi_0 - \phi_\infty) \tag{3.3-16}$$

For reaction (3.3-3) it follows that

$$kt = \ln (\phi_0 - \phi_\infty)/(\phi - \phi_\infty) \tag{3.3-17}$$

Thus when log ($\phi - \phi_\infty$) is plotted against time, a linear plot is obtained with a slope equal to ($-k/2.3$). Thus the knowledge of ϕ_0 is not necessary, but that of ϕ_∞ is.

When the physical quantity measured (ϕ) is a linear function of the concentration of a reaction product (P) formed in reaction (3.3-3), then

$$kt = \ln [P]_\infty / ([P]_\infty - [P]) \qquad (3.3-18)$$

where $[P]_\infty$ is the concentration of product P at $t = \infty$. For

$$[P] = \text{const}(\phi - \phi_0) \qquad (3.3-19)$$

and

$$[P]_\infty = \text{const}(\phi_\infty - \phi_0) \qquad (3.3-20)$$

combination with equation (3.3-18) gives

$$kt = \ln (\phi_\infty - \phi_0)/(\phi_\infty - \phi) \qquad (3.3-21)$$

Thus the plot of log ($\phi_\infty - \phi$) versus time is linear with a slope equal to $-k/2.3$.

In this treatment it is assumed that the value of ϕ_0 remains unchanged during the reaction and is additive with ϕ. Similar expression can be derived for the case when the value of ϕ_0 depends on concentration of the reactant A (eq. 3.3-3) and changes with time in the course of reaction.

In cases where the initial concentration of the reactants is unknown or is inaccessible, it is possible [3-7] to start the handling of data with the differential equation

$$\frac{dx}{dt} = k(c_0 - x)^n \qquad (3.3-22)$$

Through elimination of ($c_0 - x$) from this equation and its integrated form it is possible to obtain the expression

$$\log \frac{dx}{dt} = 0.4343kt + \log c_0 k \qquad (3.3-23)$$

for $n = 1$ and

$$\left(\frac{dx}{dt}\right)^{(1 - n)/n} = (n - 1)k^{1/n} t + k^{(1-n)/n} c_0^{(1-n)} \qquad (3.3-24)$$

for n different from one.

From the time change of concentration the value of dx/dt_i is determined as a tangent for various values of t_i and the value of $\log(dx/dt_i)$ or of $(dx/dt_i)\exp(1-n)/n$ is plotted against t_i (Fig. 3-8). For $n = 1$ the slope M/P of the plot is $0.4343k_1$; for $n \neq 1$ the slope becomes $(n-1)k^{1/n}$. The intercept U for $n = 1$ equals $\log c_0k$; for $n \neq 1$, the intercept equals $k^{(1-n)/n} c_0^{1-n}$.

From the slope of the dependence in Fig. 3-8, it is thus possible to determine the value of the rate constant k_1, while from the intercept U it is possible to verify the value of n and of the initial concentration c_0.

An alternative approach starts from the integrated form transformed (for $n \neq 1$) into

$$x = c_0 - [c_0^{1-n} - (1-n)kt]\,\frac{1}{1-n} \tag{3.3-25}$$

which shows a linear relationship between functions x and t. This expression does not contain initial concentrations of the reactants. Thus, for $n = 2$

$$x = c_0 - \frac{c_0}{1 + c_0kt} = \frac{c_0^2kt}{1 + c_0kt} \tag{3.3-26}$$

and after rearrangement

$$\frac{t}{x} = \frac{1}{c_0^2 k} + \frac{1}{c_0}t \tag{3.3-27}$$

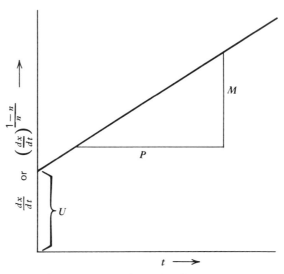

FIG. 3-8. Establishment of reaction rate equation, order of reaction, and estimate of the value of the rate constant from measurement of reaction rates.

When t/x is plotted as a function of time (t) the slope of the linear plot $M/P = 1/c_0$. After the initial concentration c_0 has been verified in this way, it is possible from the intercept $U = 1/c_0^2 k$ to determine the value of k.

In a similar way it is possible for first-order reactions to obtain a relationship $x = c_0(1 - e^{-kt})$ which is unsuitable for a simple graphical treatment. For calculations of rate constants in such cases it is possible to use the Guggenheim and Roseveare methods, discussed in Section 3.5.1.

For certain types of reactions it is possible to adopt a procedure that circumvents the tedious "trial and error" approach. This procedure, which makes it possible to choose the best empirical equation for reaction rate and to determine the reaction order, is denoted as the method of dimensionless parameters [3-5].

The dimensionless parameters introduced are $\alpha = c/c_0$, which is the relative concentration, and $\tau = kc_0^{n-1}t$, which is the time parameter. Inserting into

$$\ln c = \ln c_0 - kt \tag{3.3-28}$$

for first-order reaction gives

$$\ln \alpha = -\tau \tag{3.3-29}$$

For all other total reaction orders inserting the dimensionless parameters into (3.3-30):

$$\frac{1/c^{n-1} - 1/c_0^{n-1}}{n-1} = kt \tag{3.3-30}$$

gives

$$(\alpha^{1-n} - 1) = (n - 1)\tau \tag{3.3-31}$$

For each total order n the dependence of α on τ has a characteristic shape (Fig. 3-9).

If it is possible to assume that the reaction is of the type a A $\longrightarrow$ P or if it is possible to arrange the reaction conditions in such a way that concentrations of all reacting species are equal, it is possible to calculate α from the measured concentration changes and plot α against the values of corresponding log t (the value of τ is not directly accessible experimentally). Because

$$\log \tau = \log t + \log kc_0^{n-1} \tag{3.3-32}$$

the curves α–log t and α–log τ are merely shifted along the time axis by log kc_0^{n-1}. Comparison of the α–log t curve with the shape of theoretical α–log τ curves makes it possible to decide which empirical equation for reaction rate applies and to which total reaction order the experimental α–log t curve corresponds. In practice

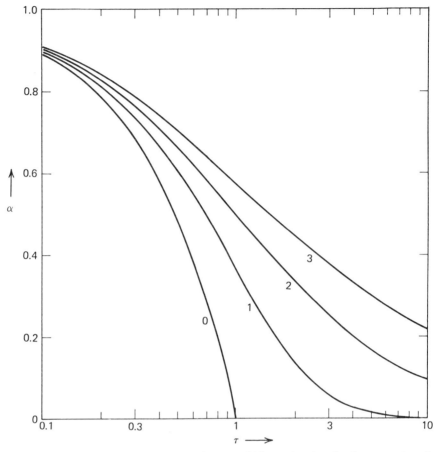

FIG. 3-9. Dependence of relative concentration $\alpha = C/C_0$ as a function of a time parameter τ for (0) zeroth; (1) first; (2) second; and (3) third order reactions (from A. A. Frost and R. G. Pearson, *Kinetics and Mechanism*, 2nd ed., Wiley, New York, 1961, p. 15).

it proved simplest to plot the experimental α–log t curve on transparent paper (using the same scale as in Fig. 3-9) and slide this paper along the time axis until the experimental points overlap one of the theoretical curves.

3.3.2.   Complex Reactions

3.3.2.1.   General Considerations

For complex reactions it can be difficult to obtain an integrated form of the RRE. Cases exist when it is necessary to use the differential form of the RRE. Examples of applications of both types together with the transformation most suitable for experimental verification will be discussed here. First another possible treatment leading to the obtaining of the RRE for complex reactions will be discussed.

This alternative involves determination of initial velocity of complex reactions at varying initial concentrations of the reactants. Such values make it possible to find the RRE. The initial velocities are derived from experimental data for the time dependence of concentration by graphical or numerical methods.

In graphical methods it is possible to estimate the tangent (d) to the concentration–time dependence at $t = 0$ either by means of a ruler (Fig. 3-10), by means of a front-surfaced plane mirror, or a movable protractor with a prism at a center. In other instances when the measurement can be carried out within a short period t_1 (relative to the half-time) it is possible to assume that $x_1/t_1 \approx (dx/dt)_0$. The accuracy of this approach is greater, the shorter the period t_1. However, the situation is complicated by the fact that at very short times the measurement of the value of x becomes less accurate and, if carried out before the solution is completely mixed, can lead to completely erroneous values. The shortest time t_1 that can be chosen for a given reaction to achieve a sufficiently accurate determination of the ratio x_1/t_1 depends on the analytical method used for determination of x.

Alternatively [3-6], it is possible to express concentration of the product P as a polynomial in time t

$$[P] = bt + ct^2 + dt^3 + \cdots \tag{3.3-33}$$

Initial rate $(d[P]/dt)_{t=0} = v_0 = b$. Neglecting higher terms

$$[P] \approx bt + ct^2 \tag{3.3-34}$$

and hence

$$[P]/t \approx v_0 + ct \tag{3.3-35}$$

Thus a plot of $[P]/t$ versus t must be linear with an intercept equal to v_0, as long as eq. (3.3-34) is a satisfactory approximation. When the concentration of P is calculated for reaction $nA \rightarrow P$ using expression

$$[P] = [A]_0 (1 - c^{-kt})^n \tag{3.3-36}$$

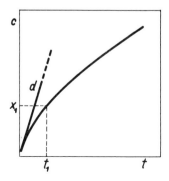

FIG. 3-10. Determination of the initial velocity (v_0) using a tangent (d) and the ratio x_1/t_1.

and compared with the value for [P] calculated from equation (3.3-34) for the first 10% of conversion, the value of [P] obtained from (3.3-34) is lower by 0.2% for $n = 1$, by 1.2% for $n = 2$, and by 3.3% for $n = 3$.

Whenever possible, experimental data should be compared with the integrated form of the RRE. The equation chosen for comparison is usually based on preliminary knowledge of the course of the reaction, on determination of reaction order in individual reactants, and on the stoichiometry of the reaction. A survey of the integrated forms of RRE most suitable for experimental verification for some most commonly encountered types of complex reactions follows. Reactions are classified as:

Type

1. Parallel reactions

$$C \longleftarrow A + B \longrightarrow D + E$$

2. Consecutive reactions

$$A + B \longrightarrow C \longrightarrow D$$

3. Competitive reactions

$$A + B \longrightarrow P \longleftarrow A + C$$

$$A + B \begin{array}{c} \nearrow Y \\ \searrow Z \end{array}$$

4. Reversible reactions $A \rightleftarrows B + C$

Some complex reactions can be classified into more than one group.

3.3.2.2. Parallel Reactions

The parallel reactions can be subclassified into

(a) Reactions of the same order (first or higher):

$$A \begin{array}{c} \nearrow U \\ \longrightarrow V \\ \searrow W \end{array}$$

(b) First-order reactions with common product:

$$\begin{array}{c} A \searrow \\ \;\; C \\ B \nearrow \end{array}$$

(c) First and second order:

$$A + B \longrightarrow C$$
$$A \longrightarrow D$$

(a) PARALLEL REACTIONS OF THE SAME ORDER. If for reaction of the type:

$$A \begin{array}{c} \overset{k_1}{\nearrow} U \\ \overset{k_2}{\rightarrow} V \\ \overset{k_3}{\searrow} W \end{array}$$

the analytical method used allows only determination of concentration change of A, than the plot of $N' = \log([A]_0 - x)$ as a function of time is linear (Fig. 3-7) with slope $M/P = k/2.3$.
Hence

$$[A] = [A]_0 e^{-kt} \tag{3.3-38}$$

where

$$k = k_1 + k_2 + k_3 \tag{3.3-39}$$

If only $[A] = f(t)$ is measured, it is impossible to separate values of k_1, k_2, and k_3, and only their sum can be determined. To be able to determine all individual rate constants for the above mentioned three competitive reactions, it is necessary, in addition to measurement of $[A] = f(t)$, to measure the time dependence of concentration of two or more products, U and V, V and W, or U and W. For those the following relations are valid:

$$[U] = [U]_0 + \frac{k_1[A]_0}{k} (1 - e^{-kt}) \tag{3.3-40}$$

$$[V] = [V]_0 + \frac{k_2[A]_0}{k} (1 - e^{-kt}) \tag{3.3-41}$$

$$[W] = [W]_0 + \frac{k_3[A]_0}{k} (1 - e^{-kt}) \tag{3.3-42}$$

Solution of two simultaneous equations for known k yields constants k_i; as $k = k_1 + k_2 + k_3$, the third value of k_i can be obtained as difference. Alternatively, it is possible to plot the concentrations of individual products as a function of $(1 - e^{-kt})$ (Fig. 3-11). The slopes of these linear plots have the values $k_i[A]_0/k$ from which the values of constants k_i can be calculated.

If, in addition to the condition that the competing side reactions are of the same reaction order, the initial concentration of products equals zero (i.e., when $t = 0$,

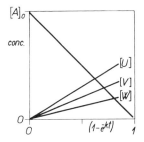

FIG. 3-11. Graphical treatment of competitive first-order reactions.

$[U]_0 = [V]_0 = [W]_0 = 0$), there is another possibility to determine the value of the constant k_i which does not involve determination of the concentration change of products. In this particular case it is sufficient to determine the ratio of yields of individual products after a chosen time interval. The ratios of yields (or concentrations) at $t = t_1$ are related to the rate constants by equations:

$$\frac{[U]_1}{[V]_1} = \frac{k_1}{k_2} \qquad (3.3\text{-}43)$$

$$\frac{[U]_1}{[W]_1} = \frac{k_1}{k_3} \qquad (3.3\text{-}44)$$

and

$$\frac{[V]_1}{[W]_1} = \frac{k_2}{k_3} \qquad (3.3\text{-}45)$$

By the combination of two of these equations with $k = k_1 + k_2 + k_3$ it is possible to separate the values of individual rate constants k_i.

An example of a reaction treated in this way is the cleavage of β-ketols in alkaline media [3-8] which follows two paths:

$$\begin{array}{ccc}
 & \overset{k_1}{\nearrow} & \text{RCOCH}\!\!=\!\!\text{CHR}' + \text{OH}^- + \text{H}^+ \\
\underset{\displaystyle \overset{|}{\text{OH}}}{\text{RCOCH}_2\text{CHR}'} & & \\
 & \overset{k_2}{\searrow} & \text{RCOCH}_2^- + \text{R}'\text{CH}\!\!=\!\!\text{O} \\
 & & \text{H}^{\cdot} \Big\Vert \\
 & & \text{RCOCH}_3
\end{array} \qquad (3.3\text{-}46)$$

By measuring the ratio of concentrations of the α,β-unsaturated ketone (RCOCH = CHR') and of the aldehyde (R'CHO) or the ketone (RCOCH₃) after a chosen time period (e.g., 30 min) polarographically or spectrophotometrically it was possible to determine the ratio of k_1 and k_2. Because it was also possible to follow the change in ketol $(RCOCH_2CH(OH)R')$ concentration with time, the value of $k = k_1 + k_2$ was accessible and hence the separation of k_1 and k_2 was possible.

(b) TWO PARALLEL REACTIONS WITH A COMMON PRODUCT. Reactions of the type

$$A \xrightarrow{k_1} C \xleftarrow{k_2} B \qquad (3.3\text{-}47)$$

can be easily examined, if it is possible to follow time changes of both starting materials A and B. Nevertheless, a solution is possible even in systems with comparable values of k_1 and k_2 where the analytical method used makes it possible only to follow concentration of the product C. The expression:

$$\log ([C]_\infty - [C]) = \log ([A]_0 e^{-k_1 t} + [B]_0 e^{-k_2 t}) \qquad (3.3\text{-}48)$$

is used. In the presence of a single component (either A or B) in the reaction mixture or if $k_1 = k_2$, the plot of $\log ([C]_\infty - [C])$ against time is linear. If the reaction rates of the two components A and B are comparable, the dependence of $\log ([C]_\infty - [C])$ on t is curved.

An example of a nonlinear dependence has been observed in the course of hydrolysis of a compound which was assumed to be diethyl tert-butyl carbinyl chloride. Since $k_1 \neq k_2$ it is possible to attribute the faster reaction to k_1 and hence $k_1 > k_2$. The plot of $\log ([C]_\infty - [C])$ against time approaches linearity after a certain time interval, that is, when $t > t_p$. For $t < t_p$ the plot shows significant deviations with a trend (Fig. 3-12) in such direction that the observed values of $\log ([C]_\infty - [C])$ are higher than would correspond to a linear plot. This indicates that the reaction mixture contains two components, the more reactive A, reacting faster with rate constant k_1, and slower reacting species B, corresponding to rate constant k_2. After the time interval t_p, when the more reactive component A is practically removed from the solution, the expression simplifies to

$$\log ([C]_\infty - [C]) = \log [B] = \log [B]_0 - \frac{k_2 t}{2.3} \qquad (3.3\text{-}49)$$

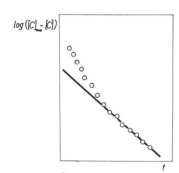

FIG. 3-12. Graphical treatment of two competitive first-order reactions when the initial concentrations of reacting species are not known.

From the slope of this linear portion of the plot it is hence possible to determine the rate constant k_2. The intercept on the concentration axis (of $t = 0$) equals log $[B]_0$. The above equation makes it thus possible to calculate the value of $[B]$ at each time t (Fig. 3-13). The value of $[A]$ can then be obtained from the difference

$$[A] = [C]_\infty - [C] - [B] \tag{3.3-50}$$

at all time intervals t and from a plot of log $[A]$ against t (Fig. 3-13) it is possible to determine values of $[A]_0$ and k_1. The experimental results were interpreted for the given system as evidence that the reaction mixture contained not one species but two isomeric reactants. The described analysis of the kinetic data indicated the presence of about 35% of compound A and 65% of compound B in the solution before hydrolysis. Compounds A and B were assumed to be isomers formed in the synthesis of the tertiary chloride.

(c) PARALLEL REACTIONS OF FIRST AND SECOND ORDER. Such reactions, sometimes denoted as heterocompetitive, follow a general scheme:

$$A + B \xrightarrow{k_1} C \tag{3.3-51}$$
$$A \xrightarrow{k_2} D \tag{3.3-52}$$

The discussion here will be restricted to the special case

$$A + B \xrightarrow{k_1} C + D \tag{3.3-53}$$
$$A \xrightarrow{k_2} D + E \tag{3.3-54}$$
$$E + B \xrightarrow[\text{fast}]{} C \tag{3.3-55}$$

where the products of the first- and second-order reactions (C and D) are identical. Such reactions are found for cases of halide hydrolysis which follow both S_N1 and S_N2 mechanisms.

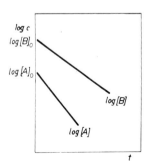

FIG. 3-13. Determination of initial concentrations of two species participating in competitive first-order reactions.

The integrated form of the RRE is relatively complicated and unsuitable for experimental verification. For this purpose the differential form of RRE (3.3-56) is used:

$$\frac{dx}{dt}\left(\frac{1}{a - x}\right) = k_1 + k_2 (b - x) \tag{3.3-56}$$

From experimental data the dependence $x = f(t)$ is found and the rate dx/dt obtained from tangents at individual times for which the values of $(a - x)$ and $(b - x)$ are known. From the slope of the plot (Fig. 3-14) of $(dx/dt)/(a - x)$ against $(b - x)$ it is possible to obtain the rate constant $k_2 = M/P$ and from the intercept, the value of k_1. In such a way, for example, the hydrolysis of primary 1-chloro-1-butene was followed.

Generally, for a system of parallel reactions, the RRE has the simplest form in the case when it is possible to prepare the reaction mixture in such a way that the common starting material at the beginning of the reaction is in stoichiometric ratio to each of the products. For reactions of the type

$$2A + B \longrightarrow A_2B \tag{3.3-57}$$

$$A + C \longrightarrow C \tag{3.3-58}$$

it is advantageous if the reaction mixture can be prepared so that $[A]:[B]:[C] = 1:\frac{1}{2}:1$.

If the orders of parallel reactions differ by one, then the dependence of the concentration of the product $[P]$ on the concentration of the starting material $[A]$ has the shape of a logarithmic curve (Fig. 3-15) such that lines drawn as equidistanced parallels with the x-axis intersect the curve at points where the corresponding x-coordinates are in a constant ratio $a/b = b/c = c/d = $ const. If the difference between the orders of the compared reactions is two, the curve showing dependence of the concentration of the product $[P]$ on the concentration $[A]$ is an arcustangent.

The tangent of the dependence of concentration of the product $[P]$ on initial concentration $[A]_0$ in the case that the reactions differ by one order is given by relation $1/\{1 + (k_1/k_2)[A]_0\}$ and can be used for determination of the ratio k_1/k_2.

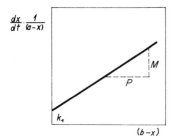

FIG. 3-14. Graphical treatment of a system of consecutive reactions (3.3-53) to (3.3-55) using the differential form of the reaction rate equation (3.3-56).

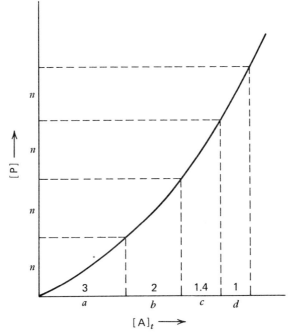

FIG. 3-15. Dependence of concentration of the product P on concentration of the starting material A for two parallel reactions. The order of these two reactions differs by one and the curve is logarithmic.

Finally, if it is possible to measure the rates of formation of both products in reactions

$$nA \xrightarrow{k_1} B \tag{3.3-59}$$

$$n'A \xrightarrow{k_2} C \tag{3.3-60}$$

then from differential equations

$$\frac{d[B]}{dt} = k_1 (a - x)^n = v_1 \tag{3.3-61}$$

and

$$\frac{d[C]}{dt} = k_2 (a - x)^{n'} = v_2 \tag{3.3-62}$$

it follows that

$$v_1 = k_1 \left(\frac{v_2}{k_2}\right)^{n/n'} \tag{3.3-63}$$

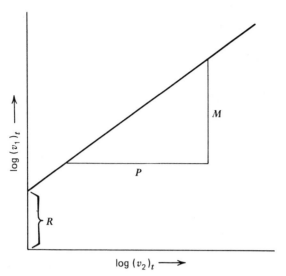

FIG. 3-16. Determination of the ratio of reaction orders and of rate constants of parallel reactions from correlation of reaction rates. $M/P = n/n'$; $R = \log k_1 - n/n' \log k_2$.

and

$$\log v_1 = \log k_1 - \frac{n}{n'} \log k_2 + \frac{n}{n'} \log v_2 \qquad (3.3\text{-}64)$$

A plot of $\log (v_1)_t$ against $\log (v_2)_t$ (Fig. 3-16) gives the ratio n/n' from the slope M/P. From the intercept it is possible to obtain the ratio k_1/k_2. If the reaction orders in both parallel reactions are equal ($n = n'$), the $\log v_1 - \log v_2$ plot shows unit slope ($M/P = 1$).

3.3.2.3. Consecutive Reactions

(a) FIRST-ORDER CONSECUTIVE REACTIONS. For a system of two, consecutive, first-order reactions

$$A \xrightarrow{k_1} B \xrightarrow{k_2} C \qquad (3.3\text{-}65)$$

it is possible to derive the equations (3.3-66) to (3.3-68):

$$[A] = [A]_0 \, e^{-k_1 t} \qquad (3.3\text{-}66)$$

$$[B] = \frac{[A]_0 \, k_1}{k_2 - k_1} \, (e^{-k_1 t} - e^{-k_2 t}) \qquad (3.3\text{-}67)$$

$$[C] = [A]_0 \left\{ 1 + \frac{1}{k_2 - k_1} \left(k_2 e^{-k_1 t} - k_1 e^{-k_2 t} \right) \right\} \qquad (3.3\text{-}68)$$

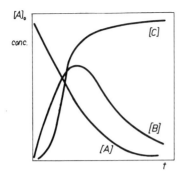

FIG. 3-17. Dependence of concentration of the starting material (A), intermediate (B), and product (C) on time for a first-order consecutive reaction following equation (3.3-65).

The time dependence of individual components for the case in which the values of the constants k_1 and k_2 are comparable and $[B]_0 = [C]_0 = 0$ is shown in Fig. 3-17.

At every instant in the course of reaction the sum of concentrations is equal to the initial concentration of the reactant:

$$[A]_0 = [A] + [B] + [C] \qquad (3.3\text{-}69)$$

For the time t_{max} when concentration of the intermediate B reaches its maximum value $[B]_{max}$ it is possible to show that

$$t_{max} = \frac{1}{k_2 - k_1} \ln \frac{k_2}{k_1} \qquad (3.3\text{-}70)$$

Like the value of t_{max}, that of $[B]_{max}$ depends on the ratio of the rate constants k_1 and k_2. For $k_2 \ll k_1$ the value of $[B]_{max}$ approaches the value of $[A]_0$ and is reached shortly after the beginning of the reaction (i.e., t_{max} is short) (Fig. 3-18). In such cases the rate of the conversion of A into C depends almost solely on the reaction with the constant k_2. With increasing value of the ratio k_2/k_1 the maximum attainable concentration $[B]_{max}$ decreases and the value of t_{max} is shifted towards larger values. For the other extreme case, when $k_2 \gg k_1$, the intermediate B is immediately transformed into the final product C and the overall rate is governed by the rate of the reaction with constant k_1.

An example illustrating the various cases mentioned in the preceding paragraph is the cleavage of tropenone methoiodide, which takes place in two steps: Hoffmann degradation of the quaternary salt and elimination reaction of the resulting Mannich base [3-9]:

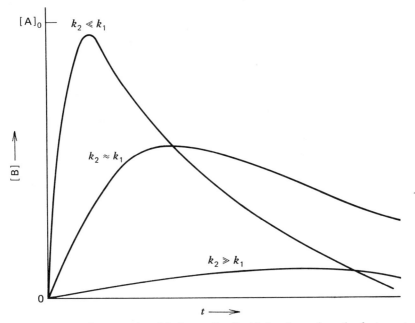

FIG. 3-18. Change of concentration of the intermediate B with time for varying ratio of rate constants k_1 and k_2 [equation (3.3-65)].

If the tropenone methoiodide is placed in a Britton–Robinson buffer of pH 9 and polarographic waves of the original compound, intermediate Mannich base, and the tropone formed as final products are recorded, the time dependence (Fig. 3-19) follows the same pattern as shown in Fig. 3-17. When pH is decreased by one unit, to pH 8, the wave of the intermediate B is barely visible and the reaction resembles A $\longrightarrow$ C. On the other hand, if pH is increased to pH 10.5, the concentration of [B] increases almost immediately, approaches the value of $[A]_0$, and only then slowly decreases.

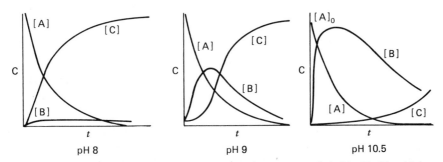

FIG. 3-19. Effect of pH on changes of concentrations of tropenone methoiodide (A), Mannich base intermediate (B), and tropone (C) with time in alkaline cleavage of tropenone methoiodide according to equation (3.3-71).

Hence

      at pH 8 $k_2 > k_1$ (and an unstable intermediate is formed)

      at pH 9 $k_2 \approx k_1$

      at pH 10.5 $k_2 < k_1$ (an intermediate is rapidly formed)

The difference between the ratios $k_2 : k_1$ at individual pH-values is due to base catalysis: Both reaction steps are general base catalyzed, but the rate of the Hoffmann-degradation with constant k_1 increases more with increasing pH than the rate of Mannich base elimination with constant k_2.

The value of k_1 can be obtained directly from a measurement of the dependence of [A] on time. Preliminary information about the value of the constant k_2 can be obtained from intersections of the [A]–t plot with the [B]–t and [C]–t plots. For the times at which concentrations of the two components are equal, equations (3.3-72) and (3.3-73) apply:

$$t_{[A] = [B]} = \frac{2.3}{k_1 - k_2} \log \frac{2k_1 - k_2}{k_1} \tag{3.3-72}$$

$$t_{[A] = [C]} = -\frac{2.3}{k_2} \log \frac{k_2 - k_1 + k_1 \, ([A]/[A]_0)}{k_1} \tag{3.3-73}$$

To decide whether the studied reaction follows the scheme (3.3-65), it proved useful [3-10] to introduce dimensionless parameters—relative concentrations α, β, and γ, relative rate κ, and time constant τ.

$$\alpha = \frac{[A]}{[A]_0} \qquad \beta = \frac{[B]}{[A]_0} \qquad \gamma = \frac{[C]}{[A]_0} \qquad \tau = k_1 t \quad \text{and} \quad \kappa = \frac{k_2}{k_1}$$

The rate equations can then be expressed in forms:

$$\alpha = e^{-\tau} \tag{3.3-74}$$

$$\beta = \frac{k_1}{k_2 - k_1} (e^{-\tau} - e^{-\kappa\tau}) = \frac{1}{\kappa - 1} (e^{-\tau} - e^{-\kappa\tau}) \tag{3.3-75}$$

$$\gamma = 1 + \frac{k_1}{k_1 - k_2} \left(\frac{k_2}{k_1} e^{-\tau} - e^{-\kappa\tau} \right) = 1 + \frac{1}{1 - \kappa} (\kappa e^{-\tau} - e^{-\kappa\tau}) \tag{3.3-76}$$

The functional relationship of individual parameters can be expressed graphically (Fig. 3-20). To establish whether the experimental data correspond to the above scheme and to find an approximate value of the rate constant, the relative concentrations found are compared with those shown graphically.

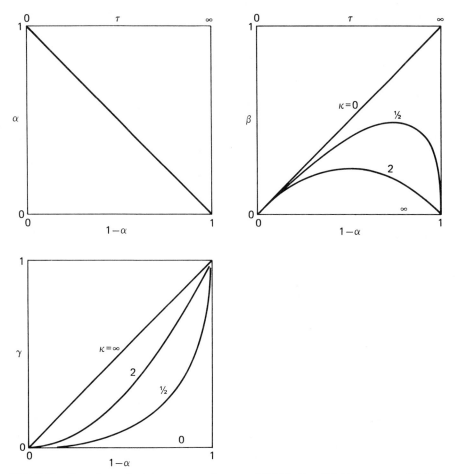

FIG. 3-20. Dependence of relative concentrations $\alpha = [A]/[A]_0$, $\beta = [B]/[A]_0$, and $\gamma = [C]/[A]_0$ for a system of first-order consecutive reactions, following equation (3.3-65) on function 1-α for varying values of the relative rate constant $\kappa = k_2/k_1$ (from A. A. Frost and R. G. Pearson, *Kinetics and Mechanism*, 2nd ed., Wiley, New York, 1961, p. 168).

(b) STEADY-STATE TREATMENT. Numerous types of more complex reactions can be dealt with only when simplifying assumptions are made. One of the most frequently used simplifications is the steady state treatment. The basic assumption for the application of this treatment is that the concentration of the intermediate (substance B) in the above scheme remains constant over most of the kinetic run. From the graphs in Fig. 3-18 it follows that for a ratio of rate constants $k_1 : k_2$ where the time-dependence of concentration of B shows a maximum, the validity of this assumption is rather limited. The assumption that [B] = const is more accurately fulfilled in systems where the intermediates are reactive (i.e., at $k_2 \gg k_1$, Fig. 3-18) and their concentration remains low during the whole kinetic run [3-11].

The steady state condition can be expressed as:

$$\frac{d[B]}{dt} = 0 = k_1[A] - k_2[B] \tag{3.3-77}$$

and hence

$$\frac{[B]}{[A]} = \frac{k_1}{k_2} \tag{3.3-78}$$

Because

$$[A] = [A]_0\, e^{-k_1 t} \tag{3.3-79}$$

it follows that

$$[B] = [A]_0 \frac{k_1}{k_2} e^{-k_1 t} \tag{3.3-80}$$

and

$$[C] = [A]_0 \left[1 - \left(1 + \frac{k_1}{k_2} \right) e^{-k_1 t} \right] \tag{3.3-81}$$

If the latter two equations are compared with the full expressions derived above, it is obvious that they are identical when $k_2 \gg k_1$ and $t \gg 1/k_2$. The first condition indicates that the intermediate under the given conditions must be considerably more reactive than the starting material A, and consequently the concentration of [B] during the kinetic run must remain low. According to the second condition the treatment can be applied in the time interval following the induction period. In most examples in which the steady state treatment has been successfully applied, the induction period is extremely short and therefore the second condition is fulfilled [3-12 – 3-14].

(c) MORE COMPLEX SYSTEMS. For the systems discussed so far it was assumed that $[B]_0 = [C]_0 = 0$. In a general treatment [3-15] of a system of consecutive reactions (3.3-82):

$$A \xrightarrow{k_A} B \xrightarrow{k_B} C \xrightarrow{k_C} D \tag{3.3-82}$$

the time-dependences of individual concentrations are described by the following differential equations

$$-\frac{d[A]}{dt} = k_A[A] \tag{3.3-83}$$

$$-\frac{d[B]}{dt} = k_B[B] - \frac{\beta}{\alpha} k_A[A] \qquad (3.3\text{-}84)$$

$$-\frac{d[C]}{dt} = k_C[C] - \frac{\gamma}{\beta} k_B[B] \qquad (3.3\text{-}85)$$

where $\alpha = n$, $\beta = n - 1$, $\gamma = n - 2$, and $\delta = n - 3$ when $(n - 1)$ is the number of reaction steps (in the above case $n - 1 = 3$ and $\alpha = 4$, $\beta = 3$, $\gamma = 2$, and $\delta = 1$). The concentration of D can be expressed by means of a similar differential equation or more simply from the relation (3.3-86):

$$[D] = S - [A] - [B] - [C] \qquad (3.3\text{-}86)$$

where S is the analytical concentration of the reacting species. Further a symbol (P,Q) is introduced expressing the fraction of compound P that is directly transformed into compound Q. Individual symbols (P,Q) for the above reaction can be expressed as follows:

$$(A,B) = \frac{\beta}{\alpha} = \frac{n-1}{n}$$

$$(A,C) = 0 \qquad\qquad (A,D) = 1 - \frac{\beta}{\alpha} = \frac{1}{n}$$

$$(B,C) = \frac{\gamma}{\beta} = \frac{n-2}{n-1} \qquad (B,D) = 1 - \frac{\gamma}{\beta} = \frac{1}{n-1} \qquad (3.3\text{-}87)$$

$$(C,D) = 1$$

From the differential equations the following definite integrals can be obtained:

$$[A] = -[A]_0 (1 - e^{-k_A t}) + [A]_0 \qquad (3.3\text{-}88)$$

$$[B] = [A]_0 \left[(1 - e^{-k_A t})(A,B) \frac{k_A}{k_A - k_B} - (1 - e^{-k_B t})(A,B) \frac{k_A}{k_A - k_B} \right]$$
$$- [B]_0 (1 - e^{-k_B t}) + [B]_0 \qquad (3.3\text{-}89)$$

Similarly, it is possible to derive expressions for [C] and [D] that are nevertheless rather complex. It is therefore suggested to use Table 3-2 in the derivation of analogous equations. From the symmetry of the equation it is obvious how it would look like for further members. The equation follows the general scheme (3.3-90):

$$[R] = [A]_0 Z_1 Y_1 + [A]_0 Z_2 Y_2 + [A]_0 Z_3 Y_3 + \cdots$$
$$+ [B]_0 (W_1 Z_1 + W_2 Z_2 + W_3 Z_3 + \cdots)$$
$$+ [C]_0 (U_1 Z_1 + U_2 Z_2 + U_3 Z_3 + \cdots) \qquad (3.3\text{-}90)$$

TABLE 3-2
Coefficients for Evaluation of Differential Equations for Consecutive Reactions

Concentration	Exponential Term	$[A]_0$	$[B]_0$	$[C]_0$
[R]	Z_1	Y_1	W_1	U_1
	Z_2	Y_2	W_2	U_2
	Z_3	Y_3	W_3	U_3
	...	...	...	...
[A]	$1 - e^{-k_A t}$	-1	0	0
	$+1$	$+1$	0	0
[B]	$1 - e^{-k_A t}$	$+(A,B)\dfrac{k_A}{k_A - k_B}$	0	0
	$1 - e^{-k_B t}$	$-(A,B)\dfrac{k_A}{k_A - k_B}$	-1	0
	$+1$	0	$+1$	0
[C]	$1 - e^{-k_A t}$	$-(A,B)(B,C)\dfrac{k_A k_B}{(k_A - k_B)(k_A - k_C)}$	0	0
	$1 - e^{-k_B t}$	$+(A,B)(B,C)\dfrac{k_A k_B}{(k_A - k_B)(k_B - k_C)}$	$+(B,C)\dfrac{k_B}{k_B - k_C}$	0
	$1 - e^{-k_C t}$	$-(A,B)(B,C)\dfrac{k_A k_B}{(k_A - k_C)(k_B - k_C)}$	$-(B,C)\dfrac{k_B}{k_B - k_C}$	-1
	1	0	0	0

Note: According to ref. 3-15.

In the derivation of the above expressions it is assumed that $(A,C) = 0$, that is, that the reaction does not involve a direct transformation of compound A into C.

For the completely general case of a system of substances A, B, C, D, . . . where each compound can form one or all compounds in consecutive processes, but where in no step a compound is formed which appears earlier in the reaction sequence, it is possible to use the differential equations:

$$- \frac{d[A]}{dt} = k_A[A] \tag{3.3-91}$$

$$- \frac{d[B]}{dt} = k_B[B] - (A,B)k_A[A] \tag{3.3-92}$$

$$- \frac{d[C]}{dt} = k_C[C] - (A,C)k_A[A] - (B,C)k_B[B] \tag{3.3-93}$$

This system differs from the previous one by the term $(A,C)k_A[A]$, which expresses direct transformation of A into C in a system of reactions (3.3-94) and (3.3-95):

$$A \longrightarrow B \longrightarrow C \longrightarrow D \tag{3.3-94}$$

$$A \longrightarrow C \tag{3.3-95}$$

The Table 3-2 must be extended in the section for [C] under factor $[A]_0$ as shown in Table 3-3. Such coefficients enable derivation of complete definite integrals. From the general expressions it is possible to derive equations for special systems, for example,

$$A \longrightarrow C \tag{3.3-96}$$

$$B \longrightarrow C + D \tag{3.3-97}$$

$$\text{where} \quad (A,B) = 0 \quad (A,C) = 1 \quad (A,D) = 0$$

$$(B,C) = \frac{\gamma}{\beta} \quad (B,D) = 1 - \frac{\gamma}{\beta} \tag{3.3-98}$$

$$(C,D) = 1$$

The table for [D] is available in the ref. [3-15].

It should be stressed that the discussion in this section was restricted to systems of consecutive first-order reactions. Systems involving higher order reactions are more difficult to treat. For example, a system of reactions of the type:

$$A + B \xrightarrow{k_1} C \tag{3.3-99}$$

$$C + D \xrightarrow{k_2} H \tag{3.3-100}$$

leads to an expression

$$\frac{d[H]}{dt} = k_2 \{[C]_0 + f(t) - [H]([D] - [H])\} \tag{3.3-101}$$

TABLE 3-3

Extension of the Table of Coefficients for Calculation of Differential Equations for a General System of Consecutive Equations

Concentration	Exponential Term	$[A]_0$
[C]	$1 - e^{-k_A t} - (A,B)(B,C)$	$\dfrac{k_A k_B}{(k_A - k_B)(k_A - k_C)} + (A,C) \dfrac{k_A}{k_A - k_C}$
	$1 - e^{-k_B t} + (A,B)(B,C)$	$\dfrac{k_A k_B}{(k_A - k_B)(k_B - k_C)} + 0$
	$1 - e^{-k_C t} + (A,B)(B,C)$	$\dfrac{k_A k_B}{(k_B - k_C)(k_B - k_C)} - (A,C) \dfrac{k_A}{k_A - k_C}$

According to ref. 3-15.

which cannot be integrated in the general case. The situation remains insoluble even for $[A]_0 = [B]_0$ and even in the presence of excess of [D] the solution is possible only by graphical integration. When both $[A]_0 = [B]_0$ and [D] is present in excess the solution is possible [3-15] by means of tabulated values of parameters for an incomplete Euler function.

3.3.2.4. Competitive Reactions

Competitive reactions can be classified as competitive consecutive reactions in which a product (C) of the first reaction is involved in another reaction with one of the original reactants (A):

$$A + B \longrightarrow C + D \tag{3.3-102}$$

$$A + C \longrightarrow D + E \tag{3.3-103}$$

and competitive parallel reactions in which a reactant (A) can be converted to a given product (E) by two pathways:

$$A + B \longrightarrow D + E \tag{3.3-104}$$

$$A + C \longrightarrow E + F \tag{3.3-105}$$

Examples of treatment of these two most frequently encountered types of competitive reactions are discussed here:

(a) COMPETITIVE CONSECUTIVE SECOND-ORDER REACTIONS. The following scheme will be considered:

$$A + B \xrightarrow{k_1} C + D \tag{3.3-106}$$

$$A + C \xrightarrow{k_2} D + E \tag{3.3-107}$$

An example of such a system is the hydrolysis of a symmetric diester under conditions when the system does not contain an excess of hydroxide ions and consequently $[OH^-]$ changes in the course of the reaction:

$$
\begin{array}{c}
COOR \\
(CH_2)_n + OH^- \xrightarrow{k_1} (CH_2)_n + ROH \\
COOR
\end{array}
\qquad (3.3\text{-}108)
$$

$$
\begin{array}{c}
COO^- \\
(CH_2)_n \\
COOR
\end{array}
$$

(B)   (A)   (C)   (D)

$$
OH^- + (CH_2)_n \xrightarrow{k_2} (CH_2)_n + ROH \qquad (3.3\text{-}109)
$$

(A)   (C)   (E)   (D)

For reactions of this type four possibilities can be observed:

I. $k_1 \gg k_2$;   II. $k_1 \ll k_2$;   III. $k_1 = 2k_2$;   IV. k_1 is comparable to k_2

I. If $k_1 \gg k_2$, the conversion of B with rate constant k_1 is almost quantitatively complete before the reaction of C with rate constant k_2 starts to contribute to the observed chemical changes. Each phase of the reaction can then be treated separately as a simple second-order reaction. This case has been observed, for example, for alkaline hydrolysis of diesters of oxalic acid.

II. If $k_1 \ll k_2$, the fast second step follows the slow first step. The observed change in diester concentration with time will correspond to simple second-order kinetics. Nevertheless, in the course of reaction, two moles of hydroxide ions (A) are consumed for each mole of diester (B). An example of this type is acetal hydrolysis. Hemiacetal (C) reacts with base in a considerably faster reaction. Acetal hydrolysis should be thus a second-order reaction. In aqueous solutions, water participates in this reaction as a base. It is present in large excess and therefore the reaction behaves like a kinetically first-order reaction.

III. $k_1 = 2k_2$. This condition is approximately fulfilled in glycerol diacetate hydrolysis. In this case the decrease in diester concentration follows second-order kinetics.

IV. *General treatment*, which can be applied to esters of some dibasic acids (e.g., succinic) where $k_1 > 2k_2$, leads to a complicated differential equation. A simple way to follow the kinetics is by means of dimensionless parameters [3-16], introduced as

$$
\alpha = \frac{[A]}{[A]_0} \qquad \beta = \frac{[B]}{[B]_0} \qquad (3.3\text{-}110)
$$

$$
\tau = [B]_0\, k_1 t \qquad \kappa = \frac{k_2}{k_1} \qquad (3.3\text{-}111)
$$

$$\alpha = \frac{1 - 2\kappa}{2(1 - \kappa)} \beta + \frac{1}{2(1 - \kappa)} \beta^\kappa \qquad (3.3\text{-}112)$$

At $[A]_0 = 2[B]_0$

$$\tau = \frac{1 - \kappa}{1 - 2\kappa} \int_\beta^1 \frac{d\beta}{\beta^2[1 + \beta^{\kappa-1}/(1 - 2\kappa)]} \qquad (3.3\text{-}113)$$

The dependence of the relative concentrations α and β on change in parameter τ is shown in Fig. 3-21. Calculation of the values of this parameter τ can be carried out by means of a tabulated function [3-16]. Again a plot of α or β as a function of log t is possible.

In another approach [3-17] concentrations of reactants B and C that react with compound A are measured. Using the differential equations (3.3-114) and (3.3-115):

$$\frac{d[B]}{dt} = -k_1[A][B] \qquad (3.3\text{-}114)$$

$$\frac{d[C]}{dt} = k_1[A]([B] - \kappa[C]) \qquad (3.3\text{-}115)$$

where $\kappa = k_2/k_1$ and by introducing

$$\beta = \frac{[B]}{[B]_0} \qquad \gamma = \frac{[C]}{[C]_0} \qquad (3.3\text{-}116)$$

it is possible to obtain from the ratio of the two differential equations the expression:

$$\frac{d\gamma}{d\beta} = \frac{\kappa\gamma - \beta}{\beta} \qquad (3.3\text{-}117)$$

and by integration

$$\kappa \ln \gamma = \ln (\beta + (1 - \kappa) \gamma) \qquad (3.3\text{-}118)$$

or

$$\gamma = \frac{\beta (1 - \beta^{\kappa-1})}{(\kappa - 1)} \qquad (3.3\text{-}119)$$

From the dependence of β on γ it is possible to obtain an approximate value of κ graphically. For $\kappa < 0.05$ the system simplifies to

$$\kappa \approx \frac{\ln (\beta + \gamma)}{\ln [1 + \gamma/(\beta + \gamma)]} \qquad (3.3\text{-}120)$$

A full treatment of the system has also been reported in refs. 3-18 and 3-19.

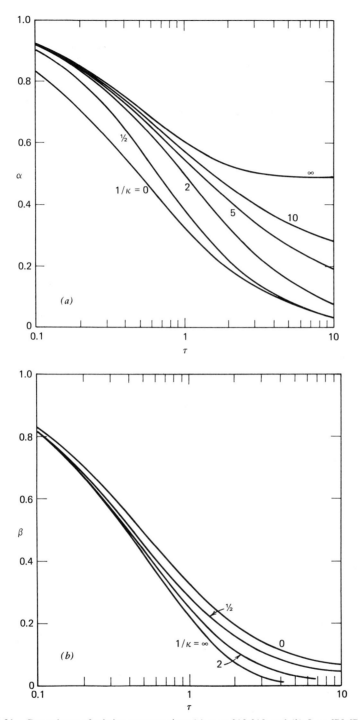

FIG. 3-21. Dependence of relative concentrations (a) $\alpha = [A]/[A]_0$ and (b) $\beta = [B]/[B]_0$ on time parameter $\tau = [B]_0 k_1 t$ for system of consecutive second-order reactions following eqs. (3.3-106) and (3.3-107) for varying values of the relative rate constant $\kappa = k_2/k_1$ (from A. A. Frost and R. G. Pearson, *Kinetics and Mechanism*, 2nd ed., Wiley, New York, 1961, pp. 182–3).

The procedure proposed by Frost and Schwemer [3-16] has been generalized [3-20] to include the case of nonstoichiometric initial ratio of the two reacting components, and a corresponding reaction scheme corresponding to hydrolysis of triesters was treated by means of approximation procedures yielding the three rate constants [3-21, 3-22]. The determination of individual rate constants is based on minimization of an appropriately chosen objective function [3-23, 3-24].

(b) COMPETITIVE PARALLEL SECOND-ORDER REACTIONS. An example of the scheme

$$A + B \xrightarrow{k_1} D + F \qquad\qquad (3.3\text{-}121)$$

$$A + C \xrightarrow[k_2]{} E + F \qquad\qquad (3.3\text{-}122)$$

is the alkaline hydrolysis of a mixture of simple esters carried out under conditions where the reaction mixture does not contain an excess of hydroxide ions.

$$R^1COOR + OH^- \xrightarrow{k_1} R^1COO^- + ROH \qquad (3.3\text{-}123)$$
$$\;\;\;(B)\qquad\quad (A)\qquad\qquad (D)\qquad\quad (F)$$

$$R^2COOR + OH^- \xrightarrow{k_2} R^2COO^- + ROH \qquad (3.3\text{-}124)$$
$$\;\;\;(C)\qquad\quad (A)\qquad\qquad (E)\qquad\quad (F)$$

This treatment can be applied for the analysis of mixtures of esters R^1COOR and R^2COOR but it is not useful for the elucidation of mechanisms, because in most instances it is possible—and simpler—to study the reaction of both esters separately.

In those cases where such treatment is needed, for example, in reaction of a mixture of isomers, it is possible to apply, after a minor modification, the procedure proposed for the previous case. In practice, this treatment has been applied to the study of oxidation of a mixture of diastereomeric diols by periodate [3-25].

Among more complicated types of competitive reactions, solutions have been found for the system

$$A + B \xrightarrow{k_1} C \qquad\qquad (3.3\text{-}125)$$
$$B + C \xrightarrow{k_2} D \qquad\qquad (3.3\text{-}126)$$
$$B \xrightarrow{k_3} P \qquad\qquad (3.3\text{-}127)$$

which was encountered in the kinetics of some coupling reactions [3-26]. Treatment has been proposed for the system (3.3-128) to (3.3-130)

$$A + B \xrightarrow{k_1} C + E \qquad\qquad (3.3\text{-}128)$$
$$A + C \xrightarrow{k_2} D + E \qquad\qquad (3.3\text{-}129)$$
$$A + D \xrightarrow{k_3} F + E \qquad\qquad (3.3\text{-}130)$$

dealt with in connection of the alkaline hydrolysis of 1,3,5-tri-(4-carboxymethoxy-phenyl)benzene [3-27]. Even for the type (3.3-131) to (3.3-134):

$$A + B \xrightarrow{k_1} C + F \qquad (3.3\text{-}131)$$
$$A + C \xrightarrow{k_2} D + G \qquad (3.3\text{-}132)$$
$$A + B \xrightarrow{k_3} E + G \qquad (3.3\text{-}133)$$
$$A + E \xrightarrow{k_4} D + F \qquad (3.3\text{-}134)$$

calculation of the rate constants is possible [3-27]. When a reagent B (which has one reactive center) reacts with a molecule A (which has two reactive centers that are not equivalent), it is necessary to consider two pairs of consecutive steps:

$$\text{1st reactive center} \qquad A + B \xrightarrow{k_1} C \qquad (3.3\text{-}135)$$
$$B + C \xrightarrow{k_2} D \qquad (3.3\text{-}136)$$
$$\text{2nd reactive center} \qquad A + B \xrightarrow{k_3} E \qquad (3.3\text{-}137)$$
$$B + E \xrightarrow{k_4} D \qquad (3.3\text{-}138)$$

Substance A can react with substance B at two reactive centers to give position isomers C and E which react with another molecule of B to give common product D.

An example of a chemical reaction where such a scheme is assumed to be operating [3-28] is the coupling of histidine (A) with p-diazobenzene sulfonic acid (B) giving two monoazoderivatives (C and E) and a bisazo compound (D).

(3.3-139)

Finally, in the reaction of a monofunctional substance B with compound A which is monofunctional in one center and n-functional in two further reaction centers, the following scheme must be considered:

$$\text{1st reactive center} \qquad A + B \xrightarrow{k_1} C \qquad (3.3\text{-}140)$$

$$nB + C \xrightarrow{k_2} D \qquad (3.3\text{-}141)$$

$$\text{further reactive center} \qquad A + nB \xrightarrow{k_3} E \qquad (3.3\text{-}142)$$

$$B + E \xrightarrow{k_4} D \qquad (3.3\text{-}143)$$

An example is the reaction of tryptophane (A) with p-diazobenzene sulfonic acid (B) where compounds C, D, and E result in reactions on amino groups or on the indole ring respectively, the latter reacting faster than the amino groups [3-29].

3.3.2.5 Reactions Involving Equilibria

Reversible reactions that will be discussed here involve first-order reactions, reactions which are first and second order, reactions which are second order in both directions, as well as, exchange reactions and more complex types of reactions.

(a) REVERSIBLE REACTIONS OF FIRST ORDER. For reaction of the type

$$A \underset{k_{-1}}{\overset{k_1}{\rightleftharpoons}} B \qquad (3.3\text{-}144)$$

if $[B]_0 = 0$, it can be shown that

$$(k_1 + k_{-1})t = \ln \frac{[A]_0 - [A]_e}{[A] - [A]_e} \qquad (3.3\text{-}145)$$

where $[A]_e$ is concentration of the species A in equilibrium. The form of this equation resembles that of an equation for an irreversible first-order reaction, if we consider as the active concentration not the analytical concentration of compound A, but rather that part of the concentration of the substance A which will undergo conversion in the reaction—that is $([A]_0 - [A]_e)$.

To show that the experimental data fit the above equation and to obtain the value of the overall rate constant characterizing establishment of the equilibrium (i.e., $k = k_1 + k_{-1}$) it is possible again to employ a graphical treatment, if $N = \log ([A] - [A]_e)$ is plotted as a function of time (Fig. 3-6).

The slope of the linear plot is $M/P = (k_1 + k_{-1})/2.3$. Separation of individual constants k_1 and k_{-1} is possible either by determination of the value of constant k_1 from the initial rate or by combination of $(k_1 + k_{-1})$ with the value of $K = k_1/k_{-1}$. When the equilibrium constant is used in the computation, it is necessary to be sure that the studied equilibrium really corresponds to the simple scheme $A \rightleftharpoons B$ and that the system does not involve more complex equilibria.

(b) REVERSIBLE REACTIONS OF FIRST AND SECOND ORDER. For reactions of the type

$$A \underset{k_{-1}}{\overset{k_1}{\rightleftharpoons}} B + C \qquad (3.3\text{-}146)$$

if $[B]_0 = [C]_0 = 0$, it is possible to derive the relation:

$$k_1 \frac{([A]_0 + [A]_e)t}{[A]_0 - [A]_e} = \ln \frac{[A]_0^2 - [A][A]_e}{([A] - [A]_e)[A]_0} \qquad (3.3\text{-}147)$$

If the following reaction variables are introduced

$$x = [A]_0 - [A] \qquad (3.3\text{-}148)$$

$$x_e = [A]_0 - [A]_e = [B]_e = [C]_e \qquad (3.3\text{-}149)$$

$$a = [A]_0 \qquad (3.3\text{-}150)$$

where x_e is equilibrium concentration of either B or C, it is possible to obtain an expression (3.3-151):

$$k_1 \frac{2a - x_e}{x_e} t = \ln \frac{ax + x(a - x_e)}{a(x_e - x)} \qquad (3.3\text{-}151)$$

If a given system follows reaction $A \rightleftharpoons B + C$, the plot of $N = \log\{[ax + x(a - x_e)]/[a(x_e - x)]\}$ as a function of time must be linear (Fig. 3-6) and the slope $M/P = (k_1/2.3)/[(2a - x_e)/x_e]$.

For example, such treatment was found successful for data obtained for hydrolysis of ethyl acetate in unbuffered media.

A similar treatment applies in the reverse case, where the reaction follows the scheme

$$B + C \underset{k_{-1}}{\overset{k_1}{\rightleftharpoons}} A \qquad (3.3\text{-}152)$$

the initial reaction is second order, provided that at the beginning of reaction $[A]_0 = 0$.

If furthermore

$$a = [B]_0 = [C]_0 \qquad (3.3\text{-}153)$$

then

$$x = [B]_0 - [B] = [C]_0 - [C] \qquad (3.3\text{-}154)$$

and

$$x_e = [B]_0 - [B]_e = [A]_e \qquad (3.3\text{-}155)$$

The integrated form of the reaction rate equation is given by (3.3-156):

$$k_1 \frac{a^2 - x_e^2}{x_e} t = \ln \frac{x_e(a^2 - xx_e)}{a^2(x_e - x)} \qquad (3.3\text{-}156)$$

The validity of this equation has been confirmed for the reaction of ethanol with acetic acid. A similar graphical treatment {for $N = [\log x_e(a^2 - xx_e)]/[a^2(x_e - x)]$}, as in the previous case, can be adopted.

(c) REVERSIBLE REACTIONS OF SECOND ORDER. For reactions of type

$$A + B \underset{k_{-1}}{\overset{k_1}{\rightleftharpoons}} C + D \qquad (3.3\text{-}157)$$

a general solution is possible, but too complicated and rather unsuitable for experimental verification. To prove that a studied reaction follows the above scheme it is recommended to study the reaction at equal initial concentrations ($[A]_0 = [B]_0$), in which case (3.3-158) applies:

$$k_1 \frac{2a(a - x_e)}{x_e} t = \ln \frac{x(a - 2x_e) + ax_e}{a(x_e - x)} \qquad (3.3\text{-}158)$$

where x_e is the equilibrium concentration of compound C or D and other symbols have the same meaning as in the previous case.

Graphical verification can be carried out in a similar way as for reactions of first and second order and $N = \log\{[x(a - 2x_e) + ax_e]/[a(x_e - x)]\}$ is plotted as a function of time; then the slope $M/P = (k_1/2.3)/[2a(a - x_e)/x_e]$.

The same general equation can be used for the reaction of the type (3.3-159):

$$2A \rightleftharpoons C + D \qquad (3.3\text{-}159)$$

All the expressions quoted in sections(b) and (c) of this chapter are used for determination of the value of the rate constant k_1 from measurements during the part of a kinetic run when the equilibrium is not yet established. If the system is sufficiently shifted from equilibrium, it is possible to determine the difference in concentrations, whereas in the vicinity of the equilibrium the difference $(x_e - x)$ becomes small, and the determination of the rate constants by most methods becomes inaccurate.

When the rate constant of the reverse reaction k_{-1} is to be calculated from the rate constant k_1 and the equilibrium constant $K = k_1/k_{-1}$ is used in this calculation, it is always necessary to check that the equilibrium is only a one-step process. For some complex reactions under conditions when $v_1 = v_{-1}$ the ratio of concentrations can be an exponential function of the rate constant. This applies, for example, to a system described by equation (3.3-160), where the reactions with rate constants k_1 and k_{-2} are rate determining and only the overall equilibrium constants $K = [C]/[A]$ is measured.

If the equilibrium is rapidly established, or if rates are to be measured in the time interval close to the establishment of the equilibrium, special techniques for the study of fast reactions described in Chapter 4 must be applied. However, this treatment allows determination of both k_1 and k_{-1}, although an independent knowledge of the equilibrium constant $K = k_1/k_{-1}$ is necessary. Potentiometric measurements are also useful in studying the rates of approach to equilibria, since a logarithmic function of concentration is involved (Section 2.4.3.10).

(d) Two Equilibrium First-Order Reactions. For the reaction

$$A \underset{k_{-1}}{\overset{k_1}{\rightleftharpoons}} B \underset{k_{-2}}{\overset{k_2}{\rightleftharpoons}} C \tag{3.3-160}$$

the expressions involved are rather complex. For concentrations of individual components it is possible [3-30] to obtain the following equations:

$$[A] = [A]_0 \left\{ \frac{k_{-1}k_{-2}}{\lambda_2\lambda_3} + \frac{k_1(\lambda_2 - k_2 - k_{-2})}{\lambda_2(\lambda_2 - \lambda_3)} e^{-\lambda_2 t} \right.$$
$$\left. + \frac{k_1(k_2 + k_{-2} - \lambda_3)}{\lambda_3(\lambda_2 - \lambda_3)} e^{-\lambda_3 t} \right\} \tag{3.3-161}$$

$$[B] = [A]_0 \left\{ \frac{k_1 k_{-2}}{\lambda_2\lambda_3} + \frac{k_1(k_{-2} - \lambda_2)}{\lambda_2(\lambda_2 - \lambda_3)} e^{-\lambda_2 t} + \frac{k_1(\lambda_3 - k_{-2})}{\lambda_3(\lambda_2 - \lambda_3)} e^{-\lambda_3 t} \right\} \tag{3.3-162}$$

$$[C] = [A]_0 \left\{ \frac{k_1 k_2}{\lambda_2\lambda_3} + \frac{k_1 k_2}{\lambda_2(\lambda_2 - \lambda_3)} e^{-\lambda_2 t} - \frac{k_1 k_2}{\lambda_3(\lambda_2 - \lambda_3)} e^{-\lambda_3 t} \right\} \tag{3.3-163}$$

where

$$\lambda_2 = \tfrac{1}{2}(p + r) \qquad \lambda_3 = \tfrac{1}{2}(p + r) \tag{3.3-164}$$

for

$$p = k_1 + k_{-1} + k_2 + k_{-2} \tag{3.3-165}$$

and

$$r = [p^2 - 4(k_1 k_2 + k_{-1}k_{-2} + k_1 k_{-2})]^{1/2} \tag{3.3-166}$$

Depending upon the relative values of rate constants, the value of [B] may or may not pass through a maximum [3-30]. Approximate values of rate constants can be obtained by modeling the dependence of [B] or [C] on time.

Solution for systems involving four components linked by equilibria is possible using Laplace transforms [3-31].

(e) COMPLEX SYSTEMS. A frequently encountered system involves a chemical equilibrium followed by an irreversible reaction:

$$A \underset{k_{-1}}{\overset{k_1}{\rightleftharpoons}} B \xrightarrow{k_2} C \tag{3.3-167}$$

The time changes of individual components can be expressed by equations (3.3-168) to (3.3-170):

$$\frac{d[A]}{dt} = k_{-1}[B] - k_1[A] \tag{3.3-168}$$

$$\frac{d[B]}{dt} = k_1[A] - k_{-1}[B] - k_2[B] \tag{3.3-169}$$

$$\frac{d[C]}{dt} = k_2[B] \tag{3.3-170}$$

Hence it is necessary to be able to follow concentration changes of at least two components of the reaction mixture, if all rate constants are to be determined.

If the initial concentrations of the intermediate B and the product C are equal to zero:

$$[B]_0 = [C]_0 = 0 \tag{3.3-171}$$

then

$$[A]_0 = [A] + [B] + [C] \tag{3.3-172}$$

at any time t and

$$\frac{d[A]}{dt} + \frac{d[B]}{dt} + \frac{d[C]}{dt} = 0 \tag{3.3-173}$$

A simplified solution on the rate equation for the system in equilibrium followed by a consecutive irreversible reaction is possible, when it can be shown that one of two conditions is valid: (1) that in the course of reaction the concentration of [B] remains small and practically constant, or (2) that an equilibrium between A and B is established in the studied solution.

The rate equation is simplified in the first case by using the relationship describing the steady-state concentration of B:

$$\frac{d[B]}{dt} = k_1[A] - k_{-1}[B] - k_2[B] = 0 \tag{3.3-174}$$

From the above

$$[B] = \frac{k_1[A]}{k_{-1} + k_2} \tag{3.3-175}$$

and hence

$$-\frac{d[A]}{dt} = \frac{d[C]}{dt} = \frac{k_1 k_2}{k_{-1} + k_2}[A] \tag{3.3-176}$$

This relationship will be considered as *the simple steady-state approximation*. The extent of deviations of $d[C]/dt$ calculated by means of this expression from the values of $d[C]/dt$ calculated by means of the exact expression has recently been discussed [3-14].

Further approximation can be obtained [3-32] by expressing the value of $d[B]/dt$ from the equation

$$[B] = \frac{k_1[A]}{k_{-1} + k_2} \tag{3.3-177}$$

that is,

$$\frac{d[B]}{dt} = \frac{k_1}{k_{-1} + k_2} \frac{d[A]}{dt} \tag{3.3-178}$$

As previously stated

$$\frac{d[C]}{dt} = -\frac{d[A]}{dt} - \frac{d[B]}{dt} \tag{3.3-179}$$

and combining with the previous expression

$$\frac{d[C]}{dt} = -\frac{d[A]}{dt} - \frac{k_1}{k_{-1} + k_2} \frac{d[A]}{dt} \tag{3.3-180}$$

Because

$$\frac{d[C]}{dt} = k_2[B] = \frac{k_1[A]}{k_{-1} + k_2} \tag{3.3-181}$$

it is possible to combine both equations and write

$$-\frac{d[A]}{dt} = \frac{k_1 k_2}{k_1 + k_{-1} + k_2} [A] \qquad (3.3\text{-}182)$$

This expression can be denoted as *the improved steady-state approximation.*

Alternatively, when an equilibrium can be established in the solution between the species A and B, then

$$-\frac{d[A]}{dt} = \frac{d[C]}{dt} = k_2 [B] \qquad (3.3\text{-}183)$$

Since at equilibrium $[B] = k_1 [A]/k_{-1}$

hence

$$-\frac{d[A]}{dt} = \frac{k_1 k_2}{k_{-1}} [A] \qquad (3.3\text{-}184)$$

This expression will be referred to as *the simple equilibrium approximation.* The equation for the steady-state approximation will become equal to equation (3.3-184) if $k_{-1} \gg k_2$. That means that establishment of the equilibrium takes place, if the reverse transformation of the intermediate product B into the starting material is considerably faster than its cleavage into C.

Similarly as in the case of steady state, also under equilibrium conditions it is possible to consider further approximation [3-32]. Using the equilibrium condition $[B] = k_1 [A]/k_{-1}$ which upon differentiation yields

$$\frac{d[B]}{dt} = \frac{k_1}{k_{-1}} \frac{d[A]}{dt} \qquad (3.3\text{-}185)$$

a combination with (3.3-173) yields

$$-\frac{d[A]}{dt} \left(1 + \frac{k_1}{k_{-1}} \right) = \frac{d[C]}{dt} = k_2 [B] \qquad (3.3\text{-}186)$$

and inserting for [B] from the equilibrium condition and rearranging gives

$$-\frac{d[A]}{dt} = \frac{k_1 k_2}{k_1 + k_{-1}} [A] \qquad (3.3\text{-}187)$$

This expression will be denoted as *the improved equilibrium approximation.*

Both improved approximations can be used for calculation of [A] and [B] provided that for the case of an unstable intermediate $k_1 \ll (k_{-1} + k_2)$ or for the case when $k_2 \ll k_{-1}$ and a slow formation of the intermediate makes the establishment of the equilibrium possible. For calculation of [C] the application of "improved approximations" is hindered only when k_1 becomes similar to k_2 and k_{-1} becomes smaller than either k_1 or k_2.

The suitability of these approximations can be demonstrated by considering several possible cases for the relative values of the rate constants in this type of reaction system. In some cases, the rate equation will simplify even further.

(a) $k_{-1} \approx k_2 \gg k_1$. The intermediate concentration remains low because the rate of its formation is slow when compared with the forward and reverse reactions of the intermediate. The steady-state approximation (3.3-176) and the improved steady-state approximation (3.3-182) can be used equally well because $k_1 \ll (k_{-1} + k_2)$. The equilibrium approximations (3.3-184) and (3.3-187) cannot be used, because the rate of the cleavage of the intermediate with constant k_2 cannot be neglected when compared with the rate of the reverse reaction with constant k_{-1} yielding the reactant A.

(b) $k_1 \approx k_{-1} \gg k_2$. The equilibrium between A and B is established much faster when compared to the rate of formation of the product C from intermediate B. The equilibrium approximation (3.3-184) thus applies, but so does the improved steady-state approximation (3.3-182), since $k_2 < (k_1 + k_{-1})$. The simple steady-state approximation (3.3-176) and improved equilibrium approximation (3.3-187) cannot be used, since the value of k_1 cannot be neglected against $(k_{-1} + k_2)$.

(c) $k_{-1} \gg k_1 \approx k_2$. A low, steady-state concentration of B is established, but since $k_1 < k_{-1}$ and $k_2 < k_{-1}$, either of the four approximations can be used.

(d) $k_{-1} \gg k_1 \gg k_2$. An equilibrium is established with a low concentration of B. Because $k_2 < k_{-1}$, the improved steady-state approximation (3.3-182), the equilibrium approximation (3.3-184) and the improved equilibrium approximation (3.3-187) all apply, but not the simple steady-state approximation (3.3-176).

(e) $k_{-1} \gg k_2 \gg k_1$. If the rate of cleavage of the unstable intermediate is greater than the rate of its formation, then the overall rate is governed by the rate of formation of the unstable intermediate. The rate equation simplifies to the equation for a simple first-order reaction:

$$\frac{-d[A]}{dt} = k_1[A] \tag{3.3-188}$$

(f) $k_2 \gg k_{-1} \gg k_1$. As in the preceding case, the rapid cleavage of the unstable intermediate occurs and the reaction rate also follows a simple, first-order rate equation (3.3-188).

(g) $k_1 \gg k_{-1} \gg k_2$. The equilibrium is shifted in favor of B, but product formation is slow. An accumulation of the intermediate B occurs early in the reaction. The simple equilibrium approximation (3.3-184) best describes the change in concentration of A.

(h) $k_1 \gg k_2 \gg k_{-1}$. In this case a relatively stable intermediate is formed, but as k_{-1} can be neglected when compared to k_1 and k_2, the treatment becomes equivalent to that of a system of first-order irreversible consecutive reactions (3.3-65) discussed in Section 3.3.2.3a).

(i) $k_2 \gg k_1 \gg k_{-1}$. As in cases e) and f), the unstable intermediate is rapidly converted into products, with the rate equation becoming that for a simple, first-order reaction (3.3-188).

Thus, with the exception of case g) involving establishment of an equilibrium between A and B followed by a slow conversion of B, in all cases, where the scheme (3.3-167) does not degenerate to simpler processes, the improved steady-state approximation (3.3-182) is the only form which can be always successfully applied. Unsuitability of the description of processes following equation (3.3-167) in terms of the rate-determining step has recently been pointed out [3-11].

In cases where application of the above approximations is impossible, in particular when the individual reaction rates become comparable, the rate equations are more complicated. In cases when an accumulation of the intermediate occurs in the course of the reaction, the possibility of monitoring concentration changes of this intermediate contributes considerably to elucidation of mechanisms of such reactions.

Concentration changes of the dimer A with time in the sequence (3.3-189) to (3.3-190):

$$A \underset{k_{-1}}{\overset{k_1}{\rightleftharpoons}} 2B \qquad (3.3\text{-}189)$$

$$B \overset{k_2}{\longrightarrow} C \qquad (3.3\text{-}190)$$

have been calculated using a method of consecutive approximations [3-33].

Another reaction system commonly encountered in practice also involves an intermediate formed in equilibrium with reactants, namely

$$A + B \underset{k_{-1}}{\overset{k_1}{\rightleftharpoons}} C \overset{k_2}{\longrightarrow} D \qquad (3.3\text{-}191)$$

where one reaction participating in equilibrium is second order and the equilibrium is followed by a first-order consecutive reaction.

Such a system may be recognized by following the time-dependence of the concentration of substance A in the presence of an excess of substance B. Each kinetic run will then correspond to a pseudo-first-order kinetics in the decrease in concentration of A with formal rate constant k'. The dependence of the measured rate constant k' on substance B concentration can correspond to either of two cases:

1. It is observed that the measured value of the constant k' is not directly proportional to concentration of the reagent in excess (B), but rather

$$k' = \frac{k_1 k_2}{k_{-1}} \frac{[B]}{1 + k_1[B]/k_{-1}} \qquad (3.3\text{-}192)$$

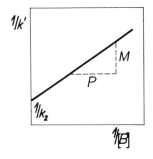

FIG. 3-22. Dependence of reciprocal value of measured pseudo first-order rate constant (k') on reciprocal concentration of the reagent in excess [B] for an equilibrium followed by a first-order reaction (3.3-191). Intercept equals $1/k_2$, $M/P = k_{-1}/k_1 k_2$.

which can be transformed into

$$\frac{1}{k'} = \frac{k_{-1}}{k_2 k_2} \frac{1}{[B]} + \frac{1}{k_2} \tag{3.3-193}$$

To prove that the reaction corresponds to the above equation, $1/k'$ obtained at varying [B] is plotted against $1/[B]$ (Fig. 3-22). The slope of the linear plot is $M/P = k_{-1}/k_1 k_2$ and the intercept at $1/[B] = 0$ is equal to $1/k_2$. Such treatment of experimental data is possible when $k_1[A][B] \gg k_{-1}[C]$ and simultaneously $k_2[C] \ll k_1[A][B] - k_{-1}[C]$. An example of such behavior is the reaction of ethylene glycol with periodic acid [3-34]

$$\begin{array}{c} CH_2OH \\ | \\ CH_2OH \end{array} + H_4IO_6^- \underset{k_{-1}}{\overset{k_1}{\rightleftharpoons}} \begin{array}{c} CH_2O \quad O \quad OH \\ \diagdown \quad \| \diagup \\ I \\ \diagup \quad \diagdown \\ CH_2O \quad O^- OH \end{array} + 2H_2O \tag{3.3-194}$$

$$\downarrow k_2$$

2.   In the second group of reactions the time change of the concentration of species A in each kinetic run in the presence of an excess of B corresponds to first-order kinetics, and simultaneously a linear relationship is found between the value of the constant k' and concentrations [B]. In such cases there are two possibilities which cannot be distinguished based solely on kinetic measurements.

(a)   Slow establishment of the equilibrium is followed by a rapid cleavage of the intermediate, so that $k_1[A][B] - k_{-1}[C] \ll k_2[C]$. If furthermore $k_1[A][B] \ll k_{-1}[C]$ the system behaves like an irreversible reaction of second order.

(b)   If the equilibrium is strongly shifted in favor of the starting material causing the concentration of the intermediate C to remain negligible, it is possible that the equilibrium is rapidly established and the cleavage of C to form D is the slow step. Under such conditions the reaction will also follow second-order kinetics. In this case the condition can be formulated as $k_1[A][B] \gg k_{-1}[C]$ and simultaneously $k_2[C] \ll k_1[A][B]$. Thus in both cases a strictly linear relationship between the experimentally accessible value of k' and concentration of B present in excess is found and hence second-order kinetics are observed and the two alternatives are kinetically indistinguishable. Such behavior has been observed [3-35, 3-36] for periodate oxidation of higher glycol homologues of the type $RCH(OH)CH(OH)R$, where either one of the steps leading to the formation of the cyclic intermediate or cleavage of the intermediate formed in a rapidly established equilibrium is involved.

Another important group of chemical reactions follows the type

$$A \underset{k_{-1}}{\overset{k_1}{\rightleftharpoons}} B + C \qquad (3.3\text{-}195)$$

$$B \overset{k_2}{\longrightarrow} D \qquad (3.3\text{-}196)$$

This class of reactions includes the solvolysis of alkyl chlorides for such alkyl groups and reaction conditions, where the first step results in an ionization of the alkyl halide and the reaction mechanism corresponds to a nucleophilic substitution of the type S_N1. This mechanism can be written as:

$$RCl \underset{k_{-1}}{\overset{k_1}{\rightleftharpoons}} R^+ + Cl^- \qquad (3.3\text{-}197)$$

$$R^+ + H_2O \overset{k_2}{\longrightarrow} ROH + H^+ \qquad (3.3\text{-}198)$$

Two conditions must be fulfilled to satisfy the application of the steady-state approximation. First, the concentration of water in the reaction mixture must be so high that it exhibits no detectable change in the course of reaction. Second, the concentration of the carbonium ion R^+ must remain small and practically constant during this time.

Hence if $[H_2O] = $ const, and $[R^+] = $ const, thus $d[R^+]/dt = 0$, the relationship (3.3-199):

$$\frac{-d[RCl]}{dt} = \frac{d[R^+]}{dt} + \frac{d[ROH]}{dt} \qquad (3.3\text{-}199)$$

can be simplified to

$$\frac{-d[RCl]}{dt} = \frac{d[ROH]}{dt} = k_2[R^+] \qquad (3.3\text{-}200)$$

Because

$$\frac{d[R^+]}{dt} = k_1[RCl] - k_{-1}[R^+][Cl^-] - k_2[R^+] = 0 \qquad (3.3\text{-}201)$$

hence

$$k_1[RCl] = k_{-1}[R^+][Cl^-] + k_2[R^+] \qquad (3.3\text{-}202)$$

and thus

$$[R^+] = \frac{k_1[RCl]}{k_{-1}[Cl^-] + k_2} \qquad (3.3\text{-}203)$$

Substituting into equation for $d[RCl]/dt$ gives:

$$\frac{-d[RCl]}{dt} = \frac{k_1 k_2[RCl]}{k_{-1}[Cl^-] + k_2} \qquad (3.3\text{-}204)$$

If we introduce $r = k_{-1}/k_2$, then

$$\frac{-d[RCl]}{dt} = \frac{k_1[RCl]}{r[Cl^-] + 1} \qquad (3.3\text{-}205)$$

Three alternatives can be considered:

1. If $v_2 \gg v_{-1}$ then, because $r[Cl^-] = v_{-1}/v_2$, $r[Cl^-] \ll 1$ and

$$\frac{-d[RCl]}{dt} = k_1[RCl] \qquad (3.3\text{-}206)$$

In this case the reaction rate measured from the decrease in [RCl] is independent of chloride ion concentration. Under such conditions it is impossible to calculate the value of r.

2. If $v_2 \ll v_1$, $r[Cl^-]$ is much larger than 1 and

$$\frac{-d[RCl]}{dt} = \frac{k_1[RCl]}{r[Cl^-]} \qquad (3.3\text{-}207)$$

Such cases can be distinguished because the reaction rate is indirectly proportional to chloride ion concentration at constant ionic strength. Nevertheless from the dependence of $d[RCl]/dt$ on $1/[Cl^-]$ it is possible to calculate only the ratio k_1/r, but it is impossible to separate k_1 and r.

3. In the transition cases, where the rate of both steps (v_2 and v_{-1}) are comparable, it is observed that the measured reaction rate of hydrolysis is a function of halide ion concentration and at a constant ionic strength this relationship becomes linear ($-d[RCl]/dt \sim 1/[Cl^-]$) at high halide concentrations.

Two methods are recommended for evaluation of the parameter r:

(a) The course of hydrolysis is followed as a function of time, that is, after selected intervals from the beginning of the reaction the concentration of the alkyl halide is determined. The equation for reaction rate can be integrated between the initial concentration $[RCl]_0$ and concentration $[RCl]_t$ after time t, and equation (3.3-208) can be used for calculation of r and k_1:

$$-k_1 t = r\left\{ [RCl]_0 \ln \frac{[RCl]_t}{[RCl]_0} + [RCl]_0 - [RCl]_t \right\} + \ln \frac{[RCl]_t}{[RCl]_0} \quad (3.3\text{-}208)$$

When experimentally the values of concentration $[RCl]_t$ are determined at two or more times t, it is possible from the pairs of equations to determine the values of k_1 and r by the method of successive approximations.

In the course of reaction an acid HX is liberated which increases the ionic strength. In order to keep the ionic strength effectively constant throughout the reaction it is necessary to add an appropriate amount of an inert electrolyte.

(b)   To determine the initial rate v_0 a series of reaction mixtures is prepared. Each contains the same initial concentration of the alkylhalide and varying concentrations of halide ions. The ionic strength is kept constant by addition of a neutral salt containing anion which does not show nucleophilic reactivity. The differential equation for the rate of hydrolysis is then transformed into (3.3-209):

$$\frac{1}{v_0} = \frac{r[Cl^-]_0 + 1}{k_1[RCl]_0} = \frac{r[Cl^-]_0}{k_1[RCl]_0} + \frac{1}{k_1[RCl]_0} \quad (3.3\text{-}209)$$

The values of r and k_1 can be determined when $1/v_0$ is plotted against $[Cl]_0$ (Fig. 3-23). The intercept on the $1/v_0$-axis at $[Cl]_0 = 0$ is equal to $1/k_1[RCl]_0$ and this enables us to calculate the value of k_1. The value of r is then obtained from the slope $M/P = r/(k_1[RCl]_0)$.

This method has a more general use than method (a) but the source of uncertainty is the determination of the initial velocity v_0. Two methods are recommended for the measurement of the values of v_0 with sufficient accuracy:

(A)   For a given reaction mixture an *apparent first-order rate constant* is defined by

$$\bar{k} = \frac{1}{t} \ln \frac{[RCl]_0}{[RCl]} \quad (3.3\text{-}210)$$

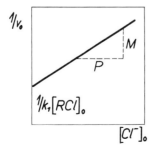

FIG. 3-23.   Graphical treatment of the dependence of reciprocal value of initial rate (v_0) on initial concentration of chloride ions for solvolysis of alkyl chlorides following S_N1 mechanism (equations 3.3-197 and 3.3-198). Intercept equals $1/k_1[RCl]_0$, $[M/P] = r/(k_1[RCl]_0)$.

This so-called "constant" is in fact not a physical constant since its value changes with time and the initial concentration of the reaction mixture. The value of $\bar{k}$ is found at several times t and from the dependence of k on time (Fig. 3-24) it is possible to extrapolate the value of $\bar{k}_0$ for each composition of reaction mixture, and then

$$\frac{1}{\bar{k}_0} = \frac{r[Cl^-]_0 + 1}{k_1} \tag{3.3-211}$$

Hence the values of r and k_1 can be determined graphically from the dependence of $1/\bar{k}_0$ on $[Cl^-]_0$ (Fig. 3-25) where $M/P = r/k_1$ and the intercept equals $1/k_1$.

(B) In this approach a number of reaction mixtures is prepared in order to keep the concentration of the halogen ion constant (i.e., using an excess of the halide). If $r[Cl^-] \ll 1$, the concentration change of the alkyl halide follows first-order kinetics with rate constant k_1'. The separation of values of k_1 and r can be carried out by means of equation (3.3-212):

$$\frac{1}{k_1'} = \frac{r[Cl^-]_0}{k_1} + \frac{1}{k_1} \tag{3.3-212}$$

using the same graphical method as in the previous case.

The actual measurements of rates of such complex reactions can be even more complicated. In practice, almost every system has its own difficult aspects. One of the fundamental problems is the control of ionic strength, since it is difficult to find electrolytes which have no specific, for example, catalytic, effects on the reaction rate. In particular, in media of low dielectric constant it is difficult to exclude the role of the change in ionic strength during the course of the reaction.

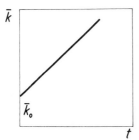

FIG. 3-24. Dependence of the "Apparent first-order rate constant" $\bar{k}$ on time for S_N1 solvolysis of alkyl chlorides, defined by equation (3.3-210). Intercept equals to extrapolated value $\bar{k}_0$.

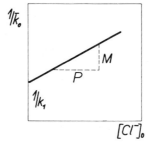

FIG. 3-25. Dependence of the reciprocal extrapolated value $\bar{k}_0$ on initial concentration of chloride ions for S_n1 solvolysis of alkyl chlorides, defined by equation (3.3-211). Intercept equals $1/k_1$, $M/P = r/k_1$.

For the system (3.3-213) to (3.3-215):

$$A \underset{k_{-1}}{\overset{k_1}{\rightleftharpoons}} B \qquad\qquad (3.3\text{-}213)$$

$$A \overset{k_2}{\longrightarrow} C \qquad\qquad (3.3\text{-}214)$$

$$B \overset{k_3}{\longrightarrow} D \qquad\qquad (3.3\text{-}215)$$

corresponding to a starting material that exists in two isomeric forms A and B in equilibrium, each of which reacts in a first-order (or pseudo-first-order) reaction to give a different product, the exact solution [3-37] was compared with approximations based on ratio of products formed and on consideration of the overall rate constant for total product formation as a time independent quantity.

(f) GENERAL CASES.   The steady-state approximation proved useful in treatment of some additional systems in which the formation of a reactive or unstable intermediate is assumed. In such cases the rate of the first step (v_1) must be smaller either than v_{-1} or v_2, but not necessarily both. A summary of some types of systems to which this treatment can be applied together with expressions for reaction rate is given in Table 3-4.

In the derivation of these expressions it was assumed that the concentration change of the intermediate X with time is numerically negligible when compared with rates of formation or cleavage of other species involved, that is, A, B, C, D, and Z.

If $v_{-1} \ll v_2$ it is possible to calculate the values of the rate constants k_1 and k_2 as for consecutive, irreversible reactions. Furthermore, if $v_1 \ll v_2$ (and not only $v_1 < v_2$ given in the original conditions), it is possible to calculate the value of k_1 as for simple first- or second-order reactions.

If, on the other hand, $v_{-1} \gg v_2$, it is possible to calculate the value of k_1/r (as in the second example in the preceding discussion of the S_N1 mechanism), but in a completely general case not the value of r alone.

If, alternatively, concentrations of C and D in reactions 3 through 7 can be chosen in such a way that the rate of the reverse reaction v_{-1} is comparable with the rate of the consecutive reaction v_2, it is possible to obtain values of both k_1 and r by plotting v_0 (the initial rate of decrease in concentration of A) against functions of initial concentrations $1/[C]_0$, $1/[D]_0$, or $[C]_0/[D]_0$, respectively.

It is possible to determine the values of the constant k_1 for reactions 1–7 (Table 3-4), if a suitable reaction can be found which will trap the intermediate X and transform it into unreactive form F. This can be achieved by adding to the reaction mixture a compound E which transforms X in an irreversible reaction, the rate of which is faster than that of reactions with rate constants k_{-1} and k_2 and consequently also than that with k_1. In such systems the decrease in concentration of A with time follows simple first- or second-order kinetics, which allows determination of k_1. From this value and the value of k_1/r, found in the reaction mixture in absence of the compound E, it is possible to calculate the value of r. Isolation of the rate

TABLE 3-4

Example of Systems of Equilibrium and Consecutive Reactions

Reaction	Mechanism	Rate $-d[A]/dt$	Constant Accessible for Evaluation[a] $r = \dfrac{k_{-1}}{k_2}$
1	$A \underset{k_{-1}}{\overset{k_1}{\rightleftharpoons}} X \xrightarrow{k_2} \text{Product}$	$\dfrac{k_1[A]}{r+1}$	$\dfrac{k_1}{r+1}$
2	$A + B \underset{k_{-1}}{\overset{k_1}{\rightleftharpoons}} X \xrightarrow{k_2} \text{Product}$	$\dfrac{k_1[A][B]}{r+1}$	$\dfrac{k_1}{r+1}$
3	$A \underset{k_{-1}}{\overset{k_1}{\rightleftharpoons}} C + X \xrightarrow{k_2} \text{Product}$	$\dfrac{k_1[A]}{r[C]+1}$	k_1 and r
4	$A + B \underset{k_{-1}}{\overset{k_1}{\rightleftharpoons}} C + X \xrightarrow{k_2} \text{Product}$	$\dfrac{k_1[A][B]}{r[C]+1}$	k_1 and r
5	$A \underset{k_{-1}}{\overset{k_1}{\rightleftharpoons}} X + D \longrightarrow \text{Product}$	$\dfrac{k_1[A]}{r[D]+1}$	k_1 and r
6	$A \underset{k_{-1}}{\overset{k_1}{\rightleftharpoons}} C + X + D \xrightarrow{k_2} \text{Product}$	$\dfrac{k_1[A]}{r[C]/[D]+1}$	k_1 and r
7	$A + B \underset{k_{-1}}{\overset{k_1}{\rightleftharpoons}} C + X + D \xrightarrow{k_2} \text{Product}$	$\dfrac{k_1[A][B]}{r[C]/[D]+1}$	k_1 and r
8	$A \underset{k_{-1}}{\overset{k_1}{\rightleftharpoons}} X \underset{k_{-2}}{\overset{k_2}{\rightleftharpoons}} Z$	$\dfrac{k_1[A] - rk_{-2}[Z]}{r+1}$	$\dfrac{k_1}{r+1}$ and $\dfrac{rk_{-2}}{r+1}$

[a]Constants, which can be evaluated, if v_{-1} is of the same order of magnitude as v_{-2} and if the decrease in [A] is measured as a function of initial composition of the reaction mixture.

Note: Table modified from G. A. Russell, Competing Reaction, in Investigation of Rates and Mechanisms of Reactions (S. L. Friess, E. S. Lewis, and A. Weissberger, Eds.), 2nd ed., Part 1, Interscience, New York, 1961, p. 352.

constant k_1 is in some cases also possible by measuring the rate of isotopically labeled compounds.

For the study of reactions of this type, it is of course most advantageous if physicochemical techniques can be applied which enable determination (preferably continuous) of the intermediate concentration in the course of reaction. Such techniques enable us not only to establish the proposed mechanism, but also to determine directly both k_1 and k_2. From an equation of the type (3.3-216):

$$\frac{-d[X]}{dt} = k_2[X] \tag{3.3-216}$$

it is then possible to determine k_2 and from known value of r to calculate the value of k_{-1}.

Two further examples of reactions involving equilibria will be discussed next. For the system

$$A + B \underset{k_{-1}}{\overset{k_1}{\rightleftharpoons}} X_1 + C \tag{3.3-217}$$

$$X_1 + D \underset{k_{-2}}{\overset{k_2}{\rightleftharpoons}} X_2 + E \tag{3.3-218}$$

$$X_2 + F \overset{k_3}{\longrightarrow} \text{Products} \tag{3.3-219}$$

it is possible to derive an equation

$$\frac{-d[A]}{dt} = \frac{v_1}{1 + v_{-1}/v_2 + v_{-1}v_{-2}/v_2 v_3}$$

$$= \frac{k_1[A][B]}{1 + r'[C]/[D] + r'[C]r''[E]/[D][F]} \tag{3.3-220}$$

where $r' = k_{-1}/k_2$ and $r'' = k_{-2}/k_3$.

The values of r', r'', and k_1 can be determined under favorable conditions by measuring the decrease in concentration of the starting material A, using the known initial concentrations of substances A, B, C, D, E, and F.

Another important type is the system

$$HA \underset{k_{-1}}{\overset{k_1}{\rightleftharpoons}} A^- + H^+ \tag{3.3-221}$$

$$A^- + B \overset{k_2}{\longrightarrow} \text{Product} \tag{3.3-222}$$

$$HA + B \overset{k_3}{\longrightarrow} \text{Product} \tag{3.3-223}$$

This type of system is assumed to be involved in one of the mechanisms proposed by Euler for the acid–base catalyzed reactions, in which the acid and base catalyzed reactions follow different patterns (Section 3.6.1.3).

If the rate v_1 is considerably faster than v_2 and v_3 then

$$-\frac{d[B]}{dt} = k_2[A^-][B] + k_3[HA][B] \qquad (3.3\text{-}224)$$

Expressing $[A^-]$ and $[HA]$ by means of $K_a = k_1/k_{-1}$ and $S_{HA} = [HA] + [A^-]$ it follows

$$-\frac{d[B]}{dt} = \left[\frac{k_2 + k_3[H^+]/K_a}{1 + [H^+]/K_a}\right] S_{HA}[B] \qquad (3.3\text{-}225)$$

This equation shows that in buffered solutions at a given pH the reaction follows second-order kinetics with the expression in brackets as the apparent rate constant k'. Change in acidity results in the change in both v_2 and v_3 and thus in a change of the value of the measured constant k'. Changing the pH-value of the buffer makes it possible to determine values of k_2 and k_3 (denoted for catalytic reactions as k_B and k_{HA}) (Section 3.6.1.4).

Similarly as in the case of simpler reactions the description of the treatment of experimental data will be closed by the description of a treatment which avoids the usual, time consuming trial-and-error method. The proposed [3-38] application of dimensionless parameters allows selection of a proper rate equation and estimate of the values of rate constants, but it is not completely general and can be applied only to some types of reactions. Nevertheless, its application can save considerable time in finding the best treatment.

This procedure can be applied to reactions the rate of which can be expressed in form

$$\frac{dx}{dt} = \kappa(1 - x)^\mu(1 - \alpha x)^\nu \qquad (3.3\text{-}226)$$

where κ, α, μ, and ν can be first assumed to be arbitrary parameters. The following examples of these types of reactions can be cited:

1. Equilibrium reactions of first order

$$(\mu = 0, \nu = 1, \alpha > 1)$$

2. Competitive reactions, one of first order, one of zeroth order

$$(\mu = 0, \nu = 1, 1 > \alpha > 0)$$

3. Competitive reactions, one of first order, one of second order

$$(\mu = 1, \nu = 1, 1 > \alpha > 0)$$

4. Numerous catalyzed reactions

($v > 0$ for catalyzed, $v < 0$ for negatively catalyzed, $\alpha > 0$ for reactions catalyzed by a reactant, $\alpha < 0$ for reaction catalyzed by an intermediate or by a product)

To determine the values of μ, v, and α, the value of n_x is found first as the slope of the dependence of log dx/dt on log $(1 - x)$. From the graph of the dependence of $(n_x - \mu)/v$ on x for various α, the values of μ, v, and α can be found by successive approximations (Fig. 3-26).

This method makes it possible to detect deviations from simple systems, indicating more complex, particularly catalytic mechanisms.

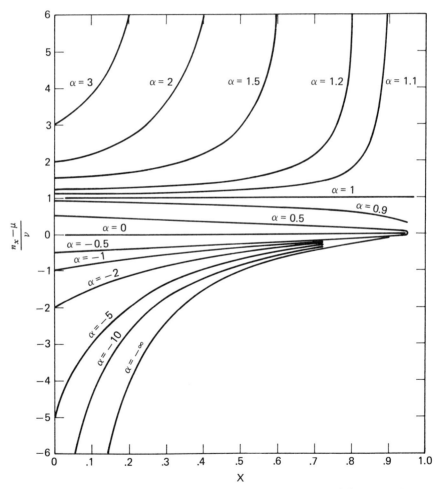

FIG. 3-26. Method of dimensionless parameters applicable to reaction types following equation (3.3-226). n_x is found from dependence of rate on log $(1 - x)$. Plot of $(n_x = \mu)/v$ against x is given for various α. (From ref. 3-38.)

3.4. DETERMINATION OF REACTION ORDER

In previous sections it has been demonstrated that determination of reaction order is frequently inseparable from finding the best reaction rate equation and sometimes follows from the equation found. A separate determination of reaction order may be carried out as part of the preliminary operations. This simplifies our task of finding the empirical rate expression. The determination of reaction order in individual components helps us to verify the form of the rate equation with respect to concentration of an individual component. It may also be used in the less demanding approaches when only the nature of the reactive species and the rate determining steps are established rather than the detailed kinetics.

Reaction order can be determined either as the overall reaction order or the order in individual components.

When the overall reaction order is determined, it is usually most convenient to choose identical initial concentrations of all components. The reaction rate equation in this way acquires the simplest form which can be verified by graphical or numerical methods described in the preceding section. The exponent n in $dx/dt = k(a - x)^n$ is then equal to the overall reaction order.

Three methods are available to determine the reaction order in individual components, in addition to the comparison with the integrated form of the rate equation (which can be carried out graphically or numerically, but is rather insensitive to distinction between smaller differences in reaction orders—e.g., between $n = 1$ and $n = \frac{3}{2}$): (a) measurement of half-times; (b) using the concentration of one or more components in excess; and (c) using differential forms of rate equations.

3.4.1. Determination of Reaction Order by Measurement of Half-Times or Fractional Times

Measurement of half-times or fractional times can be used for determination of reaction order for reactions where the rate equation has the form

$$dx/dt = k(a - x)^n \tag{3.4-1}$$

The half-time of a reaction $(\tau_{1/2})$ is defined by the time needed for conversion of half of a chosen reactant.*

For $n = 1$

$$\tau_{1/2} = \frac{\ln 2}{k} \tag{3.4-2}$$

$n \neq 1$

$$\tau_{1/2} = \frac{2^{n-1} - 1}{k(n - 1)a^{n-1}} \tag{3.4-3}$$

*It is not, as is sometimes wrongly assumed, half of the time needed for reaction to take place, since the completion time for the majority of reactions approaches infinity.

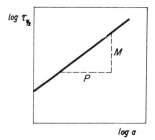

FIG. 3-27. Determination of the reaction order (n) from the dependence of the half-time ($\tau_{1/2}$) on logarithm of the initial concentration a. $M/P = 1 - n$.

For first-order reactions the value of $\tau_{1/2}$ is independent of the initial concentration of a, whereas for higher order reactions the value of $\tau_{1/2}$ depends on the initial concentration of a.

To determine the reaction order in the latter case when the measured value of the half-time $\tau_{1/2}$ is a function of the initial concentration, eq. (3.4-3) is transformed into (3.4-4):

$$\log \tau_{1/2} = \log f(k,n) - (n - 1) \log a \qquad (3.4\text{-}4)$$

Hence the dependence of $\log \tau_{1/2}$ on $\log a$ should be linear and the slope of the linear plot corresponds to $(n - 1)$ (Fig. 3-27).

This treatment, which involves measurement of values of half-times ($\tau_{1/2}$) at several initial concentrations a, can be simplified (but made also less conclusive) by carrying out the measurement of two half-times $\tau_{1/2}$ and $\tau'_{1/2}$ at two initial concentrations a and a' only (Fig. 3-28). The value of the reaction order n can then be calculated from

$$n = 1 + \frac{\log \tau'_{1/2} - \log \tau_{1/2}}{\log a - \log a'} \qquad (3.4\text{-}5)$$

This equation can be modified to apply to $\tau_{1/4}$, $\tau_{3/4}$, $\tau_{20\%}$, and so on, and generalized for τ_m.

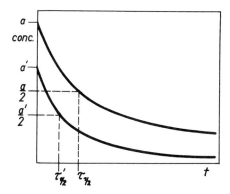

FIG. 3-28. Determination of the reaction order (n) from measurement of half-times ($\tau_{1/2}$ and $\tau_{1/2}'$) at two initial concentrations (a and a'), using equation (3.4-5).

When determining the reaction order by means of fractional times τ_m it is not necessary to carry out two sets of experiments at different initial concentrations a. Alternatively, it is possible to carry out the determination of the reaction order n from a single kinetic run, if the concentration at the end of one time interval is considered to correspond to the initial concentration of the second time interval. Simultaneously, the end of the first interval is considered to be the beginning of the second time interval (Fig. 3-29).

As another possibility, the time interval can be measured from the end of the preceding interval, but the change in concentration is measured from the beginning of the reaction. If the time (τ_m) is measured after the concentration a has decreased by a fraction given by a function of $(1 - y)$ so that

$$\tau_1 \sim a(1 - y) \tag{3.4-6}$$

$$\tau_2 \sim a(1 - y)^2 \tag{3.4-7}$$

then

$$n = 1 + \frac{\log\,[(\tau_2/\tau_1) - 1]}{\log\,[1/(1 - y)]} \tag{3.4-8}$$

The value of y can be chosen arbitrarily, for example, for $y = 0.5$ for a second-order reaction, the measured ratio $\tau_2/\tau_1 = 3$. Nevertheless, it is recommended in general to choose the value of y so that $y < 0.5$, for example, $y = 0.2$, since more accurate concentration data are usually obtained at lower conversion levels. In the case of $y = 0.2$ the values of τ_1 and τ_2 correspond to time intervals after which the initial reactant concentration has decreased to 80% and 64% respectively.

If the basic assumption that $dx/dt = k(a - x)^n$ is not fulfilled, all of the above treatments yield values of n that change with concentration of the reaction components. Whenever the value of n obtained is found to be significantly different from an integer, it can be concluded that the reaction involved corresponds to a more complex reaction rate equation.

The use of dimensionless parameters as discussed first in Section 3.3.1 represents a generalization of such treatments.

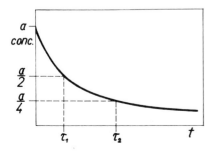

FIG. 3-29. Determination of the reaction order (n) from a single concentration–time dependence.

3.4.2. Determination of the Reaction Order Using Excess Concentration of Reactants

For reactions with rates that follow rate equations of the type (3.4-9):

$$v = k[A]^n[B]^m[C]^p[D]^r \qquad (3.4\text{-}9)$$

it is possible to use one of several approaches for determination of the reaction orders in individual components:

1. Concentration of *all* components of the reaction mixture are chosen to be high so that all of them remain practically constant during the section of the kinetic run being studied. This condition is fulfilled in the initial stages of a reaction, when the changes in concentrations are small when compared with initial concentrations. The kinetics under such conditions is zeroth order and

$$v = \frac{x_2 - x_1}{t_2 - t_1} = \text{const} = k[A]^n[B]^m[C]^p[D]^r \qquad (3.4\text{-}10)$$

If the concentration of one of the components in excess is varied (e.g., concentration of A decreased to $\frac{1}{2}$ or doubled), the rate must change by an exponent corresponding to a multiple of the reaction order in the particular component (e.g., if the rate decreased to $v^{1/2}$ or increased to v^2 as a result of the above mentioned change in concentration of A, it is possible to deduce that $n = 1$). In turn, initial concentrations of B, C, and D are similarly varied, and the individual reaction orders m, p, and r can be determined.

2. Alternatively, it is possible to make the concentrations of *all components except one* high so that their concentrations remain effectively constant in the course of one kinetic run. This is the so-called isolation method.

The reaction order n is then determined for the isolated component, as

$$v = k[A]^n[B]^m[C]^p[D]^r = k'[A]^n \qquad (3.4\text{-}11)$$

where k' is the effective rate constant incorporating all the constant terms, that is

$$k' = k[B]^m[C]^p[D]^r \qquad (3.4\text{-}12)$$

The reaction order in A (i.e., n) can be determined by means of the empirical rate equations and all other procedures described earlier in this chapter. If practical limitations such as solubilities or the scope of analytical methods used do not prevent it, the reaction orders in other components (i.e., m, p, and r) can be determined when the reaction mixtures are prepared so that initial concentrations of B, C, and D are in turn kept small when compared with the rest. This type of treatment is restricted to reactions where the concentration of products does not affect the reaction rate and which are not complicated by consecutive or side reactions.

3. In another modification, the concentrations of *all components except one*

are again kept high and the reaction rate is followed by measuring the concentration change of the only dilute component. This procedure is particularly advantageous if the concentration change of this dilute component follows first-order kinetics. The effective rate constants k' are determined for individual kinetic runs at varying concentration of one of the components present in excess. For example the initial concentration $[A]_0$ is made small when compared to $[B]_0$, $[C]_0$, and $[D]_0$, and the effective rate constant k'_1 is obtained from equation (3.4-13):

$$v_1 = k'_1 [A] \qquad\qquad (3.4\text{-}13)$$

where

$$k'_1 = k [B]_0^m [C]_0^p [D]_0^r \qquad\qquad (3.4\text{-}14)$$

Then the concentration of one of the components in excess, $[B]_0$ for example, is increased by a factor j and all other concentrations are kept constant. The corresponding rate constant k'_2 can be determined by means of the equation

$$v_2 = k'_2 [A] \qquad\qquad (3.4\text{-}15)$$

From measurements at these two concentrations of the component B it is possible to determine the reaction order in this component, when

$$j^m = \frac{k'_2}{k'_1} \qquad\qquad (3.4\text{-}16)$$

or from (3.4-17):

$$m = \frac{\log k'_2 - \log k'_1}{\log j} \qquad\qquad (3.4\text{-}17)$$

Similarly it is possible to determine the reaction orders p and r in components C and D.

The method can be modified and its reliability improved when the value of a formal rate constant k' is determined not for two but several values of $[B]_0$. If the value of $m = 1$, the plot of k' against $[B]_0$ is linear (Fig. 3-30), otherwise the value of m is best determined from the dependence of $\log k'$ on $\log [B]_0$ [3-39].

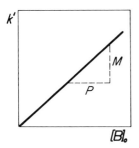

FIG. 3-30. Determination of the reaction order (m) from the dependence of the measured pseudo-first-order rate constant k' on the initial concentration of the component B present in excess.

3.4.3.   Determination of Reaction Order from Measurement
of Reaction Rates

In cases when the use of the differential form of the rate equation is preferable to that of the integrated form, the reaction rate is followed. The reaction rate of numerous reactions follows the simple equation (3.4-18):

$$v = \prod_{i=1}^{i=n} [A_i]^{\nu_i} \tag{3.4-18}$$

where $\prod$ indicates a product and $[A_i]$ represents the concentrations of all compounds which affect the reaction rate, that is, reactants, products (for equilibrium reactions), and catalysts. The exponent ν_i in this reaction can be determined by carrying out at least $(n + 1)$ measurements of initial reaction rates for set with initial concentrations $[A_i]_{0,1}$ which follow equation (3.4-19):

$$v_{0,1} = k \prod_{i=1}^{i=n} [A_i]_{0,1}^{\nu_i} \tag{3.4-19}$$

If one of these initial concentrations, $[A_j]_0$, for example, is increased by a factor g and the rest is kept unchanged, a second group of initial concentrations $[A_i]_{0,2}$ may be dealt with. For initial velocities, equation (3.4-20) applies:

$$v_{0,1} = k \, g^{\nu_j} \prod_{i=1}^{i=n} [A_i]_{0,2}^{\nu_i} = g^{\nu_j} \, v_{0,2} \tag{3.4-20}$$

By comparison of initial velocities for the two sets of initial concentration we obtain

$$g^{\nu_i} = \frac{v_{0,2}}{v_{0,1}} \tag{3.4-21}$$

and hence

$$\nu_j = \frac{\log v_{0,2} - \log v_{0,1}}{\log g} \tag{3.4-22}$$

Similarly, by varying the concentration of each component A_i, it is possible to determine the value of ν_i for each component. If possible, the value of g is usually chosen $g \geq 2$. For example, for reaction

$$A + B + C \longrightarrow Products \qquad (3.4\text{-}23)$$
$$\downarrow$$
$$D$$

it is possible to determine $v_{0,1}$ at $[A] = [A]_0$, $[B] = [B]_0$, and $[C] = [C]_0$. Then $v_{0,2}$ is determined for $[A] = 2[A]_0$, $[B] = [B]_0$ and $[C] = [C]_0$. Thus

$$v_A = \frac{\log v_{0,2} - \log v_{0,1}}{\log 2} \qquad (3.4\text{-}24)$$

Similarly, measuring $v_{0,3}$ at $[A] = [A]_0$, $[B] = 2[B]_0$, and $[C] = [C]_0$ it is possible to calculate

$$v_B = \frac{\log v_{0,3} - \log v_{0,1}}{\log 2} \qquad (3.4\text{-}25)$$

and finally measuring $v_{0,4}$ for $[A] = [A]_0$, $[B] = [B]_0$, and $[C] = 2[C]_0$ it is possible to obtain

$$v_C = \frac{\log v_{0,4} - \log v_{0,1}}{\log 2} \qquad (3.4\text{-}26)$$

It should be stressed that such simple determination of order in individual components cannot be used as the sole proof of the form of the equation for reaction rate. It is impossible to be *a priori* sure that the studied reaction follows equation (3.4-18) and not a more complicated expression. For example, when on the right-hand side of equation (3.4-18) are additive factors, the described treatment cannot be applied.

Based on equation (3.4-18) it is also possible to find the value of v_i from determined value of $(a - x)$ and measured value of v using nomograms [3-40].

When it is necessary to carry out the measurement of initial velocities, it is useful to vary the initial concentrations of all components of the reaction mixture in the widest possible concentration ranges.

Finally, it must be stressed once again that with more complex reactions the overall reaction order looses its meaning, since the reaction rates are not simple functions of concentration. In such cases, systematically planned experiments enabling verification of the complete reaction rate equation are necessary.

3.5. CALCULATION OF THE RATE CONSTANT

After the best fitting reaction rate equation is found and the reaction order in individual components determined, it is possible to attempt to find the best value of the rate constant.

Procedures used in the determination of the rate constant belong to three categories: (1) calculation of the value of the rate constant for paired time and concentration data by means of the integrated form of the rate equation, (2) graphical methods usually based on determination of the slope of a linear plot, and (3) use of computers.

3.5.1. Calculation by Means of Integrated Form

In the simplest application of this type, measured values of the concentration followed after chosen time intervals are inserted into the integrated form of the reaction rate equation. For each time interval the value of the rate constant k is calculated. For example, for a second-order reaction (at identical initial concentrations, thus following the scheme $2A \rightarrow P$) the rate constants for the intervals ending at t_1, t_2, or t_3 can be calculated by means of equations (3.5-1) to (3.5-3):

$$k'_{0,1} = \frac{1}{t_1}\left(\frac{1}{[A]_1} - \frac{1}{[A]_0}\right) \tag{3.5-1}$$

$$k'_{0,2} = \frac{1}{t_2}\left(\frac{1}{[A]_2} - \frac{1}{[A]_0}\right) \tag{3.5-2}$$

$$k'_{0,3} = \frac{1}{t_3}\left(\frac{1}{[A]_3} - \frac{1}{[A]_0}\right) \tag{3.5-3}$$

If a suitable reaction rate equation has been used, the calculated value of k' should remain practically constant for various periods t_i, apart from statistical variations.

The limitation of this method is the high statistical weight for the accuracy of the initial concentration. Inaccuracy in determination of the value of $[A]_0$ can considerably affect all calculated values of $k'_{0,i}$. If the difference $[A]_n - [A]_0$ is too small or too large, the calculated values will be inaccurate. The procedure is suitable for use in those cases in which the reaction mixture can be prepared accurately, when the time needed for mixing of the solutions is short in comparison with the first time interval (t_1) after which the measurement is carried out, and when the analytical procedure for determination of $[A]_n$ is less accurate than that for determination of $[A]_0$. Since the mean deviation of the value of initial concentration $[A]_0$ varies only slightly from the mean deviation in determination of concentration $[A]_n$, this method is rarely the most suitable method for determination of the best value of rate constant.

The reaction rate equation integrated between two suitably chosen limits is most frequently used for calculation of values of rate constants for pairs of successive measurements (so-called "Schrittformel"—stepwise equation). For example, for the above-mentioned case of a second-order reaction the individual equations are (3.5-4) to (3.5-6):

$$k_{0,1} = \frac{1}{t_1}\left(\frac{1}{[A]_1} - \frac{1}{[A]_0}\right) \tag{3.5-4}$$

$$k_{1,2} = \frac{1}{t_2 - t_1}\left(\frac{1}{[A]_2} - \frac{1}{[A]_1}\right) \tag{3.5-5}$$

$$k_{2,3} = \frac{1}{t_3 - t_2}\left(\frac{1}{[A]_3} - \frac{1}{[A]_2}\right) \tag{3.5-6}$$

If the time intervals t_i are identical or only slightly different, it is possible to show that the mean value of $\bar{k}$ can be expressed by equation (3.5-7):

$$\bar{k} = \frac{1}{nt_1}\left(\frac{1}{[A]_n} - \frac{1}{[A]_0}\right) \tag{3.5-7}$$

The mean value is used, since it is assumed that by combining successive values into pairs the contributions of individual measurements to overall error are balanced. The equation for the mean value of $\bar{k}$, nevertheless, indicates that this value is similar to that which would be obtained by utilizing only initial and final measurements in the calculation. The statistical weight of these two measurements is considerably greater than weights corresponding to values of concentration and time inside this time interval. The calculated value of the rate constant is even less reliable because these values at short and long times are usually measured with the least accuracy.

To obtain more reliable values of a rate constant than can be obtained by previously described procedures, it is necessary to choose the sets of data purposefully. The planning of such experiments is best demonstrated by an example.

Measurements of concentrations after given time intervals are divided into two groups. In the first group $(n + 1)$ measurements of concentration after times t_0, t_1, $2t_1$, $3t_1$, . . . nt_1 are carried out. The time intervals are preferably chosen in such a way that $nt_1 < \tau_{1/2}$. Then a second group of measurements of concentration is carried out, starting after time $\tau \geq \tau_{1/2}$. This second group of measurements is carried out again after time intervals which differ by a constant increase t_1. Two sets of measurements are thus obtained and by combination of one concentration obtained from the first and one from the second set it is possible to obtain values of constant k_i:

1st set	2nd set	Constant calculated
t_0	τ	k_0
t_1	$\tau + t_1$	k_1
$2t_1$	$\tau + 2t_1$	k_2
$3t_1$	$\tau + 3t_1$	k_3
.	.	.
.	.	.
.	.	.
.	.	.
.	.	.
nt_1	$\tau + nt_1$	k_n

The arithmetic mean of values of k_i calculated in this way gives a value which corresponds well to the true value of the rate constant. The advantage of this procedure is that the mean deviation of measurements is practically the same for all values of k_i.

Another procedure for the handling of experimental data is suitable in such cases when the measurements of concentration or physical quantity proportional to it at time $t = 0$ and/or $t = \infty$ present difficulties. The treatment, called the Guggenheim method [3-41], can also be used when the initial and/or final reading (or concentration) for the first-order reaction is unknown or inaccessible.

As in the previous case, two series of experiments are carried out. First, after time intervals $t_1, t_2, t_3, \ldots, t_n$ concentrations $c_1, c_2, c_3, \ldots, c_n$ (or physical quantities $\phi_1, \phi_2, \phi_3, \ldots, \phi_n$) are measured. After an interval (τ') a second series of measurements is carried out after times $(\tau' + t_1), (\tau' + t_2), (\tau' + t_3), \ldots, (\tau' + t_n)$ and concentrations $c_1', c_2', c_3', \ldots, c_n'$ are determined. The value of τ' is preferably chosen so that it is larger than t_n and it is particularly suitable when $t_n < \tau_{1/2} < \tau'$. Then

$$\ln (c_i - c_i') = \ln (c_0 - c_\infty)(1 - e^{-k\tau'}) - kt_i \qquad (3.5\text{-}8)$$

or

$$\ln (c_i - c_i') = \text{const} - kt_i \qquad (3.5\text{-}9)$$

The value of k can be then determined graphically by plotting $\ln (c_i - c_i')$ against t_i and finding the slope ($M/P = k$, Fig. 3-31).

Before the Guggenheim method is used it must be confirmed that the reaction involved is a simple irreversible reaction following first-order kinetics. It is possible

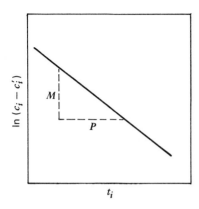

FIG. 3-31. Determination of the value of the rate constant using the Guggenheim method [3-4]. $M/P = k$.

that the log $(c_i - c_i')$ plot against time can be found to be linear for some reactions involving equilibria and competing first-order reactions, but such application would yield an erroneous result.

For reactions where it is impractical to carry out measurements beyond the $\tau_{1/2}$, another approach has been proposed [3-42–3-45]. If the description is restricted to measured physical quantities ϕ_i, then, as in the Guggenheim treatment, a set of values ϕ_1 to ϕ_n is obtained after intervals t_1 to t_n. A second set of measurements of ϕ_1' to ϕ_n' is then carried out at time intervals $\tau + t_1$ to $\tau + t_n$, where the value of τ is constant for a given series, but no limitations are involved in the choice of the value of τ.

For the first-order reaction then

$$\ln (\phi - \phi_n) = \ln (\phi_\infty - \phi_0) - kt_n \qquad (3.5\text{-}10)$$

and similarly

$$\ln (\phi_\infty - \phi_n') = \ln (\phi_\infty - \phi_0) - k(\tau + t_n) \qquad (3.5\text{-}11)$$

Division and rearrangement of these two equations yields (3.5-12):

$$\phi_n = \phi_\infty(1 - e^{k\tau}) + \phi_n' e^{k\tau} \qquad (3.5\text{-}12)$$

Thus if readings from the first series (ϕ_n) are plotted against the corresponding readings from the second series (ϕ_n'), the plot should be linear with a slope of $e^{k\tau}$. Finding the slope thus enables estimate of the value of k. Since for $t = \infty$, $\phi_n = \phi_n' = \phi_\infty$, the value of ϕ_∞ corresponds to the point on the linear plot where ϕ_n and ϕ_n' have the same value.

An alternative approach to the situation, when readings at $t = 0$ or $t = \infty$ are not available, is based on application of the method of nonlinear least squares to data obtained at arbitrary time intervals [3-46, 3-47]. This latter method yields the standard deviation in k and an estimate of the value ϕ_∞. Recently, an optimization procedure for finding the best value of ϕ_∞ was devised [3-48]. Also, three general time-log linearization procedures have been proposed [3-49].

When dealing with second-order reactions, provided that the initial concentrations of both components are equal, Roseveare [3-50] proposed to calculate the value of the rate constant by means of equation (3.5-13):

$$k = \frac{[(c_2 - c_1) - (c_3 - c_2)]^2}{2(t_2 - t_1)(c_3 - c_1)(c_2 - c_1)(c_3 - c_2)} \qquad (3.5\text{-}13)$$

where c_1, c_2, and c_3 are concentrations or quantities proportional to concentrations measured after time intervals t_1, t_2, and t_3 where $t_2 = (\Delta + t_1)$, $t_3 = (\Delta + t_2)$, and so on, for $\Delta = $ const.

Alternatively [3-51, 3-52], using the expression

$$\frac{1}{[A]} - \frac{1}{[A]_0} = kt \qquad (3.5\text{-}14)$$

or

$$[A] = [A]_0 - kt[A]_0[A] \qquad (3.5\text{-}15)$$

and relationships

$$[A] = \text{const} \, (\phi - \phi_\infty) \qquad (3.5\text{-}16)$$

and

$$[A]_0 = \text{const} \, (\phi_0 - \phi_\infty) \qquad (3.5\text{-}17)$$

it is possible to derive for $t = t$

$$\phi_n = \phi_0 - [A]_0 k t_n \, (\phi_n - \phi_\infty) \qquad (3.5\text{-}18)$$

and similarly for $t = t_n - \tau$

$$\phi'_n = \phi_0 - [A]_0 k (t_n + \tau)(\phi'_n - \phi_\infty) \qquad (3.5\text{-}19)$$

Subtraction of (3.5-19) from (3.5-18) yields

$$\phi_n - \phi'_n = [A]_0 k \tau \phi_\infty + [A]_0 k \{(t_n + \tau)\phi'_n - t_n \phi_n\} \qquad (3.5\text{-}20)$$

The time interval τ between the first measurement in first set (ϕ_n) and second (ϕ'_n) is usually chosen about equal to $\tau_{1/2}$. Values of $(\phi_n - \phi'_n)$ are then plotted against the corresponding values of $\{(t_n + \tau)\phi'_n - t_n \phi_n\}$. This plot should be linear with a slope of $[A]_0 k$. Weighting factors for the values of ϕ should be introduced using a Fortran program into the least squares method in finding the slope of the plot [3-53].

This method, like other time-lag methods, is useful when the final reading of the value ϕ is not available or is unreliable. Such a situation is encountered when long-term readings are uncertain due to consecutive reactions or long-term baseline instability, when an excessive time is required for the reaction to reach completion or when the molar absorptivity or limiting current constant of the product is not known.

In addition to arithmetic mean of the individual rate constants obtained by one of the above-mentioned procedures, two further methods are sometimes used to obtain the most reliable value of the rate constant: the method of averages and the least squares procedure.

For the method of averages the measured values are divided into two equal sets. The first set contains measurements 1 through m, the second measurements $m + 1$ through n. For a first-order reaction the following expression is used for calculation of the rate constant:

$$\bar{k} = 2.3 \frac{\sum_{i=1}^{i=m} \log c_i - \sum_{i=m+1}^{i=n} \log c_i}{\sum_{i=m+1}^{i=n} t_i - \sum_{i=1}^{i=m} t_i} \tag{3.5-21}$$

Similarly for a second-order reaction at equal initial concentrations it is possible to use the expression:

$$\bar{k} = \frac{\sum_{i=1}^{i=m} 1/c_i - [m/(n - m)] \sum_{i=m+1}^{i=n} (1/c_i)}{\sum_{i=1}^{i=m} t_i - [m/(n - m)] \sum_{i=m+1}^{i=n} t_i} \tag{3.5-22}$$

When the least squares procedure is used, the calculation of the rate constant $\bar{k}$ is carried out by means of equation

$$\bar{k} = \frac{(1/n) \sum_i t_i \sum_i (1/c_i) - \sum_i (t_i/c_i)}{(1/n) (\sum_i t_i)^2 - \sum t_i^2} \tag{3.5-23}$$

for $i = 1$ to $i = n$ for a second-order reaction at equal initial concentrations and

$$\bar{k} = 2.3 \frac{(1/n) \sum_i t_i \sum_i \log c_i - \sum_i \log c_i}{\sum_i t_i^2 - (1/n) (\sum_i t_i)^2} \tag{3.5-24}$$

for first-order reactions.

A less rigorous, but more frequently used procedure, is the least squares treatment applied to calculated values of rate constants rather than to measured concentrations.

When using statistical methods, as above, to obtain a value of the rate constant, it is important to exercise caution, because such treatments do not necessarily give the best value of the rate constant. Such treatments are namely based on the assumption of acquisition of a great number of measurements, which is rarely fulfilled in kinetic studies.

With respect to these limitations, graphical methods and the use of computers seem most promising.

3.5.2. Graphical Methods

As shown on numerous examples in the preceding sections, for application of graphical methods the reaction rate equation can be transformed in such a way that the dependence of some function of concentration on time or its function shows a linear plot. The rate constants are then determined from the slope of this plot or from the intercept on the y-axis.

It is possible to use regression analysis to find the best linear plot fitting the experimental points. The availability of linear regression treatment, even for simple calculators, makes this approach more attractive. Nevertheless, it has been proved that the human eye is able to find the best linear plot which rarely differs significantly from the calculated regression line, provided that the correlation is reasonably good (i.e., for regression coefficients $r > 0.90$).

The best line is hence frequently drawn, using a ruler made from transparent material, so that it would best fit experimental points. The line need not pass through any of the experimental points, but is drawn in such a way that the sum of deviations shown by the experimental points from the linear plot that are above the line are approximately equal to the sum of deviations below the line. Since the measurements of concentrations are usually of lower accuracy than measurements of time, the overall accuracy of the experimental points is more affected by the accuracy in the direction of concentration changes. Hence, when estimating and comparing deviations from the linear plot the emphasis should be directed towards deviations in the direction of the concentration (or its function) axis.

It is preferable to choose the scales on the two axes such that the resulting line shows an angle of about 45°. The scale of the individual axis is chosen so that the smallest unit on the axis, which still can be estimated, is of the same order of magnitude as the estimated mean deviation of measurement. If this is impossible, the graphical method will furnish only an approximate value of rate constants, but this situation is only rarely encountered.

If it is possible to estimate the uncertainty of measurements corresponding to individual points, it is preferred to indicate it by lines of appropriate length in the direction of the measured quantity centered on the particular point, namely error darts. If the uncertainty in one parameter is not considered negligible, it is recommended to indicate the estimated uncertainty by the size of circles (or squares) or more appropriately by ellipses (or rectangles). The plotted line should then either cross the circles or ellipses or pass in their close vicinity.

The graphical method is particularly useful in showing deviations of whole groups of measurements (trends) or of individual points. Isolated points of dubious accuracy, or which evidently deviate from the otherwise linear plot, are neglected in plotting the best line.

3.5.3.   General Comments on Determination of Values of Rate Constants

For first-order reactions the most accurate values are obtained after reaching about 60% conversion, for second-order reactions about 50% conversion, provided that the error in the measurement of time is small when compared with the accuracy in concentration measurement. Thus in numerical as well as graphical treatments less weight should be given to data obtained at the very beginning of the reaction and at conversions higher than 70%, 80%, or 90% (dependent on reaction rate and type of measurement).

The final reported value of a rate constant should never correspond to results obtained in a single kinetic run, even if such a value is based on a number of concentration measurements. At least three parallel kinetic runs should be measured; moreover, if the independence of the rate constant on composition of the reaction mixture (e.g., concentration of a particular component) should be established, it is recommended to follow at least three kinetic runs at three different values of the parameter involved (e.g., at least at three concentrations of the particular component).

Finally, it should be stressed that the same attention should be paid to the treatment of kinetic data as to actual measurements. Insufficient treatment of experimental data can lead to a gross misinterpretation of mechanism similar to a discussion of results not taking into consideration a part of experimental findings.

3.5.4.   The Use of Computers

In the past, computers have been used predominantly in dealing with reaction rate equations in which the differential equations either had no known exact (closed) solution or for which the treatment was too time consuming. Recently the application of computers has increased considerably, and in the future it seems that most of the data handling will be done by computers. The problems of reaction kinetics can be dealt with by both digital and analog computers.

Digital computers can be used for calculation of the best value of the rate constant, if the reaction scheme is known. Alternatively, any of the procedures described in Section 3.3 for finding the best equation for reaction rate can be carried out by means of a digital computer with a plotter. Nevertheless, the linearization of the relationships advantageous in manual use of graphical methods is not necessary when digital computers are used. Using modeling or optimization procedures it is possible to calculate the concentration–time dependence for a particular component of a reaction mixture based on the chosen reaction scheme and compare it with experimental data. If differences between the calculated and experimental data exist, it is possible to prove whether such differences can be diminished by varying variable parameters (usually values of rate constants) or whether they indicate that the proposed mechanism does not apply. Computer programs can be developed that include variations of concentration on parameters dependent on reaction medium, such as pH, ionic strength, solvent, or temperature.

Digital computers are undoubtedly the most important type of computer used in reaction kinetics. Development of programs for such applications is currently in the process of cataclysmic growth.

The interest is in particular indicated by the Computer Series in the *Journal of Chemical Education*, which recently entered its third decade [3-54], and the wealth of publications from which few were selected [3-55–3-63]. The nature of such applications depends on the type and size of the computer and the language used. We have therefore abstained here from a more detailed discussion of examples and applications, and such an attempt is postponed until permanent trends in future development are better recognized.

On the other hand, some attention will be paid in the next paragraphs to analog computers. The principles and use of such computers are less widely known. Furthermore, the discussion of their application for investigation of kinetic problems is particularly suitable from a pedagogical point of view [3-64–3-68]. For these computers the relationship between programming and reaction kinetics is particularly straightforward. It seems thus worthwhile to discuss their principles and use, as they are less easily accessible in most chemical laboratories.

The construction of analog computers is based on the fact that some physical processes follow differential equations. If the properties of the physical system are varied and some quantities characterizing the studied system are measured, it is possible to follow the relationship between the adjustable parameters of the system and the measured quantities. In the application of analog computers, it is advantageous if one of the measured variables is time. It is then possible to follow the changes in the second, dependent variable with time, and to study the effect of adjustable properties of the physical system in a way similar to that in reaction kinetics where the change in concentration with time is followed and the effect of the adjustable composition of the reaction mixture is studied.

The physical properties that can be used in the construction of analog computers can be different—for example, mechanical, hydrodynamic [3-69, 3-70], or electrical. Since analog computers based on electrical circuits are most widely used, further discussion will be restricted to this type. The discussion starts with treatment of simple reactions for which analog computers would be rarely used in practice. Nevertheless, discussion of such systems demonstrates the principles on which analog computers operate.

The simplest case can be illustrated using a circuit shown in Fig. 3-32. The relationship between current (i) and the voltage (E) in this closed circuit can be expressed by Ohm's law, $E = iR$, where R represents the resistance in the circuit. If the current is measured by the instrument M_i (with a small internal resistance when compared with R) and the voltage by the instrument M_E (with high internal resistance when compared with R), then an increase in the value of R at a given voltage E results in a decrease in the current i.

If next the value of the resistance R is chosen and the voltage applied to this resistance is continuously changed with time (e.g., by using a motor driven potentiometer), the relationship

$$k't = iR \qquad (3.5\text{-}25)$$

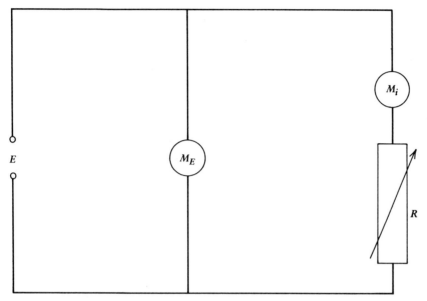

FIG. 3-32. Circuit for demonstration of the principle of an analog computer. E—source of voltage; R—variable resistor; M_i—current detector, amperemeter; M_E—voltage detector, voltmeter.

is obtained. When the change in current with time is recorded, the linear dependence shown in Fig. 3-33, with the slope k'/R, is obtained. This slope can be changed if the resistance R is varied. For comparison with a kinetic equation (3.5-25) is transformed to

$$i = \frac{k'}{R_i} t \qquad (3.5\text{-}26)$$

This form indicates that the circuit (in Fig. 3-33) behaves analogously to a reaction of zeroth order for which the equation (3.5-27)

$$x = k_1 t \qquad (3.5\text{-}27)$$

applies. To find the value of k_1 for a given chemical reaction the value of the resistance R_1 is varied until the slope of the dependence of current (i) on time (t) (or on the applied voltage (E), which is a linear function of time) has the same slope as the experimentally found dependence of the reaction variable x on time. But first it is necessary to transform both measured quantities—reaction variable x and current i—to the same scale. If the above equation is transformed into

$$\frac{i}{k'} = \frac{1}{R_i} t \qquad (3.5\text{-}28)$$

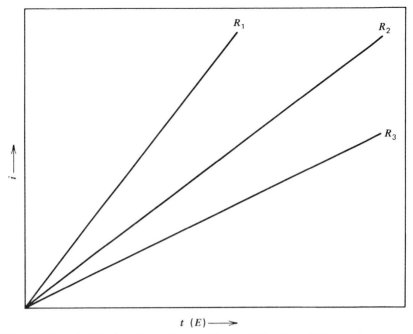

FIG. 3-33. Record of the dependence of current (i) on applied voltage (E) for varying value of the ohmic resistance (R). Because the voltage is varied linearly with time, current can be measured as a function of time.

it is obvious that it is necessary to find the proportionality constant k' from the relation $i/k' = x$. This can be done by calibration.

When more complex reactions are treated, the circuits used are consequently more complicated. The circuits must make it possible to add and subtract voltages, multiply by a constant factor, integrate, make derivatives, and so on. The construction of the used circuits can be complex, but their description can be simplified by employing block schemes, using symbols like:

		Function
$-\!\!\bigcirc\!\!-$	Potentiometer	Multiplying by $k < 1$
$-\!\!\bigcirc\!\!-\!\!\triangleright\!\!-$	Potentiometer + amplifier	Multiplying by $k > 1$
$-\!\!\triangleright\!\!-$	Inverter	Multiplying by -1
$\equiv\!\!\triangleright$	Adder	Addition and subtraction
$-\square\!\!\triangleright\!-$	Integrator	Integration
$-\boxed{\times}\!\!\triangleright\!-$	Multiplier	Multiplication by function x
$-\boxed{f(x)}\!-$	Diode function generator	$y = f(x)$

Potentiometers enable us to multiply by a given factor, inverters to change the sign, exponential function generators produce an output logarithm of the voltage applied on the input.

The principle of application becomes apparent from treatment of some simple reaction types. Hence for a first-order reaction that has the form

$$A \longrightarrow B \qquad (3.5\text{-}29)$$

where the rate equation has the form

$$\frac{d[B]}{dt} = k[A] \qquad (3.5\text{-}30)$$

it is possible to use the computer shown in Fig. 3-34.

The value of the initial concentration $[A]_0$ is adjusted by choosing a proper position of the potentiometer A_0. Then, on the potentiometer denoted as k, chosen values proportional to the values of the rate constant (k) are set, that is, k_1, k_2, k_3, ... A recorder is connected to the output of [A] or [B] and curves 1, 2, 3 for dependence of [A] on t or of [B] on t are recorded (Fig. 3-35) for $k_1 > k_2 > k_3$.

The experimentally found curve of dependence [A]–t or [B]–t is next compared with curves recorded by computer. By trial and error a setting of the potentiometer k is sought for which the recorded curve is closest to the experimentally found curve.

For a second-order reaction of a type

$$A + B \longrightarrow C + D \qquad (3.5\text{-}31)$$

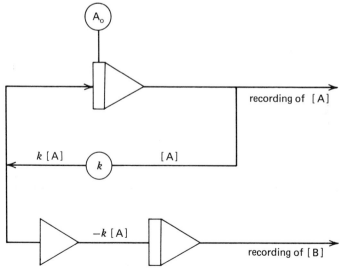

FIG. 3-34. Scheme of the analog computer for a first-order reaction following equation (3.5-30). Symbols in text.

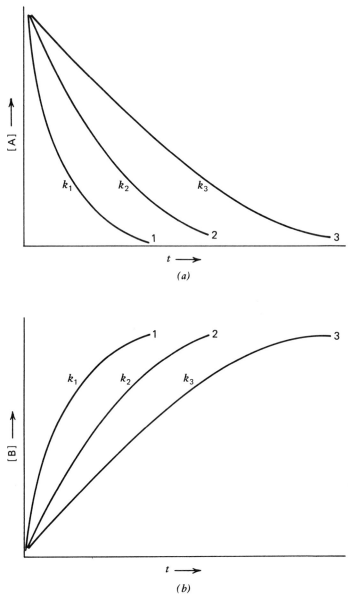

FIG. 3-35. Dependence of concentration (*a*) of compound A, and (*b*) of compound B on time recorded by analog computer shown in Fig. 3-34 for varying values of rate constants, obtained by varying position of potentiometer *k*.

where the reaction rate equation can be written as

$$\frac{d[C]}{dt} = \frac{d[D]}{dt} = k[A][B] \tag{3.5-32}$$

the circuit used is more complex (Fig. 3-36).

Again, the validity of the used rate equation and the value of the rate constant are proved by consecutive adjustment of the potentiometer 100 k. The time change of concentration of the particular species which is accessible to measurement is then recorded.

An example of the application of analog computers to the study of kinetics is the investigation of the alkaline cleavage of chalcone, accompanied by retroaldolisation fission of the ketol formed. This reaction sequence is depicted by the following equations (where acid–base equilibria are omitted):

$$C_6H_5COCH{=}CHC_6H_5 + OH^- \underset{k_{-1}}{\overset{k_1}{\rightleftharpoons}} C_6H_5COCH_2CHC_6H_5 \tag{3.5-33}$$
$$\overset{}{|} $$
$$OH$$

$$x_1 x_2$$

$$C_6H_5COCH_2{-}CHC_6H_5 \underset{k_{-2}}{\overset{k_2}{\rightleftharpoons}} C_6H_5COCH_3 + OCHC_6H_5 \tag{3.5-34}$$
$$\phantom{C_6H_5COCH_2{-}}|$$
$$\phantom{C_6H_5COCH_2{-}}OH$$

$$x_2 x_4 x_3$$

For this system it is useful to define by means of reaction variables x_1–x_4 (directly related to experimentally accessible quantities) new reaction variables x_1' – x_4' by means of equations (3.5-35) to (3.5-37):

$$x_1' = -k_1x_1 + k_{-1}x_2 \tag{3.5-35}$$
$$x_2' = -k_{-1}x_2 + k_2x_2 + k_{-2}x_3x_4 + k_1x_1 \tag{3.5-36}$$
$$x_3' = x_4' = k_2x_2 - k_{-2}x_3x_4 \tag{3.5-37}$$

For boundary (initial) conditions

$$t = 0 \qquad x_1 = (x_1)_0 \qquad x_2 = x_3 = x_4 = 0 \tag{3.5-38}$$

it follows that

$$x_3 = x_4 \qquad (x_1)_0 = x_1 + x_2 + x_3 \tag{3.5-39}$$

and the circuit shown in Fig. 3-37 can be used.

Experimentally, the reaction of chalcone in alkaline media has been followed polarographically [3-8]. The change of four waves i_1–i_4 on polarographic curves with time has been recorded. For the study of the cleavage reaction it is particularly

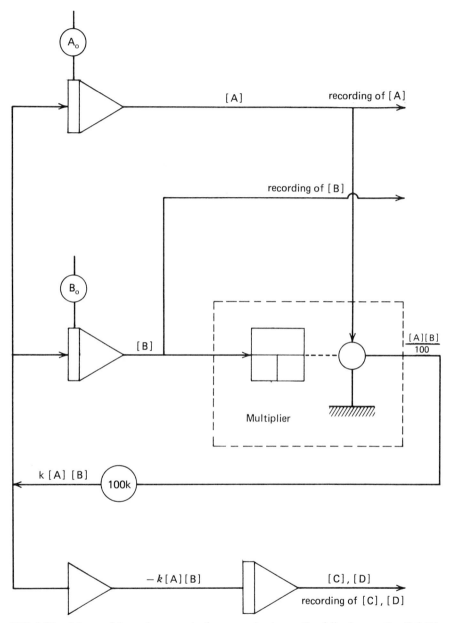

FIG. 3-36. Scheme of the analog computer for a second-order reaction following equation (3.5-32). Symbols in text.

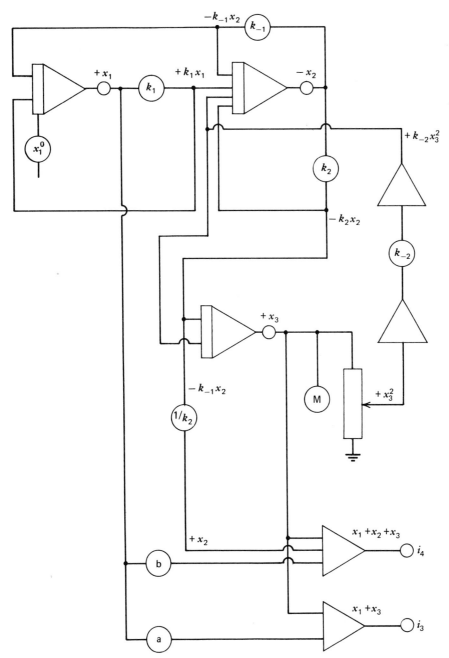

FIG. 3-37. Scheme of the analog computer for a system of reactions following equations (3.5-33) and (3.5-34). Variables x_1 to x_4 and symbols used described in text.

important to follow the time change of wave i_3, which is proportional to concentration of benzaldehyde, and wave i_4, which is proportional to the sum of concentrations of ketol and acetophenone. Therefore in the circuit in (Fig. 3-37), potentiometers (a) and (b) are included, which enable recording of i_3 and i_4 time changes. The relations used are $i_3 = a_1 x_1 + a_2 x_3$ and $i_4 = a_3 x_1 + a_4 x_2 + a_5 x_4$. The calibration of potentiometers can be carried out by comparison with the limiting current of solutions containing known amounts of benzaldehyde, ketol, and acetophenone. Since the waves of acetophenone and the ketol have approximately the same diffusion current constant, calibration can be carried out with acetophenone only. The potentiometer x_1^o is set into a position corresponding (and proportional) to an initial chalcone concentration.

From the time change of wave i_1, the value of the rate constant k_1 is determined and set on the corresponding potentiometer. Thereafter the positions of potentiometers corresponding to constants k_{-1}, k_2, and k_{-2} are varied in a series of successive approximations until the curve recorded by the computer is in agreement with the experimentally found dependence.

The principal advantage of the modeling by means of analog or digital computers is the ability to confirm simultaneously the validity of an applied reaction rate equation and to estimate the value of the rate constant. Their application makes it also possible to prove whether a difference between the assumed and observed time dependence is due to an unexpected value of one rate constant or if a completely different mechanism is operating.

3.6. EFFECT OF SOLUTION COMPOSITION ON REACTION RATE

The knowledge of the reaction rate equation frequently makes it possible to draw deductions on the reaction scheme. On the other hand it does not usually offer information on the nature of the reactive forms of the species participating in the reaction. Frequently the knowledge of the reaction rate equation or the reaction order in individual components is insufficient for drawing conclusions on the character and/or composition of the transition state. It is possible to make deductions on the mechanism from the knowledge of starting materials, products, and equation for reaction rate, but frequently several possibilities remain and a choice would be simply a guess.

To be able to use deductive reasoning rather than guessing, further information must be acquired. Important contributions to the eliminating–deducing process can be gained from the study of the effects of varying composition of reaction mixture on reaction rates. In particular, the study of the effects of pH, solvent, ionic strength, and temperature on the values of rate constants offers further diagnostic tools. Rate constants are determined by one of the procedures described in preceding sections. It is essential to check repeatedly that the change in composition of reaction mixture does not affect the kinetics and that the same reaction rate equation operates over the entire range in which the composition of solution is varied.

3.6.1. Effects of pH of Reaction Mixtures on Reaction Rates

The rate of numerous chemical reactions in solutions is affected by the change in acidity, where the rate constant depends on pH or analogous functions (when extended acidity scales are used). Such change in the value of the rate constant can be observed when:

(a) The rate-determining step is preceded by a rapidly established acid–base equilibrium which causes a change in the actual concentration of the reactive form of one of the participating reactants.

(b) The rate-determining step is the formation of an acid or a base (e.g., in reactions of carbanion-enolates or some carbonium ions).

(c) The rate-determining step involves addition of hydroxide ion as a nucleophile or a proton transfer. The solvent base or acid form is consumed or transformed in the course of the reaction and is not regenerated.

(d) The reaction is general acid–base catalyzed.

The first and third cases are sometimes described as specifically catalyzed. Alternatively, so-called generally catalyzed reactions, when studied in detail, usually involve one of the first three types as rate governing. It is recommended to retain the term of specific or general acid or base catalysis only for such systems, where the detailed sequence of reactions and the acidity sensitive step have not yet been recognized.

3.6.1.1. *Effects of Rapidly Established Acid–Base Equilibria on Reaction Rates*

In a system of reactions where the slowest step is preceded by rapidly established acid–base equilibria, either the reagent R (present in excess) or the reactant B (the concentration of which changes during the reaction) or both reagent R and reactant B can undergo change in the degree of ionization by acting as acids and bases.[*] It can easily be shown that the effect on the rate constant in the first two cases can be described by analogous expressions. Therefore the third type, corresponding to systems in which both reactant and reagent react as acids and/or bases, will be discussed separately.

Only the simplest types of rate determining reactions taking place subsequent to the acid–base equilibria are considered in the following discussion, where a rapid establishment of precedent equilibria is assumed for all systems, even when this condition is not repeatedly stated. The treatment of more complex reactions follows a similar pattern, only the rate equations needed for evaluation of the measured, formal rate constant k' (at each pH-value) are more complicated.

---

[*]The symbols RH^+ and BH^+ represent the form richer by one proton than the conjugate bases B and R; the symbols do not express actual charges of the species, which, in the present discussion, are of no consequence.

In addition to the formal rate constant k', which is a function of acidity and possibly other parameters, the following symbols are introduced:

$K_1, K_2, K_3, \ldots$	Equilibrium constant referring to rapidly established acid–base reaction
$k_1, k_2, k_{-1}, k_{-2}, \ldots$	Rate constant
R	Reagent present in reaction mixture in excess. In some cases it may undergo an acid–base reaction and form RH^+ and/or RH_2^{2+}.
A	Reactant whose concentration changes in the course of the process, but whose degree of ionization does not change with acidity in the studied acidity range
B, BH^+	Reactant whose concentration changes in the course of reaction and whose ionization changes with pH
S_R, S_A, S_B	Analytical concentration of reagent or reactant respectively

For realization of the types of reactions discussed here, series of reaction mixtures are prepared which vary in pH, but all of which contain the same analytical concentration S_R, S_A, or S_B of the reagent R and of reactants A and/or B. The change in pH can be secured either by addition of components to appropriate buffers of sufficient buffer capacity (i.e., containing buffer in concentration at least tenfold of that of reagent R), or the reagent R is used as a buffer. In the latter case, reaction mixture is prepared from R and its conjugate acid RH^+, varying the ratio $[RH^+]/[R]$ but keeping the total sum of concentration $[RH^+] + [R] = S_R$ in the given series constant. This procedure limits the use to the range of pH $= pK_{RH^+} \pm 1$, but this pH-region is frequently the most important. It is also possible to keep concentration of one of the components $[R]$ or $[RH^+]$ constant and change the pH-value by changing the concentration of the other.

To the solution, buffered by a nonreactive buffer or by a reagent present in excess, reactants A and/or B are added so that all reaction mixtures in the compared series contain the same final concentrations of reactants.

The ionic strength is preferably maintained constant in all reaction mixtures in the given series by the addition of neutral salts.

(a) ACID FORM REACTS. When only the acid form of the reagent RH^+ undergoes reaction at a measurable rate, the following scheme is followed:

$$RH^+ \xrightleftharpoons{K_1} R + H^+ \tag{3.6-1}$$

$$RH^+ + A \xrightarrow{k_1} Product \tag{3.6-2}$$

The equation for the reaction rate is:

$$-\frac{d[A]}{dt} = k_1 [RH^+][A] \tag{3.6-3}$$

When the rate is measured at various pH-values and hence the ratio $[RH^+]/[R]$ is changed, it is convenient to introduce the analytical concentration S_R defined as:

$$S_R = [RH^+] + [R] \qquad (3.6-4)$$

Then

$$K_1 = \frac{[R][H^+]}{[RH^+]} \quad \text{and} \quad [R] = \frac{K_1[RH^+]}{[H^+]} \qquad (3.6-5)$$

$$S_R = [RH^+]\left(1 + \frac{K_1}{[H^+]}\right) \qquad (3.6-6)$$

and

$$-\frac{d[A]}{dt} = k_1 \frac{[H^+]}{K_1 + [H^+]} [A] S_R \qquad (3.6-7)$$

or

$$-\frac{d[A]}{dt} = k' [A] S_R \qquad (3.6-8)$$

This is a simple rate equation valid for each pH in which the value of the determined formal rate constant k' changes with pH according to the equation (3.6-9):

$$k' = k_1 \frac{[H^+]}{K_1 + [H^+]} \qquad (3.6-9)$$

For a similar reaction scheme (3.6-10 and 3.6-11), where only the acid form of the reactant BH^+ reacts with the unprotonated form of the reagent R (and the reaction of form B can be neglected):

$$BH^+ \underset{}{\overset{K_1'}{\rightleftharpoons}} B + H^+ \qquad (3.6-10)$$

$$BH^+ + R \overset{k_1}{\longrightarrow} \text{Product} \qquad (3.6-11)$$

the rate equation is given by (3.6-12):

$$-\frac{d[B]}{dt} = k_1 \frac{[H^+]}{K_1' + [H^+]} [R] S_B \qquad (3.6-12)$$

and the formal rate constant k' follows (3.6-13):

$$k' = k_1 \frac{[H^+]}{K_1' + [H^+]} \qquad (3.6\text{-}13)$$

Equation (3.6-13) is formally similar to equation (3.6-9), but the value K_1' now corresponds to dissociation of the acid form of reactant BH^+ whereas in equation (3.6-9) K_1 corresponds to the dissociation of the reagent RH^+.

In both cases the plot of the measured formal rate constant k' against pH possesses the form given in Fig. 3-38 (to exclude the arbitrariness in the choice of the scale, values of k'/k_1 are plotted instead of k' making the graph generally valid). Equations for k' (3.6-9 and 3.6-13) and Fig. 3-38 illustrate the change in the value of the formal rate constant k' with pH when $(pK_1 - 2) < pH < (pK_1 + 2)$.

Alternatively, when the value of k' is determined at $pH \ll pK_1$ (or pK_1'), the value of K_1 (or K_1') in equations (3.6-9 and 3.6-13) can be neglected against $[H^+]$ and $k' = k_1$; the value of the rate constant determined in this pH-range is hence identical with the true value of k_1.

When, oppositely, the formal rate constant k' is determined at $pH \gg pK_1$ (or pK_1'), the value of $[H^+]$ can be neglected against K_1 (or K_1') and

$$k' = \frac{k_1[H^+]}{K_1} \qquad (3.6\text{-}14)$$

The determination of rate constants at $pH \gg pK_1$ is hence ill-suited for distinguishing the type of the reaction system involved. For this purpose measurements in the pH-region between $(pK_1 - 2)$ and $(pK_1 + 2)$ are most suitable.

The determination of the rate constant k_1 is simplest when the reaction rate can be measured in acidic media at $pH < (pK_2 - 2)$. It is nevertheless possible to determine the value of k_1 from measured values of k' in other pH-regions. In the medium pH-range, at $(pK_1 - 2) < pH < (pK_1 + 2)$, it is preferable to determine

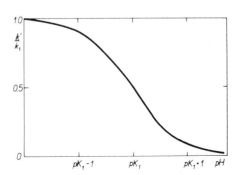

FIG. 3-38. Dependence of the relative values of rate constants k'/k_1 on pH for an antecedent protonation and reaction of the acid form following equations (3.6-1) and (3.6-2), or (3.6-10) and (3.6-11).

the value of K_1 by an independent method before using the equation for k'. At pH $>$ ($pK_1 + 2$) an independent determination of K_1 is necessary, otherwise only the ratio k_1/K_1 can be computed. Determinations at pH $>$ ($pK_1 + 2$) are carried out usually for those cases in which the reaction at lower pH-values is too fast to be followed.

An example of a reaction with the conjugate acid form is illustrated by the acid decomposition of dialkyldithiocarbamates [3-71] (3.6-15 to 3.6-17):

$$
\underset{H^+}{\underbrace{\underset{R^2}{\overset{R^1}{>}}N\!-\!C\overset{S}{\underset{S^{(-)}}{<}}}} \overset{K_1}{\rightleftharpoons} \underset{R^2}{\overset{R^1}{>}}N\!-\!C\overset{S}{\underset{S^{(-)}}{<}} + H^+ \tag{3.6-15}
$$

$$
\underset{H^+}{\underbrace{\underset{R^2}{\overset{R^1}{>}}N\!-\!C\overset{S}{\underset{S^{(-)}}{<}}}} \overset{k_1}{\longrightarrow} R^1R^2NH + CS_2 \tag{3.6-16}
$$

$$
R^1R^2NH + H^+ \rightleftharpoons R^1R^2NH_2^+ \tag{3.6-17}
$$

and of nitrosamines [3-72]:

$$
\underset{H^+}{\underbrace{\overset{R^1-N-R^2}{\underset{NO}{|}}}} \overset{K_1}{\rightleftharpoons} \overset{R^1-N-R^2}{\underset{NO}{|}} + H^+ \tag{3.6-18}
$$

$$
\underset{H^+}{\underbrace{\overset{R^1-N-R^2}{\underset{NO}{|}}}} \overset{k_1}{\longrightarrow} \text{Product} \tag{3.6-19}
$$

(b) BASE FORM REACTS. When only the base form of the reagent in excess R reacts by a measurable rate, the corresponding scheme is:

$$
RH^+ \overset{K_1}{\rightleftharpoons} R + H^+ \tag{3.6-20}
$$

$$
R + A \overset{k_2}{\longrightarrow} \text{Product} \tag{3.6-21}
$$

From the equation for the reaction rate

$$
-\frac{d[A]}{dt} = k_2 [R] [A] \tag{3.6-22}
$$

and expression for analytical concentration of the reagent after introducing $[RH^+] = [R][H^+]/K_1$, it follows:

$$S_R = [R] \left(1 + \frac{[H^+]}{K_1} \right) \tag{3.6-23}$$

then

$$-\frac{d[A]}{dt} = k_2 \frac{K_1}{K_1 + [H^+]} [A] S_R \tag{3.6-24}$$

$$k' = k_2 \frac{K_1}{K_1 + [H^+]} \tag{3.6-25}$$

Hence again the time change of $[A]$ follows simple kinetics at a given pH and the measured formal rate constant k' increases with pH in accordance with equation (3.6-25) as shown in Fig. 3-39. Similar kinetics is observed and similar expression for k' obtained when the base form of the reactant B is considerably more reactive towards the reagent R than the acid form BH^+ and the dissociation constant K_1 again corresponds to a dissociation of the reactant B rather than of the reagent R.

At pH $\ll$ pK_1 the value of K_1 can be neglected against $[H^+]$ and the last equation simplifies to:

$$k' = \frac{k_2 K_1}{[H^+]} \tag{3.6-26}$$

showing an inverse proportionality between k' and $[H^+]$. Oppositely at pH $\gg$ pK_1 the measured formal rate constant k' becomes identical with k_2. The alkaline media at pH $>$ (pK_1 + 2) are hence most suitable for determination of k_2 values.

For conclusive evidence that the reaction follows this particular scheme, the measurement of constants k' in the pH-range pK_1 $\pm$ 2 is most useful.

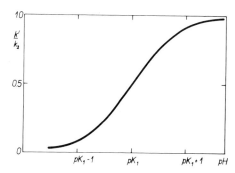

FIG. 3-39. Dependence of the relative values of rate constants k'/k_2 on pH for an antecedent dissociation and reaction of the conjugate base following equations (3.6-20) and (3.6-21).

Examples of reactions following this scheme are reactions of amines with carbon disulfide [3-71]:

$$NR_2H_2^+ \underset{K_1}{\rightleftharpoons} NR_2H + H^+ \tag{3.6-27}$$

$$NR_2H + CS_2 \xrightarrow{k_2} \underbrace{R_2N-C\overset{\displaystyle S}{\underset{\displaystyle S^{(-)}}{\big\langle}}}_{H^+} \tag{3.6-28}$$

$$\underbrace{R_2N-C\overset{\displaystyle S}{\underset{\displaystyle S^{(-)}}{\big\langle}}}_{H^+} \underset{K_D}{\rightleftharpoons} R^2N-C\overset{\displaystyle S}{\underset{\displaystyle S^{(-)}}{\big\langle}} + H^+ \tag{3.6-29}$$

or elimination reactions of Mannich bases (at pH $\ll$ pK_N) [3-73]:

$$C_6H_5COCH_2CH_2\overset{(+)}{\underset{H}{N}}\langle\bigcirc\rangle O \underset{K_1}{\rightleftharpoons} C_6H_5COCH_2CH_2N\langle\bigcirc\rangle O + H^+ \tag{3.6-30}$$

$$C_6H_5COCH_2CH_2N\langle\bigcirc\rangle O \xrightarrow{k_2} C_6H_5COCH=CH_2 + NH\langle\bigcirc\rangle O \tag{3.6-31}$$

$$NH_2^+\langle\bigcirc\rangle O \underset{K_N}{\rightleftharpoons} NH\langle\bigcirc\rangle O + H^+ \tag{3.6-32}$$

Another example is the hydrolysis of acyl cyanides [3-74], which follows reactions (3.6-33)

$$ArCOCN + OH^- \underset{pK_1 \approx 2.5}{\overset{k_1}{\rightleftharpoons}} Ar\overset{\displaystyle OH}{\underset{\displaystyle O^-}{C}}CN \tag{3.6-33}$$

$$Ar\overset{\displaystyle OH}{\underset{\displaystyle O^-}{C}}CN \xrightarrow{k_2} ArC\overset{\displaystyle OH}{\underset{\displaystyle O}{\big\langle}} + CN^- \tag{3.6-34}$$

(c) BOTH ACID AND BASE FORMS REACT. If both acid (RH^+) and base (R) forms of the reagent react with the reactant A at similar rates (i.e., when the values of k_1 and k_2 in the following scheme are comparable; when this condition is not fulfilled and one of the rate constants is considerably greater than the other, the present case degenerates into one of the above-mentioned types), the scheme (3.6-35 to 3.6-37) applies:

$$RH^+ \stackrel{K_1}{\rightleftharpoons} R + H^+ \tag{3.6-35}$$

$$RH^+ + A \stackrel{k_1}{\longrightarrow} Product_1 \tag{3.6-36}$$

$$R + A \stackrel{k_2}{\longrightarrow} Product_2 \tag{3.6-37}$$

and the rate can be described by equation (3.6-38):

$$-\frac{d[A]}{dt} = k_1 [RH^+][A] + k_2 [R][A] \tag{3.6-38}$$

or

$$-\frac{d[A]}{dt} = k_1 \frac{[H^+]}{K_1 + [H^+]} [A] S_R + k_2 \frac{K_1}{K_1 + [H^+]} [A] S_R \tag{3.6-39}$$

$$-\frac{d[A]}{dt} = \frac{k_1[H^+] + k_2 K_1}{K_1 + [H^+]} [A] S_R \tag{3.6-40}$$

and

$$k' = \frac{k_1[H^+] + k_2 K_1}{K_1 + [H^+]} \tag{3.6-41}$$

The following cases can be considered:

1. When the rate constants are determined at pH $\ll$ pK_1, K_1 can be neglected against [H$^+$] and, because $k_1 \approx k_2$, the value of $k_2 K_1/[H^+]$ can be neglected against k_1. The formal rate constant k' becomes equal to k_1.

2. When the reaction is followed in alkaline solutions with pH $\gg$ pK_1, [H$^+$] can be neglected against K_1, $k_1[H^+]/K_1$ can be neglected against k_2, and the formal rate constant k' equals k_2.

3. If the rate is determined at pH $=$ pK_1, the equation for formal rate constant k' becomes $k' = (k_1 + k_2)/2$.

4. In the region pH $=$ (p$K_1 \pm 2$), the measured value of k' changes with pH in the form of a dissociation curve. Two possibilities exist, namely that $k_1 > k_2$ (Fig. 3-40a) or $k_1 < k_2$ (Fig. 3-40b).

Analogous results can be obtained when both the acid (BH$^+$] and base (B) forms of the reactant show a comparable reactivity against the reagent R.

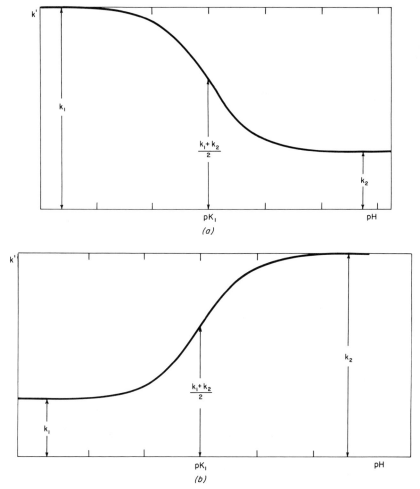

FIG. 3-40. Dependence of the measured formal rate constant k' on pH for reactions preceded by a rapidly established acid–base equilibrium, in which both acid and base forms react according to equations (3.6-35) to (3.6-37). (a) $k_1 > k_2$; (b) $k_1 < k_2$.

(d) BIVALENT ACID FORM REACTS. When the reagent is a bivalent acid that reacts with the substance A only in its fully protonated form RH_2^{2+}, the scheme (3.6-42 to 3.6-44) is to be considered:

$$RH_2^{2+} \xrightleftharpoons{K_1} RH^+ + H^+ \tag{3.6-42}$$

$$RH^+ \xrightleftharpoons{k_2} R + H^+ \tag{3.6-43}$$

$$RH_2^{2+} + A \xrightarrow{k_1} \text{Product} \tag{3.6-44}$$

The rate equation is defined as:

$$- \frac{d[A]}{dt} = k_1 [RH_2^{2+}][A] \tag{3.6-45}$$

The analytical concentration of the reagent is given by

$$S_R = [RH_2^{2+}] + [RH^+] + [R] \tag{3.6-46}$$

which, when $[RH^+]$ and $[R]$ are expressed by means of dissociation constants K_1 and K_2, becomes

$$S_R = [RH_2^{2+}] \left(1 + \frac{K_1}{[H^+]} + \frac{K_1 K_2}{[H^+]^2} \right) \tag{3.6-47}$$

Then

$$- \frac{d[A]}{dt} = \frac{k_1}{1 + K_1/[H^+] + K_1 K_2/[H^+]^2} S_R [A] \tag{3.6-48}$$

and

$$k' = \frac{k_1}{1 + K_1/[H^+] + K_1 K_2/[H^+]^2} \tag{3.6-49}$$

For $pK_1 < (pK_2 - 2)$, the value of $K_1 K_2/[H^+]^2$ can be neglected against $(1 + K_1/[H^+])$; the equation for the formal rate constant k' degenerates into $k' = k_1[H^+]/(K_1 + [H^+])$; and the value of the formal rate constant shows a pH-dependence depicted in Fig. 3-38.

For $pK_1 \geq (pK_2 - 2)$ the course of the k'–pH-dependence corresponds to a steeper, unsymmetric curve (Fig. 3-41, curves 1–3). The effect of the second dissociation step can be detected only for such a ratio of the dissociation constants.

(e)   MONOBASIC FORM OF A BIVALENT ACID REACTS.   When the reagent is a bivalent acid that reacts with the substance A only in the monoprotonated form RH^+, the scheme (3.6-50 to 3.6-52) is involved:

$$RH_2^{2+} \overset{K_1}{\rightleftharpoons} RH^+ + H^+ \tag{3.6-50}$$

$$RH^+ \overset{K_2}{\rightleftharpoons} R + H^+ \tag{3.6-51}$$

$$RH^+ + A \overset{k_2}{\longrightarrow} Product \tag{3.6-52}$$

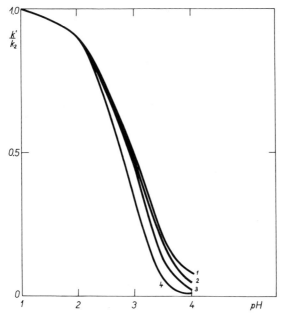

FIG. 3-41. Dependence of the relative values of rate constants k'/k_2 on pH for a reaction of the acid form of a bivalent acid following equations (3.6-42) to (3.6-44).

The rate equation is given by:

$$- \frac{d[A]}{dt} = k_2 \, [RH^+] \, [A] \qquad (3.6\text{-}53)$$

Using the expression for analytical concentration of the reagent (S_R) and definitions of K_1 and K_2, it follows that

$$- \frac{d[A]}{dt} = \frac{k_2}{1 + [H^+]/K_1 + K_2/[H^+]} \, S_R \, [A] \qquad (3.6\text{-}54)$$

and experimentally determined k' is equal to:

$$k' = \frac{k_2}{1 + [H^+]/K_1 + K_2/[H^+]} \qquad (3.6\text{-}55)$$

The pH-dependence of the formal rate constant k' for $pK_1 \ll (pK_2 - 3)$ possesses a shape shown in Fig. 3-42. The pH-dependence for $pK_1 \gg (pK_2 - 3)$ is shown in Fig. 3-43 for three values of the ratio K_1/K_2. For reasons given above, the values of k'/k_2 are plotted against pH in these two figures.

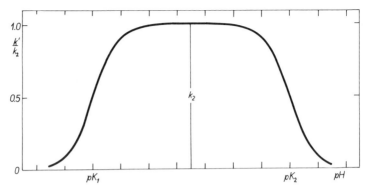

FIG. 3-42. Dependence of the relative values of rate constants k'/k_2 on pH for a reaction of the monovalent anion of a bivalent acid following equation (3.6-50) to (3.6-52), when $pK_1 = (pK_2 - 7)$.

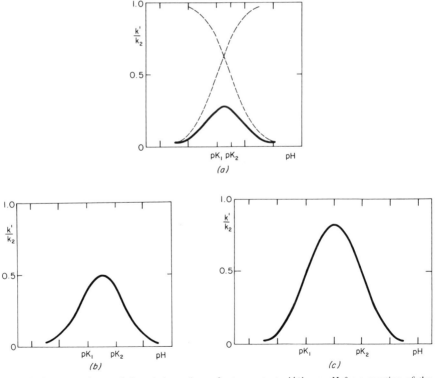

FIG. 3-43. Dependence of the relative values of rate constants k'/k_2 on pH for a reaction of the monovalent anion of a bivalent acid following equations (3.6-50) to (3.6-52) for: (a) $pK_1 = (pK_2 - 0.5)$, (b) $pK_1 = (pK_2 - 1.0)$, (c) $pK_1 = (pK_2 - 2.0)$.

An example of the type given in Fig. 3-42 is the reaction of periodic acid with ethyleneglycol [3-34, 3-75]:

$$H_5IO_6 \overset{K_1}{\rightleftharpoons} H_4IO_6^- + H^+ \tag{3.6-56}$$

$$H_4IO_6^- \overset{K_2}{\rightleftharpoons} H_3IO_6^- + H^+ \tag{3.6-57}$$

$$H_4IO_6^- + (CH_2OH)_2 \overset{k_2}{\longrightarrow} Product \tag{3.6-58}$$

(the reaction is actually complicated by acid–base equilibria of an intermediate formed).

An example corresponding to the type given in Fig. 3-43 is the hydrolysis of pyridoxal-5-phosphate [3-76]:

$$Py\ CH_2OPO(OH)_2 \overset{K_1}{\rightleftharpoons} Py\ CH_2OPO(OH)O^- + H^+ \tag{3.6-59}$$

$$Py\ CH_2OPO(OH)O^{(-)} \overset{K_2}{\rightleftharpoons} Anion + H^+ \tag{3.6-60}$$

$$Py\ CH_2OPO(OH)O^{(-)} \overset{k_2}{\longrightarrow} Py\ CH_2OH + H_2PO_4^{(-)} \tag{3.6-61}$$

(where Py =

) $\tag{3.6-62}$

(f) DIBASIC FORM OF A DIVALENT ACID REACTS.   When the reagent is a bivalent acid that reacts with substance A only in the completely dissociated form R, the following scheme (3.6-63 to 3.6-65) is to be considered:

$$RH_2^{2+} \overset{K_1}{\rightleftharpoons} RH^+ + H^+ \tag{3.6-63}$$

$$RH^+ \overset{K_2}{\rightleftharpoons} R + H^+ \tag{3.6-64}$$

$$R + A \overset{k_3}{\longrightarrow} Product \tag{3.6-65}$$

The rate equation is given by:

$$- \frac{d[A]}{dt} = k_3\ [R][A] \tag{3.6-66}$$

As in the previous two cases, equation (3.6-66) is transformed into form (3.6–67):

$$- \frac{d[A]}{dt} = \frac{k_3}{1 + [H^+]/K_2 + [H^+]^2/K_1K_2}\ S_R\ [A] \tag{3.6-67}$$

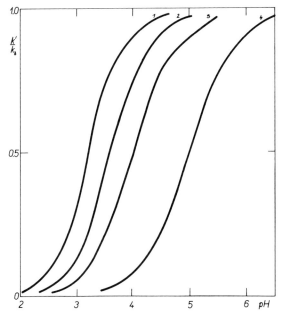

FIG. 3-44. Dependence of the relative values of rate constants k'/k_2 on pH for a reaction of a dianion of a bivalent acid following equations (3.6-63) to (3.6-65) for: (1) $pK_1 = pK_2$, (2) $pK_1 = (pK_2 - 0.5)$, (3) $pK_1 = (pK_2 - 1.0)$, (4) $pK_1 = (pK_2 - 2.0)$.

Hence

$$k' = \frac{k_3}{1 + [H^+]/K_2 + [H^+]^2/K_1K_2} \qquad (3.6-68)$$

For $pK_1 \ll pK_2$ the latter equation degenerates to the form $k' = k_3K_2/(K_2 + [H^+])$. For $K_1 = K_2$ the dissociation curve is twice as steep as that given by equation for monobasic acid (Fig. 3-44, curve 1). For comparable values of K_1 and K_2 asymmetric dissociation curves (Fig. 3-44, curves 2–4) are obtained.

(g)   ALL THREE FORMS OF A DIBASIC ACID REACT.   The case when all three possible forms of a dibasic acid react with comparable rates with compound A is considered next. The system is described by the following reactions:

$$RH_2^{2+} \xrightleftharpoons{K_1} RH^+ + H^+ \qquad (3.6-69)$$

$$RH^+ \xrightleftharpoons{K_2} R + H^+ \qquad (3.6-70)$$

$$RH_2^{2+} + A \xrightarrow{k_1} Product_1 \qquad (3.6-71)$$

$$RH + A \xrightarrow{k_2} Product_2 \qquad (3.6-72)$$

$$R + A \xrightarrow{k_3} Product_3 \qquad (3.6\text{-}73)$$

The rate equation is in the form (3.6-74):

$$-\frac{d[A]}{dt} = k_1[A][RH_2^{2+}] + k_2[A][RH^+] + k_3[A][R] \qquad (3.6\text{-}74)$$

The concentration of the various ionized forms is then expressed using the analytical concentration of the reagent (S_R):

$$[RH_2^{2+}] = \frac{S_R}{1 + K_1/[H^+] + K_1 K_2/[H^+]^2} \qquad (3.6\text{-}75)$$

$$[RH^+] = \frac{S_R}{1 + [H^+]/K_1 + K_2/[H^+]^2} \qquad (3.6\text{-}76)$$

and the rate equation can be written in the form:

$$-\frac{d[A]}{dt} = \left(\frac{k_1}{1 + K_1/[H^+] + K_1 K_2/[H^+]^2} + \frac{k_2}{1 + [H^+]/K_1 + K_2/[H^+]^2} \right.$$
$$\left. + \frac{k_3}{1 + [H^+]/K_2 + [H^+]^2/K_1 K_2} \right) S_R [A] \quad (3.6\text{-}77)$$

The pH-dependence of the measured formal rate constant k' is described by equation (3.6-78):

$$k' = \frac{k_1[H^+]^2 + k_3 K_1 K_2}{[H^+]^2 + K_1[H^+] + K_1 K_2} + \frac{k_2[H^+]K_1}{[H^+]^2 + K_1[H^+] + K_1 K_2/[H^+]} \qquad (3.6\text{-}78)$$

The cases when $K_1 \gg K_2$ are considered first. The limiting cases are:

$$\text{For} \quad [H^+] \gg K_1 \qquad k' = k_1$$
$$\text{For} \quad K_1 \gg [H^+] \gg K_2 \qquad k' = k_2$$
$$\text{For} \quad [H^+] \ll K_2 \qquad k' = k_3$$

and for $[H^+] = K_1$ is $k' = (k_1 + k_2)/2$ and for $[H^+] = K_2$ is $k' = (k_2 + k_3)/2$.

The course of the k'–pH dependence is shown in Fig. 3-45 for: (a) $k_1 > k_2 > k_3$; (b) $k_2 > k_1 > k_3$; (c) $k_1 > k_3 > k_2$; (d) $k_2 > k_3 > k_1$; (e) $k_3 > k_1 > k_2$; (f) $k_3 > k_2 > k_1$.

When the value of K_1 is comparable to that of K_2, the course of the pH-dependence of the formal rate constant k' is more complicated. Some examples are given in Fig. 3-46 for the same ratio of rate constants as above.

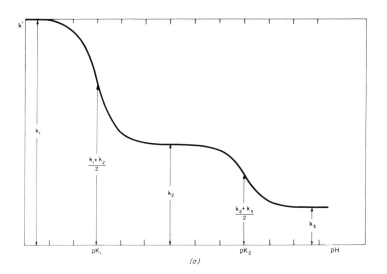

(a)

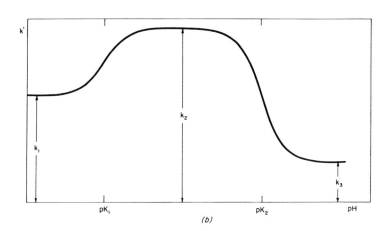

(b)

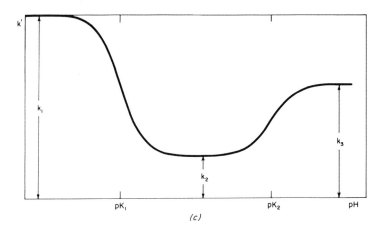

(c)

160

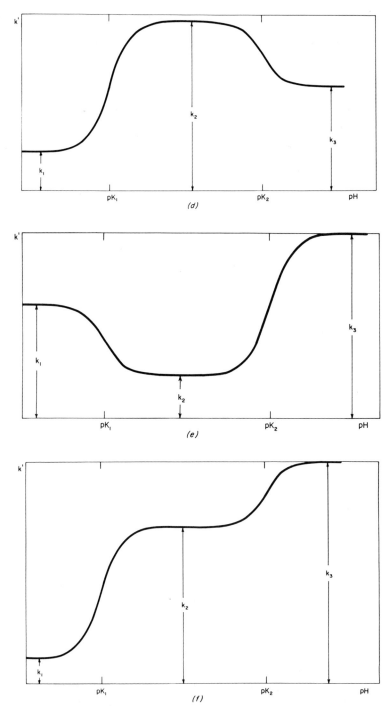

FIG. 3-45. Dependence of the measured formal rate constant k' on pH for a reaction in which all three forms of a bivalent acid react according to equations (3.6-69) to (3.6-73), if $pK_1 \leq (pK_2 - 3)$ for: (a) $k_1 > k_2 > k_3$, (b) $k_2 > k_1 > k_3$, (c) $k_1 > k_3 > k_2$, (d) $k_2 > k_3 > k_1$, (e) $k_3 > k_1 > k_2$, (f) $k_3 > k_2 > k_1$.

161

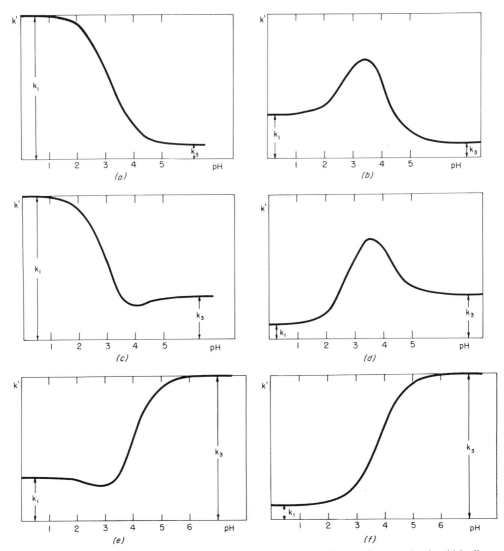

FIG. 3-46. Dependence of the measured formal rate constant k' on pH for a reaction in which all three forms of a bivalent acid react according to equations (3.6-69) to (3.6-73) and pK_1 is comparable with pK_2 (e.g., $pK_1 = (pK_2 - 1)$), for: (a) $k_1 > k_2 > k_3$, (b) $k_2 > k_1 > k_3$, (c) $k_1 > k_3 > k_2$, (d) $k_2 > k_3 > k_1$, (e) $k_3 > k_1 > k_2$, (f) $k_3 > k_2 > k_1$.

The case when only two of the three species react with a measurable velocity can be derived from the present treatment for all three possible combinations of pairs of reacting species by simply setting the value of the appropriate rate constant equal to zero.

Thus far, systems of reactions were discussed in which the formal rate constant computed at a given pH using a proper rate equation depends on pH due to a rapidly established acid–base equilibrium preceding the rate determining step, involving the acid–base properties of one reactant only. In the following paragraphs reactions will be discussed in which two reaction partners exhibit acid–base properties.

(h) BASE FORM OF THE REAGENT REACTS WITH THE ACID FORM OF THE REACTANT.   When base form of the reagent R reacts with the acid form of the reactant BH^+ the reaction scheme is described by eqs. (3.6-79 – 3.6-81). Alternatively, when the acid form of the reagent RH^+ reacts with the base form of the reactant, the reaction scheme corresponds to eqs. (3.6-79), (3.6-80), and (3.6-82):

$$RH^+ \underset{}{\overset{K_1}{\rightleftharpoons}} R + H^+ \tag{3.6-79}$$

$$BH^+ \underset{}{\overset{K_B}{\rightleftharpoons}} B + H^+ \tag{3.6-80}$$

$$R + BH^+ \overset{k_1}{\longrightarrow} Product_1 \tag{3.6-81}$$

$$RH^+ + B \overset{k_2}{\longrightarrow} Product_2 \tag{3.6-82}$$

The sequence of equations (3.6-79 to 3.6-81) corresponds to the rate equation

$$\frac{-d[B]}{dt} = k_1[RH^+][B] \tag{3.6-83}$$

and using the following expressions for analytical concentrations S_R and S_B

$$S_R = [RH^+]\left(1 + \frac{K_1}{[H^+]}\right) \tag{3.6-84}$$

$$S_B = [B]\left(1 + \frac{[H^+]}{K_B}\right) \tag{3.6-85}$$

rate equation can be expressed in the form:

$$\frac{-d[B]}{dt} = \frac{k_1 S_R S_B}{(1 + K_1/[H^+])(1 + [H^+]/K_B)} \tag{3.6-86}$$

and the formal rate constant k' (relative to k_1) is given by:

$$\frac{k'}{k_1} = \frac{K_B[H^+]}{(K_1 + [H^+])(K_B + [H^+])} \tag{3.6-87}$$

When the sequence of the first two acid–base equilibria, (3.6-79 and 3.6-80), followed by the reaction of RH^+ with B (3.6-82) is considered, the rate equation is given by:

$$\frac{-d[B]}{dt} = \frac{k_2 S_R S_B}{(1 + [H^+]/K_1)(1 + K_B/[H^+])} \tag{3.6-88}$$

and the formal rate constant k' is given by:

$$\frac{k'}{k_2} = \frac{K_1 [H^+]}{(K_1 + [H^+]) (K_B + [H^+])} \qquad (3.6\text{-}89)$$

Only when the value of K_1 or K_B (or both) can be determined by an independent experiment, is it possible to decide whether reaction of R with BH^+ or reaction of RH^+ with B is the rate determining step. The course of the pH-dependence of the measured formal rate constant k' for the former case of reactions (3.6-79 to 3.6-81) will be discussed in some detail.

(a)   We shall first consider the system when $K_B \gg K_1$. The following limiting cases can be considered:

	k'/k_1
$[H^+] \gg K_B \gg K_1$	$K_B/[H^+]$
$[H^+] = K_B; [H^+] \gg K_1$	0.5
$K_B \gg [H^+] \gg K_1$	1.0
$K_B \gg [H^+]; [H^+] = K_1$	0.5
$K_B \gg K_1 \gg [H^+]$	$[H^+]/K_1$

Under these conditions, the value of k' is comparable to k_1 in the whole pH-range and the course of the pH-dependence of k' is bell-shaped, as shown in Fig. 3-47 (right scale for k'/k_1).

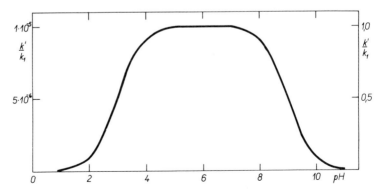

FIG. 3-47.   Dependence of the relative values of rate constants k'/k_1 on pH for a reaction in which the base form of the reagent reacts with the acid form of the reactant following equations (3.6-79) to (3.6-82). For $pK_B = 3$, $pK_1 = 9$ use the scale on right-hand side; for $pK_B = 9$, $pK_1 = 3$ the scale on the left-hand side.

(b) If on the other hand $K_1 \gg K_B$ the limiting conditions are as follows:

	k'/k_1
$[H^+] \gg K_1 \gg K_B$	$K_B/[H^+]$
$[H^+] = K_1; [H^+] \gg K_B$	$K_B/2[H^+]$
$K_1 \gg [H^+] \gg K_B$	K_B/K_1
$K_1 \gg [H^+]; [H^+] = K_B$	$[H^+]/2K_1$
$K_1 \gg K_2 \gg [H^+]$	$[H^+]/K_1$

In all these cases k' is only a very small fraction of k_1; the ratio k'/k_1 achieves its highest value in the medium pH-range where it is equal to K_B/K_1. Hence, if $K_B/K_1 = 10^{-3}$, the maximum value of k' is 0.1% of k_1; if the ratio $K_B/K_1 = 10^{-5}$, the maximum value of k' is only 0.001% of k_1. The shape of the pH-dependence is similar as in the previous case (Fig. 3-47, left scale), only the scale is much smaller. When the reaction (3.6-81) is not a very fast reaction, the reaction rate at higher values of the ratio K_B/K_1 may be too small to be measurable and the reaction can be considered as not taking place.

(c) Finally when values of K_1 and K_B are comparable, that is, when $(pK_B - 2) \leq pK_1 \leq (pK_B + 2)$, the course of the k'–pH dependence for several selected ratios K_1/K_B is given in Fig. 3-48. It is obvious that the maximum value reached by the formal rate constant k' is only a small fraction of the value of the rate constant k_1. The shape of the k'–pH dependence does not depend on whether $K_1 < K_B$ or alternatively $K_B < K_1$. From the shape of the k'–pH dependence it is possible to estimate the two pK-values of the two acid–base reactions involved. On the other hand it is not possible to distinguish which of the measured constants belongs to dissociation of the reactant BH^+ and which to the reagent RH^+. It can be seen from Fig. 3-48 that the same shape of the k'–pH plot will be obtained for $pK_1 = (pK_B - 1)$ as well as for $pK_B = (pK_1 - 1)$. To correctly ascribe the value of pK_1 or pK_B it is necessary to determine at least one of the values of the equilibrium constants by an independent method.

An example of this type is the oxidation of 2-propanol by chromic acid. It has been suggested earlier that the chromacidium ion is the reactive species and that the reaction rate is affected by the protonation of the alcohol. It has been, nevertheless, shown [3-77] that the dependence of the measured formal rate constants on the H_0-function corresponds to the reaction scheme (3.6-90 to 3.6-92):

$$H_3CrO_4^+ \underset{}{\overset{K_1}{\rightleftharpoons}} H_2CrO_4 + H^+ \qquad (3.6\text{-}90)$$

$$ROH_2^+ \underset{}{\overset{K_B}{\rightleftharpoons}} ROH + H^+ \qquad (3.6\text{-}91)$$

$$H_3CrO_4^+ + ROH \overset{k_1}{\longrightarrow} Product \qquad (3.6\text{-}92)$$

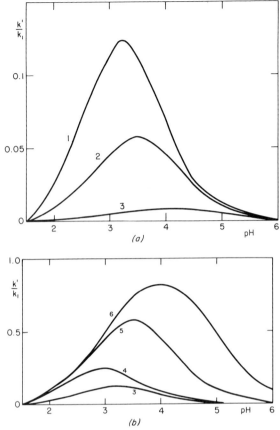

FIG. 3-48. Dependence of the relative values of rate constants k'/k_1 on pH for a reaction in which the base form of the reagent reacts with the acid form of the reactant following equations (3.6-79) to (3.6-82) for comparable values of K_B and and K_1. Part (a): $K_1 > K_B$, $K_1 = 10^{-3}$, K_B varies: (1) 3×10^{-4}; (2) 1×10^{-4}; (3) 1×10^{-5}. Part (b): $K_1 \leq K_B$, $K_B = 1 \times 10^{-3}$, K_1 varies: (3) 1×10^{-5}; (4) 1×10^{-4}; (5) 1×10^{-3}; (6) $K_B = 3 \times 10^{-4}$, $K_1 = 1 \times 10^{-3}$. Observe the change of the scale for k'/k_1 for parts (a) and (b).

Figure 3-49 shows that experimental points fit well the theoretical curve computed for the scheme (3.6-90 to 3.6-92) using equation (3.6-93):

$$\frac{k'}{k_1} = \frac{K_B[H^+]}{(K_1 + [H^+])(K_B + [H^+])} \tag{3.6-93}$$

together with the following set of constants: $k_1 = 516$ liter mol^{-1} s^{-1}; $K_1 = 2.16 \times 10^5$ mol $liter^{-1}$; $K_B = 1.51 \times 10^4$ mol $liter^{-1}$. The same agreement can be obtained when reaction (3.6-94):

$$H_2CrO_4 + R\overset{\cdot}{O}H^+ \overset{k_2}{\longrightarrow} \text{Product} \tag{3.6-94}$$

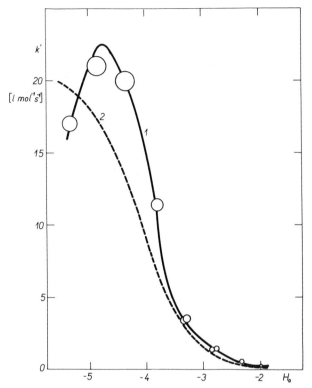

FIG. 3-49. Dependence of the measured formal rate constant k' (experimental points) on acidity function H_0 for oxidation of 2-propanol by chromic acid. Curve (1) corresponds either to sequence (3.6-90) to (3.6-92) for $k_1 = 516$ liter mol^{-1} s^{-1}, $K_1 = 2.16 \times 10^5$ mol liter^{-1}, $K_B = 1.51 \times 10^4$ mol liter^{-1} or to sequence (3.6-90), (3.6-91), and (3.6-94) for $k_2 = 35.9$ liter mol^{-1} s^{-1}, $K_1 = 1.51 \times 10^4$ mol liter^{-1}, $K_B = 2.16 \times 10^5$ mol liter^{-1}. Curve (2) is calculated under consideration of protonation only of one component for $kK = 20.5$ s^{-1} and $K = 1.26 \times 10^4$ mol liter^{-1}.

is considered as the rate-determining step using equation (3.6-95):

$$\frac{k'}{k_2} = \frac{K_1[H^+]}{(K_1 + [H^+]) (K_B + [H^+])}$$
(3.6-95)

and the values of constants $k_2 = 35.9$ liter mol^{-1}s^{-1}; $K_1 = 1.51 \times 10^4$ mol liter^{-1}; $K_B = 2.16 \times 10^5$ mol liter^{-1}. There are no reliable data available for the value K_1.

The value of K_B from a kinetic study [3-78] is 1.6×10^3 mol liter^{-1}, from Raman spectra* [3-79] in hydrochloric acid solution 5.2×10^4 mol liter^{-1}, and in perchloric acid solution 1.4×10^5 mol liter^{-1}. The spread of the data thus excludes

---

*The kinetic data were obtained [3-78] for concentrations of 2-propanol varying between 0.13 and 0.5 M, the Raman spectra were recorded in about 0.8 M 2-propanol. Effect of alcohol concentration on the value of the dissociation constant, reported recently for dissociation of ethanol [3-80] cannot be excluded.

the possibility of choice between the use of equations (3.6-93 and 3.6-95) until the value of K_1 or a more reliable value of K_B is available.

When the theoretical curve was computed for the scheme (3.6-91) and (3.6-94) suggested in literature [3-81] considering the acid–base properties of only one reactant and using values $kK_B = 20.5$ and $K_B = 1.26 \times 10^4$ a much poorer agreement is observed (Fig. 3-49). The present treatment [3-77] differs from the earlier [3-81] one in the evidence for presence or absence of one additional acid–base equilibrium, and leads to a striking difference in the value of the rate constant of the rate-determining step. Whereas the present treatment [3-77] gives $k_1 = 5.16 \times 10^2$ liter mol^{-1} s^{-1}, that given earlier [3-81] gave $k = 1.6 \times 10^{-3}$. This is of particular importance if effects of temperature or structure on reaction rates were studied, since only the k_1 value should be used in such studies. A simple comparison of formal rate constants for oxidation of various alcohols at one sulphuric acid concentration cannot be recommended, since their values are also affected by the value of the constant K_B in addition to structural effects on k_1.

(i) TWO ACID OR TWO BASE FORMS REACT. These cases can be described by the schemes involving (3.6-96 to 3.6-98) or the first two equilibria (3.6-96) and (3.6-97) and the reaction of R and B (3.6-99) respectively:

$$RH^+ \overset{K_1}{\rightleftharpoons} R + H^+ \qquad (3.6\text{-}96)$$

$$BH^+ \overset{K_B}{\rightleftharpoons} B + H^+ \qquad (3.6\text{-}97)$$

$$RH^+ + BH^+ \overset{k_3}{\longrightarrow} \text{Product} \qquad (3.6\text{-}98)$$

$$R + B \overset{k_4}{\longrightarrow} \text{Product} \qquad (3.6\text{-}99)$$

For the reaction of two acid forms, corresponding to reactions (3.6-96 to 3.6-98) the equation for the reaction rate is given by (3.6-100):

$$-\frac{d[B]}{dt} = k_3\,[RH^+][BH^+] \qquad (3.6\text{-}100)$$

and can be transformed using

$$S_R = [RH^+]\left(1 + \frac{K_1}{[H^+]}\right) \qquad S_B = [BH^+]\left(1 + \frac{K_B}{[H^+]}\right) \qquad (3.6\text{-}101)$$

into:

$$-\frac{d[B]}{dt} = \frac{k_3}{(1 + K_B/[H^+])\,(1 + K_1/[H^+])}\,S_B\,S_R \qquad (3.6\text{-}102)$$

The ratio of the formal (k') and proper (k_3) rate constants thus possesses the form (3.6-103):

$$\frac{k'}{k_3} = \frac{[H^+]^2}{(K_B + [H^+]) (K_1 + [H^+])} \qquad (3.6\text{-}103)$$

1. When considering first that $K_1 \gg K_B$ (or $K_B \gg K_1$, which is equivalent in this case, inserting K_B for K_1 and vice versa in the following relations) then:

	k'/k_3
$[H^+] \gg K_1$	1.0
$[H^+] = K_1; K_1 \gg K_B$	0.5
$K_1 \gg [H^+] \gg K_B$	$[H^+]/K_1$
$[H^+] = K_B; K_1 \gg K_B$	$0.5\, K_B/K_1$
$K_1 \gg K_B \gg H^+$	$[H^+]^2/K_1 K_B$

For $pK_1 < (pK_2 - 2)$ the pH-dependence of the ratio k'/k_3 possesses the form of a normal dissociation curve (Fig. 3-50, curve 1) with $pK = pK_1$ for measured formal rate constants k' of magnitude comparable with that of the rate constant k_3. When another method is available allowing determination of rate constants that are smaller than k_3 by several order of magnitudes, another sudden decrease of the formal rate constant k' can be observed in the pH-region comparable with the value of pK_B (Fig. 3-50, curve 2).

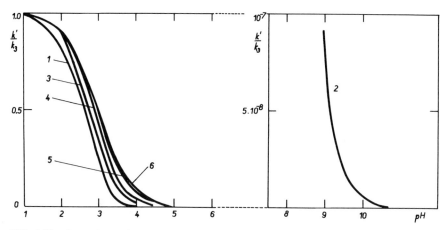

FIG. 3-50. Dependence of the relative values of rate constants k'/k_3 on pH for a reaction of the acid form of the reagent with the acid form of the reactants following equations (3.6-96) to (3.6-98) for $pK_1 = 3.0$ and varied pK_2: (1) and (2) 10.0, (3) 5.0, (4) 4.0, (5) 3.5, (6) 3.0.

2.   The course of the pH-dependence of the ratio k'/k_3 for those cases in which the value of pK_1 is comparable with that for pK_B is given in Fig. 3-50 (curves 3–6) (for selected values of the ratio K_1/K_B). The shape of these curves is unsymmetrical and for $K_1 = K_B$ they are twice as steep as for $pK_1 \ll pK_B$. Nevertheless their shape differs from that found for an antecedent dissociation of a dibasic acid.

For the reaction of two base forms (3.6-99), following the two acid–base equilibria (3.6-96 and 3.6-97), the rate equation (3.6-104) can be derived in a similar manner:

$$- \frac{d[B]}{dt} = \frac{k_4}{(1 + [H^+]/K_B)(1 + [H^+]/K_1)} S_B S_R \qquad (3.6\text{-}104)$$

and the ratio of the formal (k') and proper (k_4) rate constants possesses the form:

$$\frac{k'}{k_4} = \frac{K_1 K_B}{(K_B + [H^+])(K_1 + [H^+])} \qquad (3.6\text{-}105)$$

1.   For $K_1 \gg K_B$ (and similarly for $K_B \gg K_1$) the limiting values are

	k'/k_4
$[H^+] \gg K_1 \gg K_B$	$K_1 K_B/[H^+]$
$[H^+] = K_1; K_1 \gg K_B$	$K_B/2 K_1$
$K_1 \gg [H^+] \gg K_B$	$K_B/[H^+]$
$[H^+] = K_B; K_1 \gg K_B$	0.5
$K_1 \gg K_B \gg H^+$	1.0

2.   For the case of comparable values of K_1 and K_B the shape of the pH-dependence of the ratio k'/k_4 is a mirror image of that shown in Fig. 3-50.

(j)   REACTION OF A DIBASIC ACID WITH A MONOBASIC ACID.   Among the possible combinations, only one will be discussed here, namely the reaction in which the base form of the reactant reacts with different rates with all three ionized forms of the dibasic acid according to the scheme (3.6-106 to 3.6-111):

$$RH^+ \xrightarrow{K_1} R + H^+ \qquad (3.6\text{-}106)$$

$$BH_2^{2+} \xrightarrow{K_{B_1}} BH^+ + H^+ \qquad (3.6\text{-}107)$$

$$BH^+ \xrightarrow{K_{B_2}} B + H^+ \qquad (3.6\text{-}108)$$

$$BH_2^{2+} + R \xrightarrow{k_1} Product_1 \qquad (3.6\text{-}109)$$

$$BH^+ + R \xrightarrow{k_2} Product_2 \qquad (3.6\text{-}110)$$

$$B + R \xrightarrow{k_3} Product_3 \qquad (3.6\text{-}111)$$

The rate equation is:

$$-\frac{d[B]}{dt} = k_1 [BH_2^{2+}][R] + k_2 [BH^+][R] + k_3 [B][R] \qquad (3.6\text{-}112)$$

Introducing:

$$S_B = [BH_2^{2+}] + [BH^+] + [B] \qquad (3.6\text{-}113)$$

and transforming

$$[BH_2^{2+}] = \frac{S_B}{1 + K_{B_1}/[H^+] + K_{B_1} K_{B_2}/[H^+]^2} \qquad (3.6\text{-}114)$$

$$[BH^+] = \frac{S_B}{1 + K_{B_2}/[H^+] + [H^+]/K_{B_1}} \qquad (3.6\text{-}115)$$

$$[B] = \frac{S_B}{1 + [H^+]/K_{B_2} + [H^+]^2/K_{B_1} K_{B_2}} \qquad (3.6\text{-}116)$$

and

$$S_R = [RH^+] + [R] \qquad (3.6\text{-}117)$$

resp.

$$[R] = \frac{S_R}{1 + [H^+]/K_1} \qquad (3.6\text{-}118)$$

inserting into expression for reaction rate we obtain:

$$-\frac{d[B]}{dt} = \left\{ \frac{k_1}{(1 + K_{B_1}/[H^+] + K_{B_1} K_{B_2}/[H^+]^2)(1 + [H^+]/K_1)} \right.$$
$$+ \frac{k_2}{(1 + [H^+]/K_{B_1} + K_{B_2}/[H^+])(1 + [H^+]/K_1)}$$
$$\left. + \frac{k_3}{(1 + [H^+]/K_{B_2} + [H^+]^2/K_{B_1} K_{B_2})(1 + [H^+]/K_1)} \right\} S_R S_B \qquad (3.6\text{-}119)$$

The value in the brackets corresponds to the formal rate constant k'. It would be possible to examine the above equation for various ratios of K_1, K_{B_1}, K_{B_2}, k_1, k_2, and k_3 similarly as in the preceding sections. We shall restrict ourselves, however, to one example of that type found in the reaction of pyridoxal with hydroxylamine [3-82].

Hydroxylamine can take part in the condensation reaction only in the base form (R):

$$NH_3OH^+ \rightleftharpoons NH_2OH + H^+ \qquad (3.6\text{-}120)$$

$$(RH^+) \qquad (R)$$

$$pK_R = 6.0$$

Pyridoxal on the other hand can exist in three ionized forms: The totally protonated form BH_2^{2+}, the unprotonated or zwitterionic form BH^+, and the totally dissociated form B

$$(3.6\text{-}121)$$

The kinetic measurements were carried out in the presence of an excess of hydroxylamine and, at a given pH, first-order kinetics with the constant k' were observed. The measured formal rate constants k' change with pH as shown by points on Figs. 3-51a and b.

An attempt was made first to interpret this dependence as due to the reaction of only totally protonated form BH_2^{2+} and of the totally dissociated form B of pyridoxal with the free amine form R of hydroxylamine. For the formal rate constant k' the following expression was derived (the quadratic terms being neglected):

$$\frac{k'}{S_R} = \frac{k_1}{(1 + [H^+]/K_R)(1 + K_{B_1}/[H^+])}$$

$$+ \frac{k_3}{(1 + [H^+]/K_R)(1 + [H^+]/K_{B_2})} \qquad (3.6\text{-}122)$$

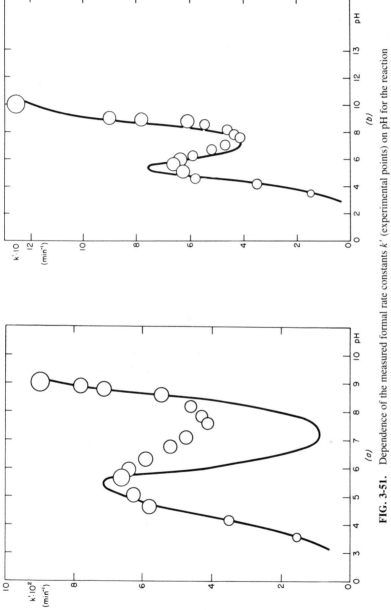

FIG. 3-51. Dependence of the measured formal rate constants k' (experimental points) on pH for the reaction of pyridoxal with hydroxylamine. (*a*) Theoretical curve calculated using equation (3.6-122), when only the reactions of the fully protonated and fully dissociated forms of pyridoxal were considered. (*b*) Theoretical curve calculated by means of equation (3.6-124) considering reactions of all three forms of pyridoxal with NH_2OH for $K_R = 10^{-6}$ mol liter^{-1}, $K_{B_1} = 5 \times 10^{-5}$ mol liter^{-1}, $K_{B_2} = 2 \times 10^{-9}$ mol liter^{-1}, $k_1 = 5 \times 10^2$ liter mol^{-1} min^{-1}, $k_2 = 3.5$ liter mol^{-1} min^{-1}, $k_3 = 12.5$ liter mol^{-1} min^{-1}, and $[NH_2OH]_0 = 0.01$ *M*.

The value k' (at analytical concentration S_R kept constant) calculated using equation (3.6-122) shows systematic deviations (Fig. 3-51a). The magnitude of these deviations shows a bell-shaped pH-dependence resembling that for distribution of the BH^+ species. Hence it was deduced that the observed deviations are due to the fact that contributions of the reaction (3.6-110) of the BH^+ species with hydroxylamine (with the constant k_2) to the total reaction rate are not negligible. It was necessary to consider the following reaction scheme:

$$BH_2^{2+} \rightleftharpoons BH^+ \rightleftharpoons B \qquad (3.6\text{-}123)$$

$$\overset{+}{R} \qquad\qquad \overset{+}{R} \qquad\qquad \overset{+}{R}$$

$$\downarrow k_1 \qquad\qquad \downarrow k_2 \qquad\qquad \downarrow k_3$$

Equation (3.6-124) for k'/S_R was used after neglecting the quadratic terms in the first and third fraction:

$$
\frac{k'}{S_R} = \frac{k_1}{(1 + K_{B_1}/[H^+])(1 + [H^+]/K_1)}
$$
$$
+ \frac{k_2}{(1 + [H^+]/K_{B_1} + K_{B_2}/[H^+])(1 + [H^+]/K_1)}
$$
$$
+ \frac{k_3}{(1 + [H^+]/K_{B_2})(1 + [H^+]/K_1)} \qquad (3.6\text{-}124)
$$

To find the best-fitting set of rate constants k_1, k_2, and k_3, the approximate value of k_1 was found from values of k' at pH < 4.5, neglecting the terms for k_2 and k_3. Similarly, the value for k_3 from values of k' was obtained at pH 8–10, neglecting the terms of k_1 and k_2. The approximate value of k_2 was deduced from the plot of differences between the theoretical curve predicted by equation for k'/S_R, which did not include the term with k_2, and experimental data. These approximate values were inserted into complete equation (3.6-124) for k'/S_R and the course of the k'–pH dependence (at given S_R value) was computed. Utilizing method of trial and error the values of the particular rate constants were found that exhibited the best agreement with the experimental data in the entire pH-range. For $K_R = 10^{-6}$ mol; $K_{B_1} = 5 \times 10^{-5}$ mol; $K_{B_2} = 2 \times 10^{-9}$ mol; $k_1 = 500$ liter mol^{-1} min^{-1}; $k_2 = 3.5$ liter mol^{-1} min^{-1}, and $k_3 = 12.5$ liter mol^{-1} min^{-1}, the agreement is shown in Fig. 3-51b.

Hence all three forms of pyridoxal are reactive, but show a marked difference in reactivity towards hydroxylamine. The role of the formation of a hydrated intermediate [3-83], [3-84] of the type $> C(OH)NHOH$ was not detected. General acid–base catalysis causes only a minor perturbance of the observed rate constants. For example, at pH 5.8 a twofold increase in acetate buffer concentration causes a change in the value of k' of about 15%. The hemiacetal formation, dehydration of the geminal diol form, and ring-opening reactions are too fast to affect the rate constants involved. Hence the major factors affecting the pH-dependence of pyridoxal oxime formation are thus the antecedent acid–base equilibria, which change the protonation of hydroxylamine and pyridoxal.

It is impossible to judge—based on the kinetic measurements only—whether a reaction of BH^+ and RH^+ is involved in the first step. Such possibility can be excluded on the basis of the known lack of reactivity of the ammonium form of hydroxylamine; the assumption of reactivity of BH_2^{2+} with R was also verified from equilibrium measurements in acidic solutions.

3.6.1.2. Reactions in Which Protons or Hydroxide Ions Participate in the Rate-Determining Step

In some reactions, with pH-dependent rates, the rate-determining step can be the dissociation of a proton or addition of a proton, or a nucleophilic attack by a hydroxide ion or cleavage resulting in formation of a hydroxide ion.

A well-known example belonging to the first group involving proton abstraction is bromination of ketones [3-85, 3-86], which for example, for acetone, follows the scheme:

$$CH_3COCH_3 + B \underset{k_2}{\overset{k_1}{\rightleftharpoons}} CH_3COCH_2^- + BH^+ \qquad (3.6\text{-}125)$$

$$CH_3COCH_2^- + Br_2 \xrightarrow[\text{fast}]{} CH_3COCH_2Br + Br^- \qquad (3.6\text{-}126)$$

The rate of establishment of the equilibrium between the keto and the carbanion form has been confirmed to be the slow step by several observations: (a) identical rates for bromination and iodination, (b) independence of the rate on bromine concentration, (c) identical rates for bromination and base-catalysed deuterium exchange, and (d) identical rates of bromination and racemization of optically active ketones of the type $RCOCHR'R''$, which also involves formation of a carbanion of the type $RCO\bar{C}R'R''$.

Another example of proton abstraction is the isomerization (3.6-127) of 2-methylbicyclo(2.2.1)hepta-2,5-diene to 5-methylenebicyclo(2.2.1)hept-2-ene in DMSO, the rate constant of which, in the presence of [18]-crown 6-ether, is a linear function of the activity of tertiary butoxide ions [3-87].

$$\text{(3.6-127)}$$

Examples of addition of hydroxide ion are hydration of α,β-unsaturated ketones followed by protonation and aldolization fission [3-8]:

$$RCOCH = CHR' + OH^- \xrightarrow{k_{OH^-}} RCO\overset{(-)}{C}HCHR' \qquad (3.6\text{-}128)$$
$$\underset{OH}{|}$$

and hydrolysis of coumarins [3-88]:

$$(3.6\text{-}129)$$

Slow protonation is assumed to occur for example in hydrolysis of alkyl vinyl ethers in dilute solutions of hydrochloric acid [3-89]:

$$CH_2 = CHOR + H^+ \longrightarrow CH_3\overset{+}{C}HOR \xrightarrow{H_2O} CH_3\underset{OH}{CHOR} + H^+ \qquad (3.6\text{-}130)$$

$$CH_3\underset{OH}{CH}\!-\!OR \rightleftharpoons CH_3CHO + ROH \qquad (3.6\text{-}131)$$

In such cases the rate constant is a linear function of concentration of H^+ or OH^- and (if the reaction with H_2O as acid or base can be neglected) the k'–$[OH^-]$ or k'–$[H^+]$ plot goes through the origin.

It is not easy to distinguish reactions of this type from reactions where a rapidly established acid–base equilibrium precedes the slow step at pH $<$ $(pK_a - 2)$ (for reactions in which the acid form of reactant or reagent participates) or pH $>$ $(pK_2 + 2)$ (for reactions in which the base form participates). In both cases the formal rate constant k' is a linear function of concentration of hydrogen or hydroxide ions. To distinguish between the two alternatives it is recommended to follow the pH-dependence of the formal rate constant k' in the widest possible range to detect a deviation from the linear k'–$[H^+]$ or k'–$[OH^-]$ plot, which would indicate that the observed dependence is a part of an exponential curve.

The concept of the hydroxide ion as being a nucleophilic reagent is well established. Alternatively, hydrogen ion can be considered as an electrophilic reagent, although in some cases the attack of the lone pair in $:OH_3^+$ on the electron-deficient center in the substrate cannot be excluded.

Since the reactions of the type discussed in this chapter consume hydrogen or hydroxide ions and these ions are not regenerated, it is considered preferable not to denote such reactions as acid–base catalyzed.

In the specific case of strongly acidic solutions it is possible to distinguish two types of system: Reactions in which the hydrogen ion is consumed and reactions in which a molecule of water participates. To find a distinction between these two alternatives the values of the measured formal rate constants k' are plotted against a suitable acidity function (e.g., H_0), as well as the molar concentration of the acid (c_{acid})(Fig. 3-52). If the k'–H_0 plot is linear, the participation of hydrogen ions can

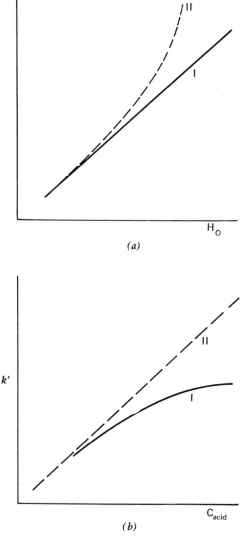

FIG. 3-52. Reactions in strongly acidic media. Dependence of the measured formal rate constant k' on: (a) acidity function H_0; (b) on concentration of the acid. I reaction involving consumption of hydrogen ion; II reaction in which a molecule of the strong acid participates.

be assumed. If the k'–log c_{acid} is linear, it can be concluded that the water molecule participates in the reaction.

3.6.1.3. Acid–Base Catalyzed Reactions

Reactions of organic substances for which the rate is affected by the presence of acids or bases, and which have not been studied in sufficient detail for inclusion in one of the aforementioned categories, are called catalyzed.

A catalyst, according to W. Ostwald [3-90], is a substance that alters the reaction rate without affecting the energy difference between the initial state and the final product of the reaction. R. P. Bell [3-91] defines a catalyst of a homogeneous reaction as a substance for which the exponent at the concentration in the reaction rate equation is higher than corresponds to an overall stoichiometric equation.

Hence, whereas the presence of the catalyst need not be indicated in the stoichiometric equation, it is necessary to express it in the description of mechanism, at least the dependence of the measured rate on the kind and concentration of the catalyst.

These definitions do not involve explicitly the condition that the catalyst must be regenerated in the course of reaction, but this is nevertheless tacitly assumed. Such regeneration can be concluded without a direct proof, if a catalyst is present and effective at a low concentration, as a trace. Nevertheless, in acid–base catalyzed reactions the process is frequently followed under conditions where the solution contains an excess of acid or a base or is well buffered. Since in such cases the concentration of the catalyst remains practically constant in the course of reaction, it is difficult to disprove the regeneration of the catalyst, unless H^+, OH^-, or acid or base is found to be present as a structural part of the product.

For reversible reactions the presence of the catalyst should not affect the position of the equilibrium. Hence the catalyst should increase the rate of the reverse reaction by the same factor as the rate of the forward reaction.

When dealing with irreversible reactions, the situation can be more complex. In some multistep reactions it is possible that the slowest step in the catalyzed reaction will be different from the slowest step governing the rate of the reaction in the absence of catalyst. For example, for the oxidation of ethylene glycol by periodate [3-34] the cleavage of the intermediate is the slowest step and the reaction is not catalyzed by acids and bases, whereas in the oxidation of butan-2,3-diol and homologues [3-35], [3-36] the reaction is acid–base catalyzed and the rate of formation of the intermediate becomes the slowest step.

If a reaction can follow competitive reaction paths, it cannot be ruled out that another path will be preferred in the presence of the catalyst than in its absence. It is thus necessary to prove analytically that the products formed in uncatalyzed and catalyzed reactions are identical. For example, for reactions where both dehydration and hydrogenation can take place, some catalysts will favor one reaction path and some catalysts the other. The composition of the reaction products can differ considerably according to the catalyst used.

It is sometimes necessary to explain why a reaction which could take place in one step (3.6-132):

$$S + R \longrightarrow \text{Products} \tag{3.6-132}$$

takes place faster in two consecutive steps (3.6-133) and (3.6-134):

$$S + HA \rightleftharpoons SH^+ + A^- \tag{3.6-133}$$

$$SH^+ + R \longrightarrow \text{Products} + H^+ \tag{3.6-134}$$

followed by

$$H^+ + A^- \rightleftharpoons HA \qquad (3.6\text{-}135)$$

This can occur when the species SH^+ is more reactive towards reagent R than the form S. The classical publications of W. Ostwald and S. Arrhenius have shown that the rate of succrose inversion is proportional to hydrogen ion concentration. Such reactions are denoted as acid catalyzed. Similarly S. Arrhenius demonstrated that saponification of esters in alkaline media depends on hydroxide ion concentration and called such reactions base catalyzed.

After the introduction of a more encompassing theory of acids and bases, Brönsted [3-91–3-95] divided acid and base catalyzed reactions into two groups: The first group includes reactions which have rates that are affected by the presence and concentration of all acids and/or bases present in the solution. Such reactions are called general acid or base catalyzed. The second group consists of reactions where the catalytic activity on reaction rate is caused only by the acid or conjugate base of the solvent (i.e., in water, H_3O^+, and OH^-). Such reactions are called specific acid or base catalyzed.

The dependence of the measured rate constant k' on acidity can be expressed for a general acid–base catalyzed reaction (i.e., reaction catalyzed both by acids and bases) in the form:

$$k' = k_0 + k_{H_3O^+} [H_3O^+] + k_{OH^-} [OH^-]$$
$$+ \sum_i k_{B_i} [B_i] + \sum_n k_{HA_n} [HA] \qquad (3.6\text{-}136)$$

Here k_0 is the rate constant for the reaction in pure solvent, $k_{H_3O^+}$ for hydrogen, k_{OH^-} for hydroxide ion catalyzed reaction, k_{B_i} stands for all rate constants catalyzed by all bases B_i, k_{HA_n} for rate constants of reactions catalyzed by all acids HA_n.

For a general acid catalyzed reaction, the expression simplifies to

$$k' = k_0 + k_{H_3O^+} [H_3O^+] + \sum_n k_{HA_n} [HA_n] \qquad (3.6\text{-}137)$$

for a general base to

$$k' = k_0 + k_{OH^-} [OH^-] + \sum_i k_{B_i} [B_i] \qquad (3.6\text{-}138)$$

for a specific acid to

$$k' = k_{H_3O^+} [H_3O^+] \qquad (3.6\text{-}139)$$

and for a specific base to

$$k' = k_{OH^-} [OH^-] \qquad (3.6\text{-}140)$$

(a) BRÖNSTED EQUATION [3-91–3-95]. For general acid or base catalyzed reactions it is of importance to compare the activity of individual catalysts. These comparisons have been made for a number of acids and bases and it was found that the stronger the acid or the base, the greater the value of the specific catalytic constant. Since the acid acts as proton donor, both when it catalyzes a reaction rate and when it reacts in an equilibrium reaction with the solvent, it seems logical to assume a parallel between the catalytic efficiency and the values of dissociation constants of the catalyst. Such deductions led Brönsted [3-92], [3-93] to derive a relationship between the values of specific rate constants (k_{HA}, k_B) and pK_a values of corresponding acids.

A decade later it was shown [3-96] that for a group of structurally related compounds a linear correlation between logarithms of rate and equilibrium constants can exist. This type of relationship is rather general and has been found for widely varying types of compounds in numerous types of reactions. Such relationships bear the general description of *linear free energy relationship* or *extra thermodynamic relationships* [3-97–3-103]. The Brönsted relationship for the dependence of the specific catalytic rate constant on values of corresponding rate constants is a special case of linear free energy relationships.

If the comparison is restricted to one type of acid or base (e.g., a series of substituted benzoic acids, of aliphatic carboxylic acids, of substituted anilines or pyridine derivatives, or of primary amines, etc.) and if no attempt is made to compare diverse compounds (e.g., catalytic activity of primary and tertiary amines) it is possible to show empirically the validity of the Brönsted equation:

$$- \log k_{HA} = \alpha \, pK_{HA} - \log G \qquad (3.6\text{-}141)$$

or

$$k_{HA} = G \, K_{HA}^{\alpha} \qquad (3.6\text{-}142)$$

The expression (3.6-141) is similar to a linear free energy relationship of the Hammett equation type. The value of the coefficient α (or β for base catalyzed reaction using K_B^{β}) almost always shows values between 0 and 1 corresponding to reaction constant (ρ). The value of the coefficient α expresses, for the given type of acid, the susceptibility and sensitivity of the given catalyzed reaction to the change in strength of the catalytically active acid. The value of log G—where $G = (K_{HA})_0 / (K_{HA}^{\alpha})_0$—is an additive term which corresponds to the ratio of values of the rate constant of the catalyzed reaction to that of the dissociation constant of the catalyzing acid for an acid selected as a standard in the given group of acids. The effect of the structure of the acid on the value of log G has been rarely discussed.

Similarly, it is possible to derive for base catalyzed reactions the relationship

$$- \log k_B = \beta p K_B - \log G \qquad (3.6\text{-}143)$$

where β has analogous meaning and shows similar values as α (usually $0 \leq \beta \leq 1$).

Also for base catalyzed reactions the validity of the Brönsted relation is restricted to groups of bases of similar structure. Thus, for rates of base catalyzed nitramide cleavage, which is catalyzed both by anions of carboxylic acids and by substituted anilines [3-104], two linear plots are observed with different values of both β and G.

If there are several sites in the catalytically active acid where the proton transfer is possible, the equation must be modified to the form (3.6-144) proposed by Brönsted [3-93]:

$$\frac{k_{HA}}{p} = G \left(\frac{q}{p} K_{HA} \right)^{\alpha} \qquad (3.6\text{-}144)$$

where p is the number of equivalent protons on the catalyst, and q the number of sites on the conjugate base of the catalyst that are able to bind a proton. The rate of the catalyzed reaction increases namely with the number of available protons, whereas the value of the equilibrium constant increases with the number of available protons p, but decreases with the number of sites q on the conjugate base. For instance for ions NH_4^+ as an acid catalyst $p = 1$ and $q = 1$, whereas for oxalic acid $p = 2$ and $q = 4$ (4 oxygen atoms equally available to bind a proton).

The basis of the correlation between the rate constant k_{HA} and the dissociation constant K_{HA} can be demonstrated on one possible mechanism of general acid catalysis which involves the reaction of the acid form of the reactant (BH^+) with the anion of the catalyzing acid (A^-):

$$BH^+ \underset{\text{fast}}{\overset{K_1}{\rightleftharpoons}} B + H^+ \qquad (3.6\text{-}145)$$

$$HA \underset{\text{fast}}{\overset{K_{HA}}{\rightleftharpoons}} A^- + H^+ \qquad (3.6\text{-}146)$$

$$RH^+ + A^- \underset{\text{slow}}{\overset{k}{\longrightarrow}} \text{Products} \qquad (3.6\text{-}147)$$

The rate of such reactions is described by

$$-\frac{d[BH^+]}{dt} = \frac{k \, S_B \, S_{HA}}{(1 + K_1/[H^+])(1 + [H^+]/K_{HA})} \qquad (3.6\text{-}148)$$

When $pK_1 \ll pH \ll pK_{HA}$, which is usually fulfilled for general acid catalyzed reactions,

$$-\frac{d[BH^+]}{dt} = \frac{k \, K_{HA}}{K_1} S_B \, S_{HA} \qquad (3.6\text{-}149)$$

Hence the measured $k_{HA} = k \, K_{HA}/K_1$ and $G = k/K_1$ when $\alpha = 1$.

The Brönsted equation also indicates that the difference between the general and specific catalysis is not qualitative but merely quantitative. The general acid (or base) catalysis is difficult to observe when the constant α (or β) shows values close to zero or one. For 0.1 M buffers, which are commonly used, it is not possible to recognize general catalysis if $\alpha > 0.8$, but such catalysis can be detected if 1.0 M buffers are used. The difficulty in detecting general catalysis for $\alpha < 0.1$ is based on the competitive effects of the added acid and the solvent. When very strong acids are used, it is possible to recognize the general character even of such reactions.

This discussion can be illustrated by the case of an acid catalyzed reaction for which the rate in a solution containing 0.1 M acetic acid and 0.1 M sodium acetate is proportional to concentration of H_3O^+, H_2O, and CH_3COOH. The dependence of the contribution for individual values of α can be seen from the following table:

| | Contribution to Catalysis | | |
Exponent α	H_3O^+	H_2O	CH_3COOH
0.1	0.002%	98%	2%
0.5	3.6%	0.01%	96.4%
1.0	99.8%	$5 \times 10^{-12}\%$	0.2%

If $\alpha = 0.1$, the reaction takes place predominantly by the collision of the substrate with the solvent molecule. Such reactions are considered to be uncatalyzed, since the reaction rate does not increase considerably even in solutions of strong acids.

If $\alpha = 0.5$, the effect of the undissociated acid predominates and the effect of hydrogen ions can be found experimentally. Such reactions are denoted as general acid catalyzed.

Finally, at $\alpha \approx 1.0$ the catalytic effect is practically due solely to hydrogen ions, and the contribution due to the undissociated acid or the solvent cannot be obtained from experimental facts. Such reactions were called specifically catalyzed.

The magnitude of the value α is sometimes interpreted as a measure of the degree of separation of the proton from the corresponding base in the transition state of the slowest reaction step. A large value of α, which should have a maximum value of unity, then represents a situation where the proton is to a large degree separated from the old base and attached to the new one and vice versa. However, the physical interpretation of values of α is more obscure now that recent experimental studies have shown that in some systems values outside the usual range $0 \leq \alpha \leq 1$ can be found.

Studies of the dependence of variation in values of constants α and β for general acid–base catalyzed reactions on structure of both the substrate and the acid or base form of the catalyst enabled construction of three-dimensional reaction-coordinate diagrams [3-104a]. Such structure–reactivity relationships allow conclusions to be made about the structure of the transition state and hence about the preference of certain reaction paths.

(b) THE MEANING OF THE VALUE OF CONSTANT k_0.   The rate constant k_0 is sometimes considered to be the rate constant of the uncatalyzed reaction, but is better interpreted as corresponding to the reaction catalyzed by solvent. If the solvent is water or other amphiprotic solvent, the situation is further complicated by the fact that the solvent molecule can act as an acid and as a base, if other catalytically active acids or bases are absent from the solution. If the reaction mixture contains a basic catalyst, water will act as an acid and vice versa.

For specifically catalyzed reactions it is usually impossible to determine the value of the constant corresponding to the reaction with the solvent. The occurrence of the reaction with the solvent is thus considered to be one diagnostic tool enabling us to distinguish whether the reaction is generally or specifically catalyzed. If the value of the constant k_0 is found different from zero, this is considered as evidence that the reaction is general catalyzed.

(c) DISTINCTION BETWEEN GENERAL AND SPECIFIC CATALYZED REACTIONS. A reaction is considered to be general acid or base catalyzed when its rate depends on the kind and concentration of the acids and/or bases present in the reaction mixture. To distinguish between a general and specific catalyzed reaction, two types of approaches have proved useful:

1. Reaction rate is followed in a series of buffers where the ratio of the acid and base component is kept constant, the dilution is varied, and the ionic strength kept constant. In practice this means preparation of a buffer in such a pH-range where the system is well poised and diluting it to varying degree by adding neutral salt. In such solutions the pH remains practically constant but the concentration of the acid and base component varies. If in such a series the value of the measured rate constant k' remains unchanged, the participation of the acids and/or bases that are contained in the buffer can be neglected. If the value of the formal constant k' significantly changes with buffer dilution, it is possible to deduce that general acid and/or base catalysis is operating.

2. Two or more buffers are prepared, having approximately the same pH and ionic strength, composed from different acids and bases. For example, acetate, phosphate, and pyridine buffers of pH 5.8 can be used as can borate and ammonia buffers of pH 9.3. It is, nevertheless, essential that the reactant does not enter into chemical reaction with a buffer component (e.g., forming borate complexes or Schiff bases with ammonia). The reaction is carried out in both or all buffers at the chosen pH: If the rate constants k' measured in individual buffers are different, the reaction is general catalyzed. If identical values of k' are found, it is not.

For reactions taking place in extremely acidic or alkaline conditions, such distinction is impossible. In such solutions the general acid or base character of the catalysis can be distinguished only if the rate constant (k_0) of the reaction with the solvent can be measured. If the value of k_0 is so small that it cannot be experimentally detected, it is considered to be zero and such reactions are attributed solely to fast reactions involving hydrogen or hydroxide ions.

For acid–base catalyzed reactions, the pH-dependence of the measured rate constant k' and its change with the nature and concentration of the buffer indicates participation of an acid or base in, or prior to, the rate-determining step. Various acid–base catalyzed reactions may follow different mechanisms, therefore it is necessary to try to distinguish at what stage in the reaction the interaction with the acid or base takes place. This must be done for each catalyzed reaction.

3.6.1.4. Separation of Values of Specific Rate Constants of Catalyzed Reactions

For reactions involving rapid establishment of antecedent acid–base equilibria, which are often denoted as "specifically acid/base catalyzed," the values of the constants $k_{H_3O^+}$ and k_{OH^-} can be most easily found, if measurements are carried out in region I and II (Fig. 3-53), assuming these regions are accessible to measurement.

If measurements can be carried out only in region III the measured rate constant k' shows the following dependence on $[H^+]$ or $[OH^-]$:

$$k' = k_{H_3O^+} [H_3O^+] \tag{3.6-150}$$

$$k' = k_{OH^-} [OH^-] \tag{3.6-151}$$

Values of $k_{H_3O^+}$ and k_{OH^-} can be obtained from plots of $k' = f[H_3O^+]$ or $k' = f[OH^-]$ (Fig. 3-54).

The same equations apply to reactions in which H^+ or OH^- is consumed in the rate-determining step. The linear plot goes through the origin, since k_{H_2O} is negligible, as mentioned above.

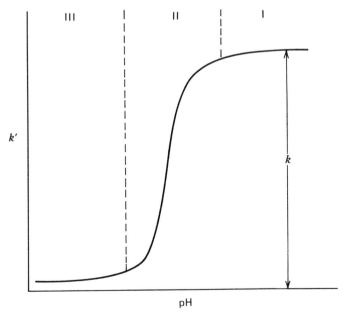

FIG. 3-53. Three pH-regions for the dependence of the formal measured rate constants k' on pH for a specifically base catalyzed reaction.

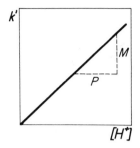

FIG. 3-54.   Determination of the rate constant for reaction with hydrogen ions from the dependence of the formal measured rate constants k' on concentration of hydrogen ions; $M/P = k_{H_3O^+}$, equation (3.6-150). Plot for reaction with hydroxide ions (3.6-151) is analogous.

Isolation of values of specific rate constants from the measured value of formal rate constant k' in the case of general catalyzed reactions is more complicated. The treatment depends on whether we are able to follow the reaction only in solution of buffers, or also in solutions of strong acids and bases.

When solutions of strong acids or bases are used and the dependence of the logarithm of the formal rate constant is plotted as a function of pH, it is possible to distinguish the character and type of the catalyzed reaction. In the most general case a plot given in Fig. 3-55 is obtained. This corresponds to a measured rate constant k' which in acidic medium is directly proportional to hydrogen ion concentration, and in alkaline media to hydroxide ion concentration. In both these pH-regions the plots log k'–pH are linear with a slope of 45°. In the region between these two is a pH-range where the value of the rate constant k' is pH-independent. This pH-range is particularly useful for the study of effects of weak acids and bases (as discussed later). The value of the measured rate constant k' can be expressed as (3.6-152), based on eq. (3.6-136):

$$k' = k_{H_2O} + k_{H_3O^+} [H_3O^+] + k_{OH^-} [OH^-] \qquad (3.6\text{-}152)$$

The rate constant in the pH-range where it remains constant is thus equal to k_{H_2O}. The values of constants $k_{H_3O^+}$ and k_{OH^-} are best determined from the dependence of k' on $[H_3O^+]$ and $[OH^-]$ (Fig. 3-56), where the intercepts correspond to k_{H_2O} and should be the same for acidic and alkaline media.

This type of dependence of the measured rate constant k' on pH is observed, for example, for mutarotation of glucose [3-105], [3-106]. This first-order reaction

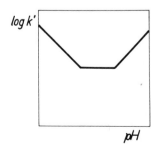

FIG. 3-55.   Dependence of the logarithm of the measured formal rate constant k' on pH for a general catalyzed reaction in solutions of strong acids and strong bases.

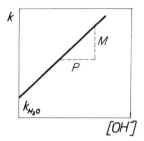

FIG. 3-56. Determination of rate constants k_{OH^-} and k_{H_2O} from the plots of values of measured formal rate constants k' as a function of concentration of hydroxide ions.

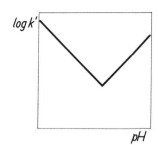

FIG. 3-57. Dependence of the logarithm of the measured formal rate constant k' on pH for a general catalyzed reaction, where the pH-independent region is not accessible.

consists of transformation of α-glucose into an equilibrium mixture of α- and β-glucose. The rate of this reaction has been followed usually polarimetrically, but it is possible to obtain identical results dilatometrically or refractometrically. The constant $k_{H_3O^+}$ for this reaction has been obtained by means of measurements in dilute solutions of hydrochloric acid, the value of k_{OH^-} in dilute solutions of hydroxides. In alkaline media it is necessary to take care particularly of the removal of carbon dioxide and of catalytic effects of the glucose anion. The experimentally obtained value of k' can be expressed as (3.6-153):

$$k' = 0.0096 + 0.258 \, [H_3O^+] + 9750 \, [OH^-] \qquad (3.6\text{-}153)$$

Alternatively, the pH-range in which the measured rate constant k' is pH-independent is experimentally inaccessible. In this case it is not possible to obtain from the log k'–pH graph (Fig. 3-57) information about the constant k_{H_2O} corresponding to the reaction with solvent. This situation occurs when $k_{H_2O} < \sqrt{K_w \, k_{H_3O^+} \, k_{OH^-}}$. The value of the rate constant at the pH where the log k'–pH plot reaches minimum is

$$k_{min} = k_{H_2O} + 2\sqrt{K_w \, k_{H_3O^+} \, k_{OH^-}} \qquad (3.6\text{-}154)$$

If the value of $k_{H_3O^+} = k_{OH^-}$, the intersection of the two linear sections with a 45° slope is at pH 7, for $k_{H_3O^+} > k_{OH^-}$ the minimum is shifted to pH < 7, for $k_{H_3O^+} < k_{OH^-}$ to pH > 7.

The values of $k_{H_3O^+}$, k_{OH^-}, and k_{H_2O} can be obtained for such reactions from plots of k' against $[H_3O^+]$ and/or $[OH^-]$ (Fig. 3-56).

Examples of behavior of this type are the halogenation of acetone [3-85] and depolymerization of dimers of dihydroxyacetone [3-106]. The latter reaction is accompanied by a volume change and can be thus followed dilatometrically. The following values of specific rate constants were found:

$$k' = 0.0025 + 1.72 \, [H_3O^+] + 4.03 \times 10^7 \, [OH^-] \qquad (3.6\text{-}155)$$

Several other types of systems can be found depending on which part of the pH-range is experimentally inaccessible.

Thus, when it is possible to follow the acid catalyzed reaction and the pH-range where the rate is pH-independent, the log k'–pH plot has the shape shown in (Fig. 3-58). From the log k'–pH plot, the value of k_{H_2O} is determined and from k'–$[H^+]$ that of $k_{H_3O^+}$.

This type of pH-dependence of the rate constant k' has been observed for hydrolysis of alkyl orthoacetates and orthocarbonates [3-107] and offers the advantages that no acid (which could react as catalyst) is formed in the course of the reaction which can be followed dilatometrically. Base catalysis is negligible.

On the other hand, in some cases it is possible to observe base catalysis and reaction with solvent (Fig. 3-59) but not acid catalysis. The slope of the rising portion is again 45°. Determination of the rate constant is analogous to the previous case: k_{H_2O} from the log k'–pH plot and k_{OH} from the k'–$[OH^-]$ plot.

Examples following this pattern are hydrolysis of β-lactones

$$
\begin{array}{c}
-\text{CH}-\text{CH}_2 \\
\mid \quad\quad \mid \\
\text{O}-\text{CO}
\end{array}
+ \text{OH}^- \xrightarrow{k}
\begin{array}{c}
-\text{CH}_2-\text{CH}_2 \\
\mid \quad\quad \mid \\
\text{OH} \quad \text{COO}^-
\end{array}
\qquad (3.6\text{-}155a)
$$

and nitroparaffin halogenation [3-91]. The latter reaction, for example, for nitromethane, is first order in the nitrocompound. The rates of chlorination and bromination are equal and independent of halogen concentration. The corresponding scheme is (3.6-156):

$$
\begin{array}{l}
\text{CH}_3\text{NO}_2 + \text{B} \\
\qquad\qquad\qquad \underset{k_2}{\overset{k_1}{\diagdown}} \\
\qquad\qquad\qquad\qquad \text{CH}_2{=}\text{NO}_2^- + \text{Br}_2 \xrightarrow{k} \text{CH}_2\text{BrNO}_2 \quad (3.6\text{-}156) \\
\qquad\qquad\qquad \underset{k_4}{\overset{k_3}{\diagup}} \quad + \text{BH}^+ \qquad\qquad\qquad + \text{Br}^- \\
\text{CH}_2{=}\text{N} \overset{\displaystyle O}{\underset{\displaystyle OH}{\Big\langle}} + \text{B}
\end{array}
$$

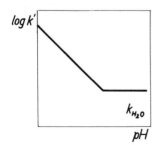 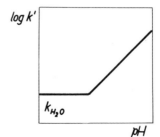

FIG. 3-58. Dependence of the logarithm of the measured formal rate constant k' on pH for a general catalyzed reaction, where the pH-range for the base catalyzed reaction is not accessible.

FIG. 3-59. Dependence of the logarithm of the measured formal rate constant k' on pH for a general catalyzed reaction, where the pH-range for the acid catalyzed reaction is not accessible.

In this system, where $K_N = k_1/k_2 = 10^{-9}$ and $K_{AC} = k_4/k_3 = 10^{-5}$, since bromination with rate constant k is faster than reactions with constants k_2 and k_4, the reaction is governed by

$$CH_3NO_2 + B \xrightarrow{k_1} CH_2{=}NO_2^- \xrightarrow[\text{fast}]{} \qquad (3.6\text{-}157)$$

and hence is base catalyzed. The acid catalysis in the available pH-range is negligible.

Finally two further types are known in which the region, where the rate is pH independent, is inaccessible (Fig. 3-60). This may be caused either by not being able to use an acid or base which is concentrated enough or by the slowness or high speed of the reaction in the critical pH-range.

An example of such a reaction is diazoacetate hydrolysis [3-108]. The rate of this reaction is very sensitive, even to traces of hydrogen ions, and measurable reaction rates are found at $[H^+] = 10^{-10}$ to 10^{-13}. Acid catalysis is thus observed even at high pH-values.

Procedures used for determination of rate constants for acids and bases other than hydrogen or hydroxide ions can be divided according to whether solutions of weak acids or buffered solutions are used. The use of buffers is more common and therefore will be discussed first in some detail.

Buffers can be used for the study of acid–base catalyzed reactions under two conditions: (1) It is possible to follow reaction rate in the pH-range where the measured constant k' is pH-independent (i.e., in the range where the plot k'–pH is parallel with the pH-axis). (2) Alternatively, it is necessary to study the reaction in a pH-range where, in addition to reaction involving acid or base (or both) buffer components, the reaction with H_3O^+ or OH^- ions participates.

1. The determination of the values of specific rate constants is simplest when it is possible to follow the reaction in the pH-range where the measured value of k' is pH-independent. The study is limited to acids and bases which have pK-values that lie in the pH-range where the value of k' remains unchanged.

First it is established whether one, or both, buffer components participate in the studied reaction. To achieve this, a buffer is prepared from the examined components so that its pH-value is within the pH-range where the rate is pH-independent. Keeping the ratio of the two buffer components constant, the concentration of the

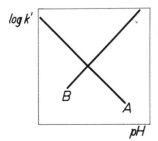

FIG. 3-60. Dependence of the logarithm of the measured formal rate constant k' on pH for a general catalyzed reaction, where (A) the pH-range for the base catalyzed reaction and the pH-independent range are not accessible, (B) the pH-range for the acid catalyzed reaction and the pH-independent range are not accessible.

buffer is varied (pH remaining practically constant). When the measured value of the rate constant of the catalyzed reaction is plotted against buffer concentration (Fig. 3-61A–C, where c stands for buffer concentration), three alternatives exist: (1) The measured rate constant k' is independent of buffer concentration (Fig. 3-61A). This indicates that none of the buffer components participate in the studied reaction. The intercept L is equal to k_0. (2) The measured rate constant k' is a linear function of the buffer concentration (Fig. 3-61B). One of the buffer components— either the acid or the base—participates in the reaction. Intercept T is equal to k_0. (3) When the measured rate constant k' is a nonlinear function of the buffer concentration (Fig. 3-61C), it can be concluded that both buffer components act as acid and/or base catalysts.

When the investigation shows that one component of the buffer participates in the reaction, a series of buffers is prepared so that concentration of the acid component of the buffer is kept constant and concentration of *the basic component of the buffer varied*. The ratio of the [acid]:[base] buffer components should remain within 10:1 to 1:10 and the pH of the buffer should remain within the pH-range where the rate constant remains pH-independent. The plot of the measured rate constant k' against the concentration of the base component of the buffer can have one of the shapes shown in Fig. 3-61A or B (where c now stands for the concentration of the basic component of the buffer). When the plot has the shape in Fig. 3-61A, the conjugate base B of the buffer does not participate in the studied reaction. The intercept L equals to k_0. Alternatively, when the plot has the shape shown in Fig. 3-61B, the conjugate base of the buffer participates in the reaction. The intercept T corresponds then to k_0 and the slope $M/P = k_B$.

The value of the constant k_{HA} can be obtained—provided that it has been established that only one buffer component reacts—following the rate constants k' in a series of buffers where the base component is kept constant and the *acid component changed*. A plot of the type shown in Fig. 3-61B is obtained (where c stands for concentration of the acid component), the intercept corresponds to k_0 and the slope $M/P = k_{HA}$. Independence of the value of k' of the base buffer component (Fig. 3-61A) proves that the acid component does not participate in the studied reaction.

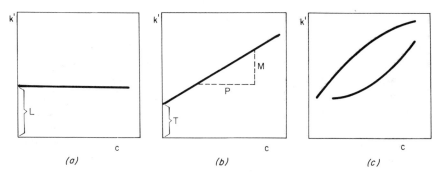

FIG. 3-61. Dependence of the values of the measured formal rate constants k' on concentration of buffer components, as described in text.

When the dependence of the measured rate constant k' on buffer concentration is nonlinear (Fig. 3-61C), the value of the constant k_{HA} is obtained in buffers by varying acid component concentration, as in the previous case. The value of k_{HA} is obtained from the slope of the k'–[HA] plot (Fig. 3-61B), but the intercept here equals to $k_0 + k_B[B]$. Similarly, the value of constant k_B is obtained by means of buffers containing constant concentration of the acid buffer component and varying concentration of the base component. The slope of the linear plot (Fig. 3-61B, where c stands for concentration of the base component of the buffer) $M/P = k_B$ and intercept is equal to $k_0 + k_{HA}[HA]$.

An example of a system that can be treated in this way is the mutarotation of glucose, [3-105, 3-106], which between pH 4 and 6 shows practically no change in the value of k' with pH. In this way it is possible to detect catalysis by anions of acetic, formic, salicylic, and amino acids.

For reactions in which only acid or only base catalysis has been found, the pH-range in which the specific constants can be determined is limited only on one side. For example, acid catalysis of ethyl orthoacetate hydrolysis [3-91, 3-101] can be studied in this way for all acids with $pK > 6$.

2.   In cases where it is not possible to carry out the investigation in a pH-range where the measured k' is independent of pH, it is necessary to use the complete equation (3.6-136). An example of application of this equation will be demonstrated on a case of base catalyzed reaction. A completely analogous procedure can be applied to acid catalyzed reactions. The first approach is to isolate values of individual specific rate constants from measurements in carbonate-free sodium (or potassium) hydroxide solutions of varying concentration. If such measurements are possible, that is, if the rate is neither too fast nor too slow, it is possible to determine values of k_{H_2O} and k_{OH^-}, as described in the previous section.

In such cases the values of k_{HA} or k_B can be obtained using the graphical treatment described below. If the rate of the reaction in solutions of sodium hydroxide catalyzed by hydroxide ions is too fast to be measured, the graphical treatment described below can be used only when the reaction is general base catalyzed. For reactions of this type which involve a general acid–base catalysis, graphical treatment is not suitable and equations (3.6-136) at varying pH-values are used to obtain values of specific rate constants by solving the system of simultaneous equations.

As in the previous case, two types of buffers are used in the study: Either the buffer capacity is varied keeping the ratio of the acid and base buffer component and hence the pH-value constant, or the acid buffer component concentration is kept constant and that of the base component is varied.

When varying the buffer concentration at constant pH, the dependence of the measured rate constant k' on buffer concentration can have one of the shapes shown in Figs. 3-61A–C (where c stands for the buffer concentration). Independence of k' of buffer concentration (Fig. 3-61A) indicates that the buffer components do not participate in the studied reaction and intercept $L = k_0 + k_{OH^-}[OH^-]$. A linear dependence of k' (Fig. 3-61B) shows that only one buffer component is involved in the base catalyzed reaction and intercept $T = k_0 + k_{OH^-}[OH^-]$. Finally, nonlinear dependence (Fig. 3-61C) indicates participation of both buffer components in the base catalyzed reaction.

When the acid component of the buffer is kept constant and the concentration of the base varied, a linear plot (Fig. 3-61B, where c stands for concentration of the base component of the buffer) is obtained. The slope $M/P = k_B + k_{OH^-} k_w/K_{HA}[HA]$ and the intercept $T = k_0$ (if only species B participates in base catalysis) or $k_0 + k_{HA}[HA]$ (when HA also acts as a base catalyst).

When concentration of the base component of the buffer is kept constant and that of the acid varied, a similar linear plot (Fig. 3-61B, where c represents the concentration of the acid buffer component) is obtained, but the intercept $T = k_0 + k_B[B]$ and the slope $M/P = k_{OH^-}$ (provided that the catalysis by the acid buffer component is negligible).

An example of such treatment is the alkaline cleavage of isothioureas [3-109] following the scheme (3.6-159):

$$\underset{\underset{R^1NR^2}{|}}{\overset{\overset{NH}{||}}{C}}-SR + B \longrightarrow \underset{\underset{R^1NR^2}{|}}{\overset{\overset{N}{|||}}{C}} + SR^- + BH^+ \qquad (3.6\text{-}159)$$

In dilute solutions of sodium hydroxide it has been proved that the measured reaction rate constant k' is linearly proportional to hydroxide ion concentration (the value of k_0 being negligibly small) and from the slope of the linear dependence, $k_{OH^-} = 7.2$ liter mol^{-1} s^{-1}. Similar values were obtained from dependences of rate constants k' measured in buffers prepared from ammonia–ammonium chloride and from sodium phosphate–sodium hydrogen phosphate.

In carbonate buffers at constant pH and varying buffer capacity, the measured rate constant k' was found to be a linear function of buffer concentration (type of curve: Fig. 3-61B). This indicates that one of the components of the carbonate buffer participates in the reaction. Linear dependence of constant k' on concentration of the hydrogen carbonate ion, when carbonate concentration was kept constant (type of curve: Fig. 3-61B), had a slope equal to k_{OH^-}. In such buffers the change in the value of the rate constant is due only to the variation in hydroxide ion concentration. Hence HCO_3^- ions practically do not affect the reaction rate. Change of rate constant k' with buffer concentration is thus due only to the role of CO_3^{2-} ions, and the specific rate constant $k_{CO_3^{2-}} = 0.0051$ liter mol^{-1} s^{-1} was obtained from the slope. Similarly it was possible to obtain the specific rate constant for the borate anion ($k_{Bor^-} = 0.0061$ liter mol^{-1} s^{-1}) by analysis of the role of composition of borate buffers on measured rate constant k'.

Hence in this particular case "catalysis" by hydroxide, carbonate, and borate anions can be detected, but no effect of H_2O, HCO_3^-, HPO_4^{2-}, PO_4^{3-}, and NH_3. It seems that nucleophilicity of the anion plays the decisive role.

Dependence on concentration of a phosphate buffer has recently been used for evaluation of specific rate constants in enolization of acetone [3-86], on concentration of acetate and TRIS buffer in carbamate decomposition [3-110], and on concentration of buffers prepared from various carboxylic acids in acid catalyzed hydrolysis of N-vinyl pyrrolidone [3-111]. Typical nucleophilic catalysis can be followed from the change in rate constants of the hydrolysis of p-nitrophenyl acetate with concentration of imidazole buffers [3-112].

To determine the values of k_{HA} for unbuffered solutions of weak acids, it must be possible to follow the reaction in a pH-range where the measured constant k' is pH independent. Furthermore, it must be possible to prepare solutions of varying concentration of the weak acid studied, the pH-values of which remain inside the pH-independent region. Effect of variation of concentration of the weak acid on the value of k' is followed. The value of k_{HA} is obtained from the linear plot of dependence of the value of k' on concentration of the weak acid, as in Fig. 3-61.

For information on mechanisms of acid and base catalyzed reactions the reader should consult references 3-84, 3-85, 3-91, 3-113, and 3-114.

3.7. EFFECT OF TEMPERATURE

Reaction rates are rather sensitive to temperature changes. An often quoted rule-of-thumb in the organic laboratory states that raising the temperature by 10°C will result in a doubling of the reaction rate. While this generalization is only occasionally accurate, it does illustrate the importance of temperature as a variable that influences reaction rates.

For most reactions, the temperature dependence of the rate constant is described within the accuracy of the experimental data by the Arrhenius equation (3.7-1):

$$k = A \, e^{-E_a/RT} \qquad (3.7\text{-}1)$$

where A is the preexponential factor, E_a is the activation energy, R is the gas constant, and T is the absolute temperature. The values of the empirical constants A and E_a are usually found graphically from the logarithmic form of the Arrhenius equation (3.7-2):

$$\ln k = \ln A - \frac{E_a}{RT} \qquad (3.7\text{-}2)$$

A graph of $\ln k$ vs. $1/T$ will give a straight line with slope $-E_a/R$ and $\ln A$ as the intercept (Fig. 3-62).

On a few occasions when very precise measurements are made, the simple Arrhenius equation may not be adequate because the assumption, that the preexponential factor is temperature independent, is not fulfilled [3-115]. In such cases, better agreement can be found by using equation (3.7-3):

$$\ln k = \ln A + \frac{C}{R} \ln T - \frac{E_a}{RT} \qquad (3.7\text{-}3)$$

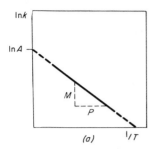
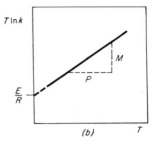

FIG. 3-62. Graphical methods for determination of values of A and E_a, using equation (3.7-2). (*a*) $\ln k = f(1/T)$: intercept equals $\ln A$, $M/P = E_a/R$; (*b*) $T \ln k = f(T)$: intercept equals E_a/R, $M/P = \ln A$.

where C is a constant that has significant magnitude for some reactions in solution. Generally, the precision of the data is not sufficient to warrant using equation (3.7-3), although this extended Arrhenius equation can be justified theoretically.

If comparisons of values of activation energies, obtained from the Arrhenius equation (3.7-1), are to be meaningful, the values of rate constants used for evaluation of activation energies by means of this equation must correspond to simple reactions. Constants k' which in fact are a product (like $k_1 K/[H^+]$) or sum (like $k_0 + k_{H^+}[H^+] + \cdots$) should not be treated in this way and a specific constant must first be isolated.

Deviations from a linear log k vs. $1/T$ plot can be caused by the fact that a complex reaction is involved which at the originally chosen temperature range corresponded to a simple reaction rate equation. A deviation from linearity of the plots of log k vs. $1/T$ indicates the need for careful revision of the reaction mechanism, so that a more complex mechanism than originally assumed can be proposed.

On the other hand, a strictly linear plot of log k as a function of $1/T$ is not in itself sufficient evidence that the measured rate constant is a simple constant, that the reaction rate equation used fully describes the observed change of concentration with time, and that the proposed mechanism is correct. Examples are known when, because of compensation of several factors, the value of log k is a linear function of $1/T$, even when it is known that the reaction involved is complex. An example is the mutarotation of glucose, in which the α-form is first converted into an open chain form which is consecutively transformed into the cyclical β-form. Whereas for calculation of rate constant k (Table 3-5) a first-order rate equation was used and the constant k used for calculation of values of constants A, E_A, and C, the studied system involves at least two equilibrium reactions (with four specific rate constants). Moreover, the value of the measured rate constant depends on pH. For rigorous treatment of mutarotation it would be necessary to separate all specific constants first.

The Arrhenius equation was derived empirically. For interpretation of the constants A and E_A in this equation two alternative formulations were offered, based on two different theoretical approaches: collision theory and a thermodynamic formulation denoted as theory of absolute reaction rates or transition state theory.

TABLE 3-5
Examples of Parameters of Equation (3.7-3)

Reaction	Temperature Range (in °C)	ln A [A]s^{-1}	$-\dfrac{C}{R}$	E_A (in kcal/mol)
$CH_3Br + H_2O \longrightarrow$ $CH_3OH + H^+ + Br^-$	17–100°	260	34.3	46.7
$CCl_3COOH \longrightarrow CO_2 + CHCl_3$	50–100°	110	10	42.9
α-glucose $\longrightarrow$ β-glucose	0– 50°	89	10.3	23.0

Note: Modified from E. A. Moelwyn-Hughes, *The Kinetics of Reactions in Solution,* 2nd. ed., Clarendon Press, Oxford, 1947, p. 67.

The collision theory of reaction rates relies on the kinetic theory of gases and leads for the dependence of rate constants on temperature to equation (3.7-4):

$$k = PZ\, e^{-E_A/RT} \qquad (3.7\text{-}4)$$

where P is the probability factor or steric factor and Z is the collision frequency, which is temperature-dependent. The energy of activation, E_A, represents the necessary minimum energy of collision required for the reaction to occur.

The interpretation based on transition state theory will be commented upon in somewhat greater detail, since collision theory can be related more directly to gas phase reactions than to reactions in solution. The transition state theory is based on the assumption that the activated complexes, $AB^{\ddagger}$, are in statistical equilibrium with the reactants. For a reaction (3.7-5):

$$A + B \underset{k_2}{\overset{k_1}{\rightleftharpoons}} AB^{\ddagger} \longrightarrow C \qquad (3.7\text{-}5)$$

it is possible to write an equilibrium constant

$$K = \frac{[AB^{\ddagger}]}{[A][B]} = \frac{k_1}{k_2} \qquad (3.7\text{-}6)$$

Applying the van't Hoff equation from classical thermodynamics to the transition state it follows that

$$\ln K^{\ddagger} = -\frac{\Delta G^{\ddagger}}{RT} \qquad (3.7\text{-}7)$$

and

$$-\frac{\Delta G^{\ddagger}}{RT} = -\frac{\Delta H^{\ddagger}}{RT} + \frac{\Delta S^{\ddagger}}{R} \qquad (3.7\text{-}8)$$

where $\Delta G^{\ddagger}$ represents the change in free energy of activation, $\Delta H^{\ddagger}$ is the change in activation enthalpy, and $\Delta S^{\ddagger}$ is the change in activation entropy.

From empirically found dependences on temperature, known from classical thermodynamics, it is possible to write expressions (3.7-9) and (3.7-10):

$$\Delta H_T^{\ddagger} = (\Delta H_{T_0}^{\ddagger} - T_0 \, \Delta C_p) + T \, \Delta C_p \qquad (3.7\text{-}9)$$

$$\Delta S_T^{\ddagger} = (\Delta S_{T_0}^{\ddagger} - \Delta C_p \ln T_0) + \Delta C_p \ln T \qquad (3.7\text{-}10)$$

where ΔC_p is the mean difference (in the given temperature interval) between specific heats of reactants and products (relative to a standard state). Combining expressions (3.7-9) and (3.7-10) with equation (3.7-7), it follows that (leaving out subscript T_0)

$$\ln K^{\ddagger} = \frac{\Delta S^{\ddagger} - \Delta C_p - \Delta C_p \ln T_0}{R} + \frac{\Delta C_p}{R} \ln T - \frac{\Delta H^{\ddagger} - T_0 \, \Delta C_p}{RT} \qquad (3.7\text{-}11)$$

Since $\ln K^{\ddagger} = \ln k_1 - \ln k_2$

$$\ln k_1 - \ln k_2 = \frac{\Delta S^{\ddagger} - \Delta C_p - \Delta C_p \ln T_0}{R}$$

$$+ \frac{\Delta C_p}{R} \ln T - \frac{\Delta H^{\ddagger} - T_0 \, \Delta C_p}{RT} \qquad (3.7\text{-}12)$$

From the extended Arrhenius equation it follows that

$$\ln k_1 - \ln k_2 = \ln A_1 - \ln A_2 + (C_1 - C_2) \frac{\ln T}{R} - \frac{E_1 - E_2}{R} \qquad (3.7\text{-}13)$$

From comparison of equations (3.7-12) and (3.7-13) it follows that

$$\ln A_1 - \ln A_2 = \frac{\Delta S^{\ddagger} - \Delta C_p - C_p \ln T}{R} \qquad (3.7\text{-}14)$$

and because $C_1 - C_2 = \Delta C_p$

$$E_1 - E_2 = \Delta H^{\ddagger} - T\Delta C_p \qquad (3.7\text{-}15)$$

Such comparison allows us to show relationships between $\ln A$ and $\Delta S^{\ddagger}$, C_p and ΔC_p, as well as E and $\Delta H^{\ddagger}$. When measurements are carried out under conditions where only limited accuracy can be reached, it is possible to use the simple Arrhenius equation. Since

$$\Delta C_p = C_1 - C_2 = 0 \qquad (3.7\text{-}16)$$

it follows that

$$\ln A_1 - \ln A_2 = \Delta S^{\ddagger}/R \qquad (3.7\text{-}17)$$

and

$$E_1 - E_2 = \Delta H^{\ddagger} \qquad (3.7\text{-}18)$$

From this relationship it follows that the constant E_A in the Arrhenius equation can be denoted as activation energy.

It is, hence, possible to write the Arrhenius equation as

$$k = A\, e^{-E/RT} \qquad (3.7\text{-}19)$$

$$k = \nu\, e^{-\Delta G^{\ddagger}/RT} \qquad (3.7\text{-}20)$$

$$k = \nu\, e^{\Delta S^{\ddagger}/R}\, e^{-\Delta H^{\ddagger}/RT} \qquad (3.7\text{-}21)$$

In these formulations ν has dimensions of a rate constant and is denoted as frequency factor.

Values of the constant E_A in the simple form of the Arrhenius equation can be due to an approximate compensation of opposite changes of ν or $\Delta S^{\ddagger}$ and $\Delta H^{\ddagger}$ with temperature. In the theory of absolute reaction rate, the value of ν is put equal to KT/h where k is the Boltzmann constant and h the Planck constant. Hence

$$\ln k = \ln kT/h + \frac{\Delta S^{\ddagger}}{R} - \frac{\Delta H^{\ddagger}}{RT} \qquad (3.7\text{-}22)$$

and for comparison, the Arrhenius equation is

$$\ln k = \ln A - \frac{E_a}{RT} \qquad (3.7\text{-}23)$$

For reactions in solution and with the assumption that $\Delta S^{\ddagger}$ is not a function of temperature, the parameters in the simple Arrhenius equation can be equated to those in transition state theory as follows:

$$E_a = \Delta H^{\ddagger} + RT \qquad (3.7\text{-}24)$$

and

$$A = \left(\frac{kT}{h}\right) e^{\Delta S^{\ddagger}/R} \qquad (3.7\text{-}25)$$

Usually, a determination of $\Delta H^{\ddagger}$ and $\Delta S^{\ddagger}$ for a single reaction offers little useful information for mechanism elucidation. Only the change of $\Delta H^{\ddagger}$ in a properly chosen reaction series (i.e., a group of related componds) can offer information on the source of such changes and hence can be related to the structure and geometry of the assumed transition state. The changes in $\Delta H^{\ddagger}$ are usually ascribed to energy changes along the reaction coordinate and may be interpreted in terms of bond strengths. Often, however, similar information is obtained from consideration of the changes in rate constants (hence $\Delta G^{\ddagger}$) in a reaction series without evaluating $\Delta H^{\ddagger}$ separately.

The value of the entropy of activation, $\Delta S^{\ddagger}$, is more often given a mechanistic interpretation. The value of $\Delta S^{\ddagger}$ indicates changes in the relative orderedness of the structures of the activated complex and the reactants, including the solvent molecules. A negative value of $\Delta S^{\ddagger}$ indicates that the transition state is more ordered than the state of the separated reactants; the increase in order can result from several sources, for example, a decrease in the number of separate molecules, formation of a cyclic transition state, steric effects, or increased solvation. Changes in solvent orientation can often dominate the value of $\Delta S^{\ddagger}$. For example, in a typical S_N1 solvolysis reaction, the reactant R-X dissociates to form two ions, R^+ and X^-, for which a positive entropy of activation would be predicted, but the overall entropy change is negative because of the increase in solvent ordering around the ions. On the other hand, solvent effects are usually minor for reactions of neutral molecules when the transition state does not involve a significant change in polarity or charge separation. Solvent effects will be discussed in more detail in Section 3.8.

The accuracy with which it is possible to obtain values of activation energy E_A depends predominantly on the accuracy of temperature control. For example, when the value of E_A is obtained from two measurements separated by a 10°C interval and when the mean variation of both temperatures involved is within $\pm$ 0.2°C, then the accuracy of the value of E_A is at best $\pm$ 2%.

Accuracy of the measured rate constant plays an important role for reactions showing a small variation of rate constants with temperature. When it is possible to determine two rate constants (at temperatures differing by 10°C) with an accuracy of $\pm$ 1%, provided that these two values are approximately in a ratio of 1:2, then the resulting E_A value is again limited to $\pm$ 2% accuracy. For reactions where the accuracy of rate constant determination is critical, its effects are added to contributions due to variations in temperature of the reaction mixture.

When it is required to obtain, in a temperature difference of 10°C, values of E_A better than $\pm$ 0.5%, variations in temperatures must be smaller than $\pm$ 0.03% and accuracy of measurement of values of k_1 and k_2 better than $\pm$ 0.3%. To achieve such accuracy in determination of rate constants it would be necessary to use analytical methods for determinations of concentration with accuracy better than $\pm$ 0.1%. This discussion indicates why it is difficult to obtain accurate values of activation energy E_A and consequently to decide if or how the value of E_A depends on temperature.

It should also be pointed out that the temperature range, over which the variation of rate constants can be followed, is limited. For the majority of reactions it is possible to follow such systems experimentally, where the half-time varies between 1 minute and 1 week. That means that the slowest reaction rate that can be followed differs from the fastest reaction rate by a factor of 10^4. If an increase of temperature by 10°C results in a doubling of the reaction rate, then change of the reaction rate by four orders of magnitude corresponds to a temperature range of 130°C. Making use of techniques for following fast reactions, it is possible to extend somewhat the range of useful temperatures. It must be considered that in extremes of the temperature range (and of the rates), the measurement is usually considerably less accurate. In practice, it is usually necessary to restrict measurements to a considerably narrower range. In practice it can be expected that most of the values of E_A or $\Delta H^\ddagger$ are accurate to ± 5–10%.

In general, with the exceptions of distinguishing cyclic from noncyclic transition states or participation of the solvent in the transition state, separation of the temperature dependent from the temperature independent term usually does not contribute more to the understanding than the values of rate constants obtained at one temperature.

3.8.   EFFECT OF SOLVENT AND IONIC STRENGTH [3-116–3-118]

Rates of chemical reactions are affected by the nature of the solvent used and by the nature and concentration of ionic species present in the reaction mixture. Changes in the solvent and nature and concentration of neutral salts added to the reaction mixture result in complex solute–solvent and solute–solute interactions. Variability of contributions of the individual types of interaction presently makes a rigorous general treatment of such effects inaccessible. Nevertheless, for some types of reactions, available treatments make it possible to deduce from the effect the nature of the solvent and from that of the ionic strength to conclude on the charge of the species participating in the reaction. Somewhat less conclusively some ideas can be deduced from such effects on the charge distribution in the transition state. Counterion effects can offer information about composition of the transition state. In some cases it is possible to obtain information about solvation of the starting materials and/or transition state.

The attempts to seek relationships between reaction rate and the nature of the solvent can be divided into three groups: treatments based on electrostatic interactions, those making use of linear free energy relationships, and finally, qualitative observations. They will be discussed separately here. The effects of ionic strength based on evaluation of coulombic interactions will be treated in the first group.

Even when the present understanding of relationships between reaction rate, nature of solvents, and ionic strength makes the evaluation of such relationships a less powerful tool in elucidation of mechanisms, the state of our knowledge should be understood by the chemist involved. Moreover, there are some examples where the study of solvent and ionic-strength effects contributed considerably to better understanding of mechanisms.

3.8.1.   Electrostatic Interactions

The involvement of electrostatic interactions in solvent and ionic-strength effects depends upon the presence of unit charges or on the charge distribution in the reacting species. Reactions between two ions, between an ion and a neutral molecule, between two neutral dipolar molecules, and between two apolar molecules will be discussed separately.

3.8.1.1. Reactions Between Two Ions

In the reactions between two ions the role of coulombic forces predominates over that of other types of neutral interactions. In such cases the agreement between the theoretical expressions and experimental data is acceptable. Better quantitative agreement is probably prevented by the lack of understanding of the behavior of ions and their precursors, polar molecules in the strong electrical fields of adjacent ions.

For the effect of the *ionic strength* on rate constants in water (at 25°C) Brönsted and Bjerrum [3-119, 3-120] derived (on the basis of the Debye–Hückel theory) the expression (3.8-1):

$$\log k/k_0 = 1.02\, z_A z_B\, \sqrt{\mu} \qquad (3.8\text{-}1)$$

where k is the rate constant at the ionic strength $\mu = \frac{1}{2}\Sigma c_i z_i^2$ (where c_i is concentration of ion i, and z the absolute value of the charge of the ion), k_0 rate constant at ionic strength equal to zero, z_A and z_B charges of ions A and B (numerical values including signs).

This relationship (3.8-1) indicates that at low concentrations (where the Debye–Hückel theory is valid) the logarithm of the rate constant should be a linear function of the square root of ionic strength. For reactions in which participating ions carry opposite charges ($z_A z_B < 0$), the reaction rate should decrease with increasing ionic strength, for reactions of two cations or two anions ($z_A z_B > 0$) the reaction rate should increase (Fig. 3-63).

Equation (3.8-1) predicts that reactions between an ion and a molecule ($z_A, z_B = 0$) are not affected by a change in the ionic strength. As will be shown in sections 3.8.1.2–4, even rates of such reactions depend on ionic strength, but to a much lesser degree than reactions between two ions.

Problems in verification of equation (3.8-1) are caused primarily by the fact that comparison with experimental data should be restricted to the range of low ionic

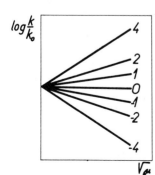

FIG. 3-63. Effect of ionic strength on values of relative rate constants (k/k_0). The values of the product of ionic charges $z_A z_B$ are given in the graph.

strengths, where the Debye–Hückel equation is valid. In practice this means that for ionic strength controlled by uni-univalent electrolytes the ionic strength should be smaller than about 0.01, for polyvalent electrolytes smaller than about 0.001. Because the concentration of the reactant is usually higher than 0.0001 M, the ionic strength can be varied only over a relatively narrow range, so that the total change of the rate constant with variation of ionic strength usually represents only 20 to 50% of its value. The use of sensitive spectrophotometric and other analytical methods, which can be used at low concentrations in solutions of low ionic strength, seems promising for future studies of ionic-strength effects.

In addition to coulombic effects causing variations of activities (considered in derivation of equation (3.8.1), specific effects of some ions can also play a role. For example, deformable anions (such as halides) can affect specific reaction rates. Interactions between these anions and reactive cations can lead to complex or adduct formation. These anions can furthermore act as nucleophilic reagents and compete with the studied reaction. Hence, cleavage of N-nitrosamines in acidic media (where the protonated form is the reactive species) depends not only on acidity, but also on concentration of chloride ions [3-72]. Some reactions depend characteristically on the kind and charge of cations. Thus some catalytic reactions are sensitive to the presence and concentration of La^{3+} and Th^{4+} ions. For some specific catalytic effects of ions NH_4^+ and NR_3H^+, it has not yet been decided whether they reflect the size of the cation or its acid–base properties. Tetraalkylammonium ions show specific effects in a number of seemingly unrelated reactions.

For some reactions between ions of equal sign, another effect plays a role: the rate of reaction of two cations can depend on concentration of the anion of an added salt, but not on the type of its cation. Oppositely, the rate of reaction involving two anions may be affected by the concentration of cation of the added salt but not on the nature of its anion. This phenomenon, treated theoretically by Scatchard [3-121], can be detected when two or more experiments are carried out at a constant ionic strength, but in the presence of different strong electrolytes.

The theory of electrolytes, proposed for interpretation of effects at higher concentrations and with polyvalent ions, necessitates introduction of further parameters related to the ionic size and repulsive forces between ions. It seems that none of the existing theories enables *a priori* prediction of ionic interactions under such conditions and, moreover, the resulting equations are complex.

The situation for general interpretation of solvent effects on reaction rate based only on electrostatic influences is even more complicated. The treatment is based on the assumption introduced by Debye and Hückel that charged particles are located in a medium that does not have internal structure and has at every point the same dielectric constant. For reaction of two ions A and B with charges z_A and z_B, Scatchard [3-121] derived for the dependence of the rate constant (k) at zero ionic strength on dielectric constant of the solvent (ε) expression (3.8.2):

$$\ln \frac{k}{k_0} = \frac{e^2 z_A z_B N_A}{RT \, r_{AB}} \left(1 - \frac{1}{\varepsilon} \right)$$

(3.8-2)

where k_0 is the rate constant in a medium with $\varepsilon = 1$, N_A is Avogadro's number and r_{AB} is the distance of closest approach which the two ions, A and B, must reach in order to react. It is assumed that $r_{AB} \approx r_A + r_B$ (where r_A, r_B are ionic radii).

Using a different electrostatic model and considering the charge distribution in the transition state led [3-122] to an expression:

$$\ln \frac{k}{k_0} = \frac{e^2 N_A}{2RT} \left(\frac{1}{\varepsilon} - 1 \right) \left[\frac{z_A^2}{r_A} + \frac{z_B^2}{r_B} - \frac{(z_A + z_B)^2}{r_\ddagger} \right] \qquad (3.8\text{-}3)$$

where $r_\ddagger$ is the radius of the transition state.

Both expressions (3.8-2) and (3.8-3) indicate linear dependence of $\ln k$ on $1/\varepsilon$. According to (3.8-2) this dependence will have a negative slope, if z_A and z_B have the same sign and a positive slope, if z_A and z_B have opposite signs.

Equation (3.8-2) expresses that increasing dielectric constant of the solvent enhances dissociation yielding oppositely charged species and association of species of the same sign of charge. Reaction between ions of equal sign will thus take place faster in a medium of a higher dielectric constant, whereas reaction of oppositely charged ions is facilitated in solvents of lower dielectric constant.

The plots of $\ln k$ vs. $1/\varepsilon$ show particularly good correlations for mixed solvents, where the dielectric constant is varied by varying proportions of solvent components. Examples of reactions studied in this way are reactions between bromoacetate and thiosulfate ions [3-122, 3-123] and between tetrabromophenolsulfophthalein and hydroxide ions [3-124].

Verification of the dependence of reaction rates on dielectric constant for ionic species is limited towards media of low dielectric constant, as in such solutions ion-pair and cluster formation become important. In binary solvent mixtures, moreover, selective solvation can also play a role.

3.8.1.2. Reactions of Neutral Molecules with Ions

Reactions between ions and dipolar molecules belong among the most frequently encountered types of organic reactions, in particular substitution and solvolytic. Unfortunately, for these reaction types our understanding of media effects is severely limited. It is evident that the rate of such reactions depends on ionic strength and nature of solvent, particularly dielectric constant. It has also been frequently shown that effects of variations of ionic strength and the value of dielectric constant are smaller for ion–molecule reactions than for ion–ion reactions.

In the treatment, based only on electrostatic interactions, for a reaction between a neutral, dipolar molecule A (with a dipole moment μ_A) and on ion B bearing a charge z_B, equation (3.8-4) was derived [3-122]:

$$\ln \frac{k}{k_0} = \frac{e^2 z_B^2 N_A}{2RT} \left(\frac{1}{\varepsilon} - 1 \right) \left(\frac{1}{r_B} - \frac{1}{r_\ddagger} \right) - \frac{N_A \, \mu_A^2 \, (\varepsilon - 1)}{RT \, r_A^3 \, (2\varepsilon + 1)} \qquad (3.8\text{-}4)$$

(where the meaning of the symbols is the same as for equations (3.8-2 and 3.8-3)) or alternatively [3-125], the expression:

$$\ln \frac{k}{k_0} = \frac{e^2 N_A}{2RT}\left(\frac{1}{\varepsilon} - 1\right)\left(\frac{1}{r_B} - \frac{1}{r_\ddagger}\right) + \frac{3e^2 N_A}{8RT}\left(\frac{2}{\varepsilon} - 1\right)\left(\frac{\mu_A^2}{r_A^3} - \frac{\mu_\ddagger^2}{r_\ddagger^3}\right) \qquad (3.8\text{-}5)$$

Alternatively, using a coulombic energy approach, equation (3.8-6) was derived [3-126]

$$\ln \frac{k}{k_\infty} = \frac{e z_B N_A \mu_A}{RT\varepsilon \, r_{AB}^2} \qquad (3.8\text{-}6)$$

where k_∞ is the rate constant in a medium with a dielectric constant of infinite magnitude and r_{AB} is the distance in which dipole A and ion B interact.

All three equations (3.8-4 to 3.8-6) indicate a linear relationship between $\ln k$ and $1/\varepsilon$. Positive slope of the plot of $\ln k$ vs. $1/\varepsilon$ indicates a positive value of the product $z_B e$, negative slope a negative value of $z_B e$. Positive slope of such plot for $S_N 2$ reactions in aprotic solvents indicates [3-127] that $r_\ddagger > r_B$, whereas negative slope of such a dependence for base catalyzed hydrolysis of esters in protic solvents requires that $r_B > r_\ddagger$. To account for this apparent contradiction, the role of specific solvation of OH^- ions by hydrogen bonding has been assumed [3-128].

In general, the possibility of using equations (3.8-4 to 3.8-6) is better for correlation of data obtained in binary solvent mixtures than for those in pure solvents of varying structure. The evidence for their use in investigation of composition of transition states is inconclusive.

3.8.1.3.   Reactions Between Two Neutral Dipolar Molecules

In the Kirkwood equation (3.8-7), the free energy of an idealized dipole located in a continuum with a dielectric constant ε is compared to the free energy in a continuum with $\varepsilon = 1$ [3-129]:

$$\Delta G = -\frac{\mu^2}{r^3}\frac{\varepsilon - 1}{2\varepsilon + 1} \qquad (3.8\text{-}7)$$

where μ is the permanent dipole moment of the solute and r its radius.

Assuming that electrostatic solvent–solute interactions resulting from dipole–dipole forces are a predominant factor in the studied solvent effects, it is possible to derive for a reaction (3.8-8) involving two dipolar molecules A and B with dipole moments μ_A and μ_B the expression (3.8-9):

$$A + B \rightleftharpoons (AB)^\ddagger \longrightarrow C + D \qquad (3.8\text{-}8)$$

$$\ln \frac{k}{k_0} = -\frac{N_A}{RT}\frac{(\varepsilon - 1)}{(2\varepsilon + 1)}\left(\frac{\mu_A^2}{r_A^3} + \frac{\mu_B^2}{r_B^3} - \frac{\mu_\ddagger^2}{r_\ddagger^3}\right) \qquad (3.8\text{-}9)$$

The reaction is assumed to involve a transition state with a dipole moment $\mu_{\ddagger}$.
An alternative electrostatic model [3-125] led to equation (3.8-10):

$$\ln \frac{k}{k_0} = \frac{3e^2 N_A}{8RT} \left(\frac{2}{\varepsilon} - 1 \right) \left(\frac{\mu_A^2}{r_a^3} + \frac{\mu_B^2}{r_B^3} - \frac{\mu_{\ddagger}^2}{r_{\ddagger}^3} \right) \qquad (3.8\text{-}10)$$

Both equations (3.8-9) and (3.8-10) indicate an increase of the reaction rate with increasing dielectric constant corresponding to a formation of a transition state which is more polar (has a larger dipole moment) than either of the reactants.

The approach based on coulombic interactions [3-126] yielded equation (3.8-11):

$$\ln \frac{k}{k_\infty} = - \frac{e^2 \mu_A \mu_B N_A}{\varepsilon \, RT \, r^3} \qquad (3.8\text{-}11)$$

It would seem possible to distinguish between the validity of equation (3.8-9), indicating linear relationship between $\ln k$ and $(\varepsilon - 1)/(2\varepsilon + 1)$, and equations (3.8-10) and (3.8-11), according to which $\ln k$ should be a linear function of $1/\varepsilon$. Nevertheless, it can be shown that the difference between $(\varepsilon - 1)/(2\varepsilon + 1)$ and $1/\varepsilon$ is usually within experimental error (e.g., for $\varepsilon = 8$ the difference is smaller than 1% [3-130]).

Reasonable fits of experimental data to linear relationship between $\ln k$ and $(\varepsilon - 1)/(2\varepsilon + 1)$ were found [3-131] for some binary solvent mixtures, where the value of ε was varied by changing the proportion of the solvent components in the mixture (e.g., benzene/ethanol, benzene/acetone, or 1,4-dioxane/acetone). Attempts to apply equation (3.8-8) to values obtained in pure aprotic and dipolar solvents showed a trend, but a very wide scatter indicated the important role of nonelectrostatic and specific solvent–solute interactions. Even poorer correlation has been found for protic solvents where hydrogen bonding plays a role. It is essential to include a sufficient number of solvents with widely differing properties in each investigation of solvent effects, since sometimes an acceptable correlation can be simulated, when the reaction is studied only in four or five solvents.

3.8.1.4. Reactions of Two Neutral Apolar Molecules

When considering cohesive forces of a solvent acting on a solute, it is possible to derive [3-121, 3-132] equation (3.8-12) for reaction of the type (3.8-8) between two neutral, apolar molecules A and B, which occurs via an isopolar transition state (AB)‡:

$$\ln \frac{k}{k_0} = \frac{1}{RT} [V_A(\delta_A - \delta_s)^2 + V_B(\delta_B - \delta_s)^2 - V_{\ddagger}(\delta_{\ddagger} - \delta_s)^2] \qquad (3.8\text{-}12)$$

where k_0 is the rate constant in a chosen standard solvent, V_i are the molar volumes, δ_i solubility parameters of the reactants and δ_s of the solvent. The values of solubility parameters δ_i are related to tabulated [3-132] values of (D_{ce}) by the relationship $\delta_i = \sqrt{D_{ce}}$. Thus the rate constant depends on the difference in molar volume between reactants and transition state, denoted as volume of activation $(\Delta V^{\ddagger})$, which is defined by equation (3.8-13):

$$\Delta V^{\ddagger} = V_{\ddagger} - V_A - V_B \qquad (3.8\text{-}13)$$

Solvents that lower the value of the volume of activation of a reaction cause an increase in the reaction rate. The rate constants depend also on relative values of cohesive energy densities (D_{ce}) of reactants, transition state, and on solvent. Cohesive energy density represents the energy needed to form a hole in a particular solvent to accommodate a solute molecule.

Applications of equation (3.8-12) in predictions of solvent effects on reactions between nonpolar molecules is limited by the fact that values for δ_i are not available for most reactive organic compounds. Attempts to correlate $\ln k$ with values of δ_s for solvents (which are tabulated [3-132]) gave at best an indication of a trend. Distinguishing between reactions with a positive and a negative value of activation volume, which would be a useful tool in the elucidation of mechanisms, is thus not reliable.

3.8.2. Semiempirical and Empirical Treatments

Experience with relationships based on evaluation of coulombic effects discussed in Section 3.8.1 strongly indicates that electrostatic interactions are often not the sole or even the predominant factor governing the solvent effects on reaction rates. In another group of attempts to correlate variations in reaction rates with the choice of solvent, empirical or semiempirical parameters are used to characterize the properties of the solvent.

In these treatments a reference process is chosen and a quantity is measured that characterizes this system. Variation of this quantity (which should be sensitive to changes in solvent composition) with the nature of the solvent is followed. Numerical values of this quantity—which can be an equilibrium constant, a rate constant, a wavelength of an absorption maximum in spectrum, or a polarographic half-wave potential—obtained for the reference process can be called the "solvent constant." Values of such solvent constants can be correlated with rate constants studied, for processes different from the reference process and the degree of correlation followed. The essential limitation of such approaches lies in the assumption that the nature of the interactions between the reactant(s), and solvent on one side and transition state and solvent on the other will remain qualitatively similar for all processes compared and differ only quantitatively.

As polar, resonance, and steric factors differ for individual processes and result in qualitative differences in solute–solvent interactions, limitations of such an approach are obvious. There is seemingly little hope for finding a set of solvent constants that would adequately describe solvent effects for all types of reactions and all substrates. The use of solvent constants is restricted to processes of a type similar to that involved in the reference process. The structure of the solute, to which such treatment should be applied, must resemble the structure of the species used in the reference process. Correlations, found between ln k and solvent constants derived from equilibrium constants or spectral data, indicate that similarity in the structure of the reactant is sufficient and need not be extended to the similarity in transition states (or products in equilibria or excited states in spectra). In such cases the chemical noncrossing rule seems to operate, indicating that solvent–solute interactions may differ for reactants and the transition state but change with the solvent structure in a more or less parallel way.

The restriction that a given set of solvent constants can be satisfactorily correlated only with rate constants for a given type of reactant in a given type of reaction can be used in mechanistic studies. Linear dependence of ln k on values of a given solvent constant indicates the possibility of a mechanism similar to that operating in the reference series, used in obtaining the solvent constants.

When an attempt is made to compare solvent effects on reactions which differ in the type of mechanism involved and on spectra of structurally different compounds, contributions of more than one type of solvent–solute interaction must be considered. Such an approach is used in multiparametric treatments, discussed in Section 3.8.2.5.

3.8.2.1. Solvent Constants from Equilibrium Constants

For equilibrium reactions in which a substrate with the ability to accept an electron pair reacts with a solvent which is an electron-pair donor, the Lewis basicity of solvents can be empirically characterized by a "Donor Number," DN_{SbCl_5} [3-133, 3-134]. Dilute solution of antimony pentachloride in 1,2-dichloroethane (as an inert solvent) is used as the electron acceptor in the reference process (3.8-14):

$$D + SbCl_5 \underset{25°C}{\overset{(CH_2Cl)_2}{\rightleftharpoons}} D\!-\!SbCl_5 \qquad (3.8\text{-}14)$$

(where D is the electron-pair donating solvent).

Enthalpy of the formation of the adduct (D-SbCl$_5$) is measured calorimetrically. Solvent donicity is then characterized by $DN_{SbCl_5} = -\Delta H_{D\text{-}SbCl_5}$[kcal mol^{-1}] and its values are tabulated [3-133]–[3-135] (Table 3-6).

Parameters derived from effects of solvents on keto–enol equilibria of 1,3-dicarbonyl compounds are much more restricted in scope of applications. For such systems K. H. Meyer proposed in 1912 [3-136] an empirical equation (3.8-15):

$$K_E = E \cdot L \qquad (3.8\text{-}15)$$

TABLE 3-6
Selected Values of Solvent Constants Derived from Equilibrium and Rate Constants

Solvent	DN_{SbCl_5} (in kcal mol^{-1})	So	Y	$N^{t\text{-BuCl}}$
Water	18.0	0	3.493	-0.26
Ethanol–water (80:20)	—	0.76^a	0.000	0.00
Methanol	19.0	1.27	-1.090	0.01
Ethanol	—	1.58	-2.033	0.09
Acetic acid	—	1.19	-1.675	-2.05
1-Propanol	—	—	—	—
1-Butanol	—	1.76^b	-2.73	—
2-Propanol	—	1.60	—	0.09
Acetonitrile	14.1	1.08	—	—
Nitromethane	2.7	—	—	—
DMSO	29.8	—	—	—
DMF	30.9	—	-3.5	—
Acetone	17.0	1.30	—	$(-0.43)^c$
Nitrobenzene	4.4	—	—	—
Pyridine	331	—	—	—
Chloroform	—	1.35	—	—
Hexamethylphosphoramide	38.8	—	—	—
Ethyl acetate	17.1	1.58	—	—
THF	20.0	—	—	—
1,4-Dioxane	14.8	1.45	—	$(-0.65)^c$
Diethylether	19.2	2.03	—	—
Benzene	0.1	1.64	—	—
Carbontetrachloride	0.0	—	—	—
n-Hexane	0.0	2.35	—	—
Cyclohexane	—	2.49	—	—

^{a}(67:33).
^{b}n-pentanol.
^{c}90%–10% H_2O.
Note: Modified from ref. 3-116.

where K_E = [Enol]/[Keto], the constant E (chosen equal to 1.0 for ethyl aceto-acetate) measures the enolization capacity of the dicarbonyl compound and solvent ("desmotropic"), and constant L is obtained from [Enol]/[Keto] for ethyl aceto-acetate in various solvents. Better correlation was found [3-137] using equation (3.8-16).

$$\log \frac{K_E^{So}}{K_E^{H_2O}} = \rho^E \, So \qquad (3.8\text{-}16)$$

where K_E^{So} and $K_E^{H_2O}$ = [Enol]/[Keto] in different solvents and water respectively.

ρ^E is a constant characteristic for the 1,3-dicarbonyl compound involved and So is a solvent constant, defined as So = $\log(K_E^{So,o}/K_E^{H_2O,o})$ where $K_E^{i,o}$ are equilibrium constants for ethyl acetoacetate in different solvents and water, respectively. Values of So obtained for 22 solvents (Table 3-6) show good correlation for both straight chain and cyclic 1,3-diketones and β-ketoacids. Nevertheless, no correlation was found between values of So and keto–enol equilibrium constants for 1,2-diketones [3-138].

Shifts of conformational equilibria of 2-isopropyl-5-methoxy-1,3-dioxan, in favor of the more dipolar axial cis-isomer with increasing polarity of the solvent, are the basis of derivation [3-139] of solvent constants D_1 (defined as equal to the change in the standard free energy associated with this equilibrium) for 17 solvents.

From solubilities, partition coefficients, and vapor pressure measurements it is possible to define solvent activity coefficients $^o\gamma^S$ [3-139a] corresponding to the changes in standard free energy observed when an electrolyte MX is transferred from a reference solvent O to another solvent S, or a free energy of transfer (ΔG_{tr}) [3-139a,b] defined as (3.8-16a):

$$\Delta G_{tr} (MX) = RT \ln {}^o\gamma_{M^+}^S \, {}^o\gamma_{X^-}^S \qquad (3.8\text{-}16a)$$

As values for individual ions cannot be measured, in order to separate the value of $\Delta G_{tr}(MX)$ into separate ionic contributions $\Delta G_{tr}(M^+)$ and $G_{tr}(X^-)$, it is necessary to make an extrathermodynamic assumption, such as:

$$\Delta G_{tr}(Ph_4As^+) \approx \Delta G_{tr}(Ph_4B^-) = [\Delta G_{tr}(Ph_4As^+) + \Delta G_{tr}(Ph_4B^-)]^{1/2}$$

This approach is based on assumption that ions Ph_4As^+ and Ph_4B^-, which are of comparable size and structure, are similarly influenced on transfer from one solvent to another. The values of $\Delta G_{tr}(M^+)$ and $\Delta G_{tr}(X^-)$ were used for correlations with kinetic data, particularly of S_N1 and S_N2 reactions [3-139c].

3.8.2.2. Solvent Constants from Rate Constants

Solvolyses of tert-butyl chloride following an S_N1 mechanism are accelerated by polar, particularly protic solvents. Because of the dependence of the rate constant (k_{So}) of the solvolysis following an S_N1 mechanism on the nature of the solvent, it is possible [3-140], [3-141] to use equation (3.8-17):

$$\ln \frac{k_{So}}{k_0} = m\,Y \qquad (3.8\text{-}17)$$

where k_0 is the rate constant obtained in the reference solvent [80% ethanol and 20% water (v/v)] and m is a coefficient dependent on the nature of the substrate which expresses susceptibility of the given substrate to variations in ionizing power of the solvent. The solvent constant Y is defined by equation (3.8-18):

$$Y = \ln \frac{k_{So}^{t\text{-BuCl}}}{k_0^{t\text{-BuCl}}} \qquad (3.8\text{-}18)$$

where $k_{So}^{t\text{-BuCl}}$ is the first-order rate constant for the solvolysis of tert-butyl chloride in solvent So at 25°C, and $k_0^{t\text{-BuCl}}$ the rate constant of the same reaction in a solution containing 80% ethanol and 20% water (v/v), chosen as the reference solvent where $Y = 0$ and $m = 1.0$. Values of solvent constants Y for pure solvents (mostly protic) and for some binary mixtures of organic solvents with water and other organic solvents are tabulated [3-142] (Table 3-6). For binary mixtures there seems to be no simple relationship between the value of Y and the solvent composition.

The rate constant of a studied solvolytic reaction k_{So} is plotted against Y and linearity of such plot indicates that the studied reaction might follow an S_N1 mechanism.

For reactions, where it is impossible to neglect the effect of the nucleophilic assistance of the solvent, a four parameter equation (3.8-19) was proposed [3-143]:

$$\ln \frac{k_{So}}{k_0} = mY + lN \qquad (3.8\text{-}19)$$

where l is a substrate dependent coefficient (similar to m) and N a solvent constant which characterizes the nucleophilicity of the solvent. Values of N were obtained [3-144] for methyl tosylate. For solvolysis of this compound, l was arbitrarily put equal to 1.00 and values of N defined by equation (3.8-20):

$$N = \ln \frac{k_{So}^{CH_3OTS}}{k_0^{CH_3OTS}} - 0.3\,Y \qquad (3.8\text{-}20)$$

where $k_{So}^{CH_3OTS}$ is the rate constant of solvolysis of methyl tosylate in solvent So, $k_0^{CH_3OTS}$ that in a mixture of 80% ethanol and 20% water (v/v), and $N = 0$ and $m = 0.3$. Values of N vary [3-144] from -5.56 for trifluoroacetic acid to 0.0 for ethanol (Table 3-6). Equation (3.8-19) shows better correlation than (3.8-17) for reactions of varying nucleophilicity and ionizing power [3-144]. It should be, nevertheless, kept in mind that introduction of another adjustable parameter (l) always results in apparent improvement in correlation, as indicated, for example, by an increased value of the correlation coefficient. Rigorous statistical evaluation of the use of the two and four parameter equations would be of importance.

Alternatively, the evaluation of the ionizing power of the solvent can be based on the study of kinetics [3-145] of anchimerically assisted solvolysis of p-methoxyneophyl tosylate at 75°C (3.8-21):

$$(3.8-21)$$

The value of $\ln k_1$ is a suitable measure of the ionizing power of the solvent even in fairly nonpolar and aprotic solvents.

Among other reactions, used as reference processes, it is possible to mention the Menschutkin reaction of tri-n-propylamine with methyl iodide at 20°C following an S_N2 mechanism [3-146], reaction of bromine with tetramethyltin in which an electrophilic aliphatic substitution occurs [3-147], and the Diels–Alder addition of cyclopentadiene to methyl acrylate [3-116], the rate of which, nevertheless, increases only slightly with increasing solvent polarity.

3.8.2.3. Solvent Constants from Electronic Absorption Spectra

Absorption spectra recorded in solvents of varying polarity vary in the wavelength of the absorption maxima, their shape, and molar absorptivity. These variations result from differences in changes in interactions between the solvent and the solute in ground state and excited state. It should be stressed that any change in spectra observed with replacement of the solvent used indicates that a different change in the interaction between the solvent and the solute in the ground state occurs than in the interaction between the solvent and the solute in the excited state. Were the energy in the excited state increased (or decreased) by the change in solvent by exactly the same amount as the energy of the ground state, no change in spectra would result. A shift of the wavelength of the absorption maximum to longer wavelengths due to a solvent effect does not offer information whether the energy of the ground state increased or decreased, only that the *difference* between the energies of the ground and excited states is decreased.

Changes in solvent used may result in a change in a proton-transfer or a charge-transfer between the solvent and solute, as reflected in solvent-dependent aggregation, ionization, complex formation, or establishment of equilibria between isomers. Such changes are usually not parallel to solvent effects on reactivity and hence will not be considered here. The great majority of data used for characterization of solvent effects was obtained from electronic spectra, hence further discussion will deal first with this type of spectra.

Interpretation of solvent effects on electronic spectra involves assumption of formation of a Franck–Condon excited state [3-148]. This is a state formed initially as a result of an electronic transition, which occurs within a time period that is short when compared with that needed for nuclear motions. In the Franck–Condon excited state, the excited solute is still surrounded by a solvent cage, resembling in size and orientation that which predominates in the ground state. By a loss of energy in the process of relaxation this initial excited state is converted into equilibrium excited state. The solvent–solute interaction in the Franck–Condon excited state, reflected in spectral changes, may differ from the solvent–solute interaction in the transition state (which is in thermal equilibrium with its environment), reflected in changes in chemical reactivity.

For a nonpolar solute in a polar or nonpolar solvent, only dispersion forces contribute to the solvation of the solute and increase somewhat with increasing polarity of the solvent. For a dipolar solute in a nonpolar solvent, both dispersion and dipole-induced dipole forces operate. If the Franck–Condon transition state is more polar than the ground state, the solvent–solute interaction increases with increasing polarity of the solvent. Both dispersion forces and dipole-induced dipole interactions increase with increasing polarity of the solvent and result in a shift of the absorption maximum to a longer wavelength. On the other hand, if the Franck–Condon transition state is less polar and hence less solvated than the ground state, the solute–solvent interaction becomes smaller with increasing solvent polarity. The dipole-induced dipole forces thus operate in a direction opposite from the dispersion forces, and a shift of the absorption maximum to shorter or longer wavelengths may result.

For a dipolar solute in a polar solvent, the dipole–dipole forces predominate (in addition to effects of hydrogen bonding discussed in Section 3.8.2.5.). The larger dipole moment of the Franck–Condon excited state results in a relatively larger stabilization due to the solvent–solute interaction than in the ground state. This is reflected by a shift to longer wavelengths with increasing solvent polarity. With so-called solvatochromic dyes (which show a pronounced shift in the wavelength and molar absorptivity of the absorption maximum with variation in the solvent), a change in the ground state dipole moment of the solute induced by surrounding solvent cage must be taken into consideration [3-149–3-151].

The majority of empirical solvent constants have been based on measurement of spectral shifts involving solvatochromic dyes. Perhaps the best known scale has been introduced by Kosower [3-152, 3-153], using spectra of solutions of 1-ethyl-

4-methoxycarbonylpyridinium iodide as the reference process. The shifts in wavelengths of the longest absorption band were measured, corresponding to an intermolecular charge-transfer transition. The stability of the electronic ground state (corresponding to an ion-pair) increases with increasing solvent polarity more than that of the first excited state (which is a radical pair). Consequently, with increasing polarity of the solvent a shift of the maximum to shorter wavelengths is observed. For example, a change from a solution of the dye in pyridine to that in methanol results in a shift of the band at longest wavelengths of 105 nm to shorter wavelengths. Solvent constant, Z, was defined as

$$Z \equiv hc \, \bar{\nu} \, N_A$$

$$(3.8\text{-}22)$$

$$Z \equiv \frac{2.859 \times 10^4}{\lambda} \, [nm]$$

where h is the Planck's constant, c the velocity of light, $\bar{\nu}$ the wave number, λ the wavelength of the photon that produces the electronic excitation, and N_A Avogadro's number. The values vary from $Z = 94.6$ for water to $Z = 54$ for benzene (Table 3-7). For aqueous solutions extrapolation of data for 1-ethyl 4-tert-butoxypyridinium iodide obtained in mixtures of water with nonpolar solvents or data for pyridine-1-oxide as a secondary standard were used. Values of Z are a measure of the transition energy (in kcal) needed to bring one mole of the standard indicator dissolved in the given solvent from ground state to the first excited state.

Similar definition of solvent constants (E_T) was based [3-152a] on changes in spectra of the pyridinium-N-phenoxide betaine dyes of the type (A), which vary from $E_T^{30} = 35.3$ for diphenylether to $E_T^{30} = 63.1$ for water (Table 3-7).

(A)

Other scales based on solvatochromic dyes involved changes in spectra of 5-dimethylamino-2,4-pentadienal [3-154], in $\pi \rightarrow \pi^*$ transition energies of mero-polymethine dyes [3-155], and in $n \rightarrow \pi^*$ transitions of N,N-dimethylthiobenza-

TABLE 3-7
Selected Values of Solvent Constants Derived from Spectral Data

Solvent	Z (in kcal mol^{-1})	E_T	S	π^*	α	β
Water	94.6	63.1	0.154	1.09[c]	1.13[c]	0.18[c]
Ethanol–water (80:20)	84.8	53.7	0.065	—	—	—
Methanol	83.6	55.5	0.050	0.60	0.98	0.62[a]
Ethanol	79.6	51.9	0.000	0.54	0.86	0.77[a]
Acetic acid	79.2	51.2	0.005	0.62[c]	1.09[c]	—
1-Propanol	78.3	50.7	−0.016	0.51[a]	0.80[a]	—
1-Butanol	77.7	50.2	−0.024	0.46[a]	0.79[a]	0.79[a]
2-Propanol	76.3	48.6	−0.041	0.46[a]	0.78[a]	0.78[a]
Acetonitrile	71.3	46.0	−0.104	0.85[a]	0.15[a]	0.31[a]
Nitromethane	71.2	46.3	−0.134	0.85[a]	0.23[a]	—
DMSO	71.1	45.0	—	1.00	0	0.76
DMF	68.4	43.8	−0.142	0.88	0	0.69
Acetone	65.5	42.2	−0.175	0.72	0.07[a]	0.48
Nitrobenzene	—	42.0	−0.218	1.01	0	0.39[b]
Pyridine	64.0	40.2	−0.197	0.87	0	0.64
Chloroform	63.2	39.1	−0.200	0.76[d]	0.34[b]	0
Hexamethylphosphoramide	62.8	40.9	—	0.87[c]	0	1.05[c]
Ethyl acetate	59.4	38.1	−0.210	0.55	0	0.45
THF	58.8	37.4	—	0.58	0	0.55
1,4-Dioxane	—	36.0	−0.179	0.55[a]	0	0.37
Diethylether	—	34.6	−0.277	0.27	0	0.47
Benzene	54	34.5	−0.215	0.59	0	0.10[a]
Carbontetrachloride	—	32.5	−0.245	0.29	0	0
n-Hexane	—	30.9	−0.337	−0.08	0	0
Cyclohexane	—	31.2	−0.324	0	0	0

[a]Less certain values.
[b]Much less certain values.
[c]Uncertain values.
[d]Doubtful values.
Note: Modified from refs. 3-116 and 3-167.

mide-S-oxide [3-156]. In an alternative approach the solvent constant S was defined [3-151] as

$$\lambda_{max}^{So} - \lambda_{max}^{EtOH} = R\,S \qquad (3.8\text{-}23)$$

using ethanol as the reference solvent ($S_{EtOH} = 0.00$) and the charge-transfer absorption of 1-ethyl-4-methoxycarbonylpyridinium iodide as the reference process ($R = 1.00$). The tabulated values of S (Table 3-7) were nevertheless not restricted to Z-values, but represent statistical averages of values obtained for a variety of reference processes.

Among other parameters used is the solvent constant Φ based on shifts of the wavelength of the $n \to \pi^*$ transition band of saturated aliphatic ketones [3-158], defined by equation (3.8-24):

$$\Phi = \frac{1}{\lambda^{So}} - \frac{1}{\lambda^{H}} \qquad (3.8\text{-}24)$$

where λ^{So} is the wavelength in the solvent So and λ^{H} that in n-hexane as reference solvent. Alternatively, the shifts of fluorescence maxima of 4-amino-N-methylphthalimide with solvent were followed [3-159].

3.8.2.4. Solvent Constants from Other Spectral Data

In infrared spectra correlations have been found [3-160] between the solvent effects on the stretching absorption of groups X = O and the X—H stretching frequency of X—H· · ·B hydrogen bonded systems, where X may be C, S, N, O, or P and B is the solvent acting as a hydrogen bond acceptor. The wave number in the studied solvent ($\bar{\nu}^{So}$) was compared to the wave number in vacuum $\bar{\nu}^{0}$ using expression (3.8-25):

$$\frac{\bar{\nu}^{0} - \bar{\nu}^{So}}{\bar{\nu}^{0}} = a\,G \qquad (3.8\text{-}25)$$

where a is the solvent coefficient and G the solvent constant. The G-values were obtained from solvent shifts of the carbonyl band of N,N-dimethylformamide and benzophenone and of the sulfonyl band of dimethylsulfoxide. The value of coefficient a was arbitrarily chosen so that for vacuum $G = 0.0$ and for dichloromethane $G = 100.0$.

Measurements of the O—D stretching band frequencies of CH_3OD in solvents that are hydrogen bond acceptors led to definition [3-161, 3-162] of a solvent constant B (3.8-26) which was considered to characterize the Lewis basicity of the solvent:

$$B = \bar{\nu}^{0}_{MeOD} - \bar{\nu}_{MeOD\cdot\cdot\cdot B} \qquad (3.8\text{-}26)$$

where $\bar{\nu}^{0}_{MeOD}$ and $\bar{\nu}_{MeOD\cdot\cdot\cdot B}$ refer to the wave number of the O—D stretching band in gas phase and aprotic solvent B, respectively. In a similar way the solvent constant B was defined using the O—H stretching frequency of phenol in the presence of B ($\bar{\nu}^{CCl_4}_{PhOH\cdot\cdot\cdot B}$), compared with the corresponding frequency of phenol ($\bar{\nu}^{CCl_4}_{PhOH}$) in carbon tetrachloride [3-163]:

$$B = \bar{\nu}^{CCl_4}_{PhOH} - \bar{\nu}^{CCl_4}_{PhOH\cdot\cdot\cdot B} \qquad (3.8\text{-}27)$$

The nitrogen hyperfine splitting constant a^{14_N} in ESR spectra of nitroxides can be used as a solvent constant [3-164], which can be determined at low concentrations of the nitroxide radical and is therefore accessible even in solvents where other indicators are not sufficiently soluble.

The ^{19}F chemical shifts in NMR spectra of 4-fluoronitrosobenzene relative to the shifts of fluorobenzene were used in the definition of solvent constants P[3-165], [3-166], the values of which varied from $P = 0.0$ in cyclohexane to $P = 2.7$ in sulfolane.

The solvent induced ^{31}P NMR chemical shifts of triethylphosphine oxide were used in the evaluation of "acceptor numbers," AN, defined as ^{31}P chemical shifts (δ_{corr}) relative to those of the 1:1 adduct of Et_3PO and $SbCl_5$ (dissolved in 1,2-dichloroethane), taken arbitrarily as 100, using equation (3.8-28):

$$AN = \frac{100\delta_{corr}}{\delta_{corr}^{Et_3PO-SbCl_5}} = 2.35\delta_{corr} \qquad (3.8-28)$$

Other types of applications of the NMR spectra, particularly of the ^{13}C NMR chemical shifts, are mentioned in Section 3.8.2.5.

3.8.2.5. Multiparameter Equations

None of the above scales attempted to include the effects of hydrogen bonding on solvent–solute interactions. The situation is further complicated by the fact that a solvent can act as hydrogen bond donor, hydrogen bond acceptor, or both.

To reflect the effect of hydrogen bonding in addition to the role of solvent polarity and at the same time to extend the use of linear solvation energy relationships to a variety of solutes, multiparameter equations were introduced. An equation that expresses the role of contributions of individual types of solvent–solute interactions for a wide variety of measured quantities—from electronic, infrared and NMR spectra to reaction rates—has been proposed by M. J. Kamlet, R. W. Taft, and their collaborators [3-167]. In general form this equation, based on linear solvation energy relationships, is given by equation (3.8-29):

$$XYZ = XYZ_0 + s\pi^* + a\alpha + b\beta \qquad (3.8-29)$$

Here XYZ is the measured quantity (wavelength or frequency, chemical shift, logarithm of equilibrium or rate constant, etc.) in the given solvent, XYZ_0 the same quantity obtained for a reference solvent. The term $s\pi^*$ describes contribution due to the dipolarity [3-168] and "polarizibility" of the solvent. The latter term expresses incompletely understood contributions of dipole–dipole interaction to solvent effects, only loosely related to the classical concept of polarizibility. The term $a\alpha$ expresses the contributions resulting from hydrogen bond formation between the solute and the solvent, in which the solvent acts as a hydrogen bond donor. Similarly the term $b\beta$ describes the contributions which result from hydrogen bond formation, in which the solvent plays the role of a hydrogen bond acceptor. As this equation was based predominantly on comparison of changes of wavelengths of absorption maxima due to solvent effects (called solvatochromic comparison method), the equation (3.8-29) and its simplified forms are called solvatochromic equations. Analogously, constants s, a, and b characterizing susceptibility of a given solute to solvent effects are called solvatochromic coefficients and constants π^*, α, and β characterizing the solvent effects are called solvatochromic constants (or parameters [3-167]).

In practical applications of equation (3.8-29) and its congeners, values XYZ must be measured for a chosen reference solvent and for a series of other solvents. Furthermore, values for solvatochromic constants π^*, α, and β must be found in tables (see ref. 3-167, Table 35) and coefficients s, a, and b must be determined.

In attempts to define values of solvatochromic constants π^*, a group of "select solvents" was chosen, which are assumed neither to form any hydrogen bonds with the solute nor to show other specific interactions with the solute. This excludes amines, hydroxylic solvents, polyhalogenated and aromatic compounds, leaving as select solvents monofunctional aprotic aliphatic compounds. For such groups of solvents equation (3.8-29) simplifies to:

$$XYZ = XYZ_0 + s\pi \qquad (3.8\text{-}30)$$

where $\pi^* = \pi$ (for select solvents).

There is no simple correlation between the values of π and those of the dielectric constant (ϵ). On the other hand there is a good correlation between values of π and the dipole moment (μ) of the solvent indicating that [3-169] dipolar effects in aprotic solvents are predominantly determined by interactions between solute and dipoles of the solvent, rather than by the bulk electrical effect of the solvent as continuous dielectric.

Attempts to extend the scale of π^* constants to protic solvents by measuring the changes in spectra were based on the use of aprotic indicators for which $b = 0$. Nevertheless, no indicator has been found which would absorb in the UV-visible range and would not act as a hydrogen bond acceptor. Values of π^* were, nevertheless, obtained from chemical shifts in the ^{13}C NMR spectra of the para-carbon in benzotrifluoride ($C_6H_5CF_3$) and phenylsulfurpentafluoride ($C_6H_5SF_5$) [3-170]. This approach is successful for select solvents and aliphatic alcohols which are strongly self-associated. For solvents, which are weaker hydrogen bond donors, the interaction between the solvent and the benzene ring of the indicator cannot be neglected. For such solvents a treatment must be used in which both the dipolar interactions and the capability of solvents to act as hydrogen bond donors are considered. To express the ability of the solvent to be a hydrogen bond donor towards a solute, indicators or solutes were chosen which themselves are not hydrogen bond donors but can act as hydrogen bond acceptors. The equation (3.8-29) then simplifies to (3.8-31):

$$XYZ = XYZ_0 + s\pi^* + a\alpha \qquad (3.8\text{-}31)$$

Evaluation of solvent constants α and π^* for such solvents is complicated particularly by self-associations of those solvents which are strong hydrogen bond donating acids. The tendency to self-associate is enhanced by amphiprotic properties of such solvents, which can also act as hydrogen bond acceptors. The self-association and amphiprotic properties render values of α and π^* obtained by means of equation (3.8-31) (Table 3-7) less reliable than values of β and π^* obtained by means of equation (3.8-32), discussed below. To find the best values of π^* and α for hydrogen bond donating solvents, a process of successive approximations was used for 16 diverse properties (including Kosower's Z-values [3-152, 3-153], Dimroth's E_T [3-152a], and Gutmann's AN [3-134, 3-135]), involving 13 indicators. These indicators were all strong hydrogen bond acceptors and were chosen in such a way that values of coefficients s and a were comparable.

For solvents, in which polar interactions and the capability of solvents to act as hydrogen bond acceptors play a role, hydrogen ion donating ability of indicators was evaluated. Using the shifts of the wavelength of absorption maxima for 4-nitroaniline, its N,N'-dimethyl derivative, 4-nitrophenol, and 4-nitroanisole as well as ^{19}F NMR shifts for 4-fluorophenol it was possible for hydrogen bond accepting solvents—in combination with data obtained for "select solvents"—to use equation (3.8-32):

$$XYZ = XYZ_0 + s\pi^* + b\beta \qquad (3.8\text{-}32)$$

By a process of successive approximation and extending the treatment to further indicators [3-171], reliable values of β as well as values of π^* for hydrogen bond accepting solvents were found and tabulated (Table 3-7, see also ref. 3-167, Table 35 and ref. 3-171).

Values obtained by means of equation (3.8-29) and its congeners correlate well with numerous solvent constants discussed in previous paragraphs. Correlations with these and other experimental data are so good that a real improvement of correlation resulting from introduction of additional parameters is strongly indicated. In some instances analysis of contributions to equation (3.8-29) enable interpretation of previously introduced solvent constants. Thus it was shown [3-172] that constant AN [3-133, 3-134] is a combined measure of solvent dipolarity-polarizibility and hydrogen donor ability, whereas constant DN is a linear function of constant β for O-bases and N-bases such as nitriles (but pyridine shows a different behavior) so that it expresses the ability of the solvent to accept hydrogen bonding.

In numerous cases it is not necessary to use the full equation (3.8-29) and reasonably good correlation can be found when the XYZ data are plotted as a function of tabulated π^* values, using an equation analogous to (3.8-30) even for solvents which are not "select." Nevertheless, in some cases—for example, for the wavelength in absorption spectra of Dimroth's betaine [3-152a] (structure A in Section 3.8.2.3), or for the rate of Menschutkin reaction [3-146]—the fit of the XYZ vs π^* plot improved considerably when the solvents were arranged into groups of chemically related compounds. Equation (3.8-30) is then extended to (3.8-33):

$$XYZ = XYZ_0 + s(\pi^* + d\delta) \qquad (3.8\text{-}33)$$

where the solvent constant δ is taken to be 0.00 for all "select solvents," 0.50 for polychlorinated aliphatic compounds (even when this value seems to depend on the structure of the indicator [3-173]) and 1.00 for all aromatic solvents. The value of the coefficient d varies from 0.0 for p-π^* electronic spectra transitions of uncharged indicators, to 0.086 for rates of Menschutkin reaction, and -0.23 for spectra of Dimroth's betaine [3-173]. The second right-hand term in equation (3.8-33) is assumed to express differences in blends of solvent dipolarity and polarizability. The uncertainty in the values of the $d\delta$ term contributes to the limited accuracy of the values of the constants α.

Even when the above treatments may seem complicated, they represent the most progressive trend in the treatment of solvent effects. In particular, this treatment distinguishes the effects of solvent (di) polarity from various hydrogen bonding effects. Such treatment clearly indicates the limitation of scales such as Z, E_T, So, DN, AN, and B.

It seems that for nonprotic solvents the values of constants π^* and β reported so far (ref. 3-167, Table 35, and ref. 3-171) are reliable and reflect only two types of effects, namely the solvent dipolarity–polarizibility and the hydrogen bonding between a donating solute and acceptor solvent, occurring at a single site. For protic solvents the mutual interactions of solvent molecules and their interplay with the solute resulting in formation of hydrogen bonds in which the solute is both donor and acceptor occur in addition to solvent polarity effects. This, together with the possibility of solutes forming hydrogen bonds at several sites and the uncertainty about the contribution of the $d\delta$ term mentioned above, all contribute to a lesser reliability of the tabulated values of α and π^* for such solvents. In this area further developments may be expected.

3.8.2.6. Some Applications of LSER

Solvatochromic equations of the type (3.8-29) will in the future undoubtedly serve well in the elucidation of mechanisms. The sign and magnitude of solvatochromic coefficients s, a, b, and d are particularly promising for deliberations dealing with the role of solvent in the transition state. Currently, nevertheless, the solvatochromic comparison method is applied predominantly to solution of spectroscopic problems, evaluation of individual types of hydrogen bonding, and comparison and evaluation of individual solvent scales. Treatment of kinetic data has so far been restricted to evidence that they are affected by similar solvent effects as spectroscopic data.

Examples of applications of solvent effects in mechanistic studies involve equation (3.8-17). Good correlations found for ln k_{S_o} and Y-values for the solvolysis rates of tert-butyl chloride, 1-adamantylbromide [3-174], 1-adamantyltosylate [3-175], and 2-adamantyltosylate [3-176] indicate that all four reactions follow the same, S_N1, mechanism, free from nucleophilic solvent participation and from rate determining elimination. Furthermore, for reactions following an S_N1 mechanism, the value of $m = 1.0$. When values of ln k_{S_o} for substrates following an S_N2 mechanism are plotted against Y, the correlation is inferior and $m = 0.25$ to 0.35.

For borderline cases such as secondary halides, the value of m is between these two extremes.

3.8.3. Qualitative Observations

In general, the rate of reactions between two uncharged molecules that result in a formation of two ions of the type (3.8-34):

$$A + B \rightleftharpoons [A \cdots B]^{\pm} \longrightarrow C^- + D^+ \qquad (3.8\text{-}34)$$

increases with increasing polarity of the solvent. A similar trend can be expected for reactions yielding products which are more polar than the starting materials.

For aliphatic nucleophilic substitutions and eliminations, considering only electrostatic interactions between solutes such as ions, dipolar molecules and transition states and solvents, Ingold and Hughes [3-177, 3-178] postulated that the rates of those reactions, in which the charge density is greater in the transition state than in reactants, increase with increasing polarity of the solvent used. When the transition state is less polar than reactants, reaction rate decreases with increasing polarity of the solvent. Finally, when there is a small difference in polarity between reactants and the transition state, solvent effects will be small. Examples of the last category are Types 2, 4, and 5 (Table 3-8) where dispersal of charge in the formation of the transition state results in a small difference in polarity when compared to the reactants.

The magnitude of the increase in rate for reactions of Type 1 can be demonstrated for solvolysis of tert-butyl chloride [3-179], for which relative rate constants are in water 335,000, in formic acid 12,200, in methanol 9, and in ethanol 1. An example of Type 2 (Table 3-8) is solvolysis of triethylsulfonium bromide [3-180] with relative rate constants equal to 1.0 in benzylalcohol and 180 in nitrobenzene. Type 3 is represented by the Menschutkin reaction of tri-n-propylamine with methyl iodide, the relative rate of which is 110,000 times faster in nitromethane than it is in n-hexane. Type 4 can be represented by the exchange reaction between methyl

TABLE 3-8
Solvent Effects Predicted for Nucleophilic Substitution Reactions

Type	Reactants	Transition State	With Increasing Solvent Polarity Reaction Rate
1. S_N1	R–X	$R^{\delta+} \cdots X^{\delta-}$	Increases strongly
2. S_N1	R–X$^+$	$R^{\delta+} \cdots X^{\delta+}$	Decreases slightly
3. S_N2	Nu + R–X	$Nu^{\delta+} \cdots R \cdots X^{\delta-}$	Increases strongly
4. S_N2	Nu$^-$ + R–X	$Nu^{\delta-} \cdots R \cdots X^{\delta-}$	Decreases slightly
5. S_N2	Nu + R–X$^+$	$Nu^{\delta+} \cdots R \cdots X^{\delta+}$	Decreases slightly
6. S_N2	Nu$^-$ + R–X$^+$	$Nu^{\delta-} \cdots R \cdots X^{\delta+}$	Decreases strongly

Note: Modified from refs. 3-177 and 3-178.

iodide and a radioactive iodide anion [3-181], the rate of which in ethanol is only 44 times faster than in water.

The second-order reaction of trimethylamine with trimethylsulfonium ion, representing Type 5, shows the anticipated relatively small decrease in rate with increasing polarity. This reaction occurs in ethanol 10 times faster than in water [3-182], indicating that initial reactants are more strongly solvated than the transition state. Finally, the reaction of trimethylsulfonium ions with hydroxide ions [3-183] shows a large decrease in rate with increasing solvent polarity expected for Type 6 (Table 3-8), with a relative rate constant of 19,600 in ethanol and 1.0 in water.

For elimination reactions, following an E1 or E2 mechanism [3-177, 3-178], the proposed changes with solvent composition are summarized in Table 3-9. β-Eliminations of Type 1 and 2 involve the same rate controlling step as S_N1 reactions and hence show similar dependence on solvent composition. Most bimolecular elimination reactions studied, following Types 3–6, particularly reactions of alkyl halides and osmium compounds in water–ethanol mixtures, show solvent effects in the expected direction and relative size.

The approach used by Hughes and Ingold [3-177, 3-178] is based on the assumption that contributions of activation entropy ($\Delta S^{\ddagger}$) to the change in free energy of activation ($\Delta G^{\ddagger}$) are either little dependent on the solvent used or are a small contribution to the value of $\Delta G^{\ddagger}$. Even when this assumption is valid for numerous reactions, exceptions are known. Furthermore, in this treatment the solvent is

TABLE 3-9
Solvent Effects Predicted for β-Elimination Reactions

Type	Reactants	Transition State	With Increasing Solvent Polarity Reaction Rate
1. E_1	$H-\overset{\mid}{C}-\overset{\mid}{C}-X$	$H-\overset{\mid}{C}-C^{\delta+}\cdots X^{\delta-}$	Increases strongly
2. E_1	$H-\overset{\mid}{C}-\overset{\mid}{C}-X^+$	$H-\overset{\mid}{C}-C^{\delta+}\cdots X^{\delta+}$	Decreases slightly
3. E_2	$Nu + H-\overset{\mid}{C}-\overset{\mid}{C}-X$	$Nu^{\delta+}\cdots H\cdots C\cdots C\cdots X^{\delta-}$	Increases strongly
4. E_2	$Nu^- + H-\overset{\mid}{C}-\overset{\mid}{C}-X$	$Nu^{\delta-}\cdots H\cdots C\cdots C\cdots X^{\delta-}$	Decreases slightly
5. E_2	$Nu + H-\overset{\mid}{C}-\overset{\mid}{C}-X^+$	$Nu^{\delta+}\cdots H\cdots C\cdots C\cdots X^{\delta+}$	Decreases slightly
6. E_2	$Nu^- + H-\overset{\mid}{C}-\overset{\mid}{C}-X^+$	$Nu^{\delta-}\cdots H\cdots C\cdots C\cdots X^{\delta+}$	Decreases strongly

Note: Modified from refs. 3-177 and 3-178.

considered to be a dielectric continuum and specific solvent–solute interactions, such as hydrogen bonding, are not taken into account.

The majority of pericyclic reactions, such as electrocyclic, sigmatropic, chele-tropic, or cycloaddition reactions, represent interaction between two uncharged molecules and involve isopolar transition states (i.e., those which are neither dipolar nor radical in nature). Such reactions exhibit only small solvent effects, because the charge distributions in the reactants and in the transition state are similar. Thus the second-order rate constants of Diels–Alder reactions increase only by a factor of 3 when reactions in nonpolar solvents are compared to those in polar media. The small variation in the rate with solvent composition can be used as one of the criteria in showing that a pericyclic mechanism is involved. Only when the cy-cloaddition proceeds through zwitterionic intermediates or dipolar ionic states, the rate constants show a considerable sensitivity to solvent effects. Thus in the reaction of n-butyl vinyl ether with tetracyanoethylene [3-184] the second-order rate constant in acetonitrile is 2600 times larger than that in cyclohexane.

Addition of uncharged electrophiles to a carbon–carbon double bond usually results in the formation of a transition state with a small, dispersed charge. Such reactions show only a small increase in reaction rate with increasing polarity of the solvent. There are, nevertheless, some additions of electrophiles where a large acceleration is observed with increasing polarity of the solvent. In such cases it can be concluded that a considerable charge is formed in the transition state.

Examples of the latter case are additions of halogens to carbon–carbon double bonds. In the addition of bromine to 1-pentene the relative rate constants increase from 1.0 in carbontetrachloride to 1.6×10^5 in methanol and 1.1×10^{10} in water [3-185, 3-186]. The mechanism of this reaction, (3.8-35 to 3.8-37), assumes a rapid establishment of an equilibrium between a charge-transfer complex, halogen, and olefin.

$$\text{>C=C< + Br}_2 \underset{\text{fast}}{\overset{\text{Br}_2}{\rightleftharpoons}} \text{C}\overset{+}{\underset{}{\cdots}}\text{C} \underset{\text{slow}}{\rightleftharpoons} \left[\begin{array}{c} \text{Br}^{\delta-} \\ | \\ \text{Br}^{\delta+} \\ \diagup \quad \diagdown \\ \text{>C}\underline{\quad}\text{C<} \end{array} \right]^{\pm} \qquad (3.8\text{-}35)$$

In the following rate-determining step a loss of halide ion occurs and a cyclic halonium intermediate is formed which then reacts with the halide ion or another nucleophile present to yield the product.

$$\left[\begin{array}{c} \text{Br}^{\delta-} \\ | \\ \text{Br}^{\delta+} \\ \diagup \quad \diagdown \\ \text{>C}\underline{\quad}\text{C<} \end{array} \right]^{\pm} + \text{So} \longrightarrow \overset{\text{Br}^+}{\underset{\diagup \quad \diagdown}{\text{>C}\underline{\quad}\text{C<}}} + \text{Br}^- \qquad (3.8\text{-}36)$$

$$\overset{\text{Br}^+}{\underset{\diagup \quad \diagdown}{\text{>C}\underline{\quad}\text{C<}}} + \text{Br}^-(\text{Nu}) \longrightarrow \overset{\text{Br} \quad \text{Br}}{\underset{| \quad |}{\text{>C}\underline{\quad}\text{C<}}} \left(\overset{\text{Br} \quad \text{Nu}}{\underset{| \quad |}{\text{>C}\underline{\quad}\text{C<}}} \right) \qquad (3.8\text{-}37)$$

Solvation of anions in protic and dipolar aprotic solvents plays an important role in nucleophilic substitutions where anions are either attacking nucleophiles or leaving groups. Even when such effects result in large changes in reaction rates, they often cannot be separated from other types of solvent effects. Hence their usefulness in mechanism elucidation has so far been limited. Similarly another aspect of anion–solvent interaction—the decrease in specific solvation of anionic bases in dipolar aprotic solvents when compared with hydroxylic solvents—which results in a dramatic increase in activity of anionic bases (usually RO^-), have so far found more synthetic than mechanistic applications.

In addition to specific solvation, the nucleophilic reactivity of anions depends also on the degree of association with the cationic counterion. Reactivity of ion pairs is usually smaller than that of the free nucleophilic anion. This was shown by the effect of the nature of the cation on reaction rate. In less polar solvents tetraalkylammonium ions are often associated with anions to a lesser degree than lithium ions. Thus, replacement of Li^+ ions by NR_4^+ ions results in an increase in reactivity of the anion. An alternative way to affect the ion-pair formation and thus the reactivity of the anion, is to bind the cations into complexes with crown ethers and similar reagents. In such species the cation is located in a hydrophilic cavity in an organic ligand with hydrophobic groups oriented towards the external solvent. Interaction of the solvent with these outer hydrophobic groups enables dissolution of inorganic salts in organic solvents where they—in the absence of complexing agents—show only a limited solubility. Resulting solutions contain non-solvated "bare," highly reactive anions, such as OH^-. This approach promises a broadening of a range of processes which can be investigated, but so far most of the studies in this area have dealt from synthetic rather than mechanistic aspects.

To clarify the participation of water molecules in a reaction, inverted micelles can be used [3-187]. Inside these micelles the concentration of water can be varied; linear dependence of the rate constant of the studied reaction on the concentration of water is an evidence of water participation in the mechanism.

3.9. STRUCTURAL EFFECTS

The relationship between the structure and reactivity of organic compounds has been the subject of numerous monographs [3-98–3-103, 3-152, 3-177, 3-188–3-192] and is treated in some detail in many advanced organic textbooks [3-193–3-199]. Therefore, only the main principles and their use in the studies of mechanisms will be mentioned here. Nevertheless, it is recognized that studies of the influence of structural effects on values of reaction rate constants and sometimes on the change in the form of the rate equation represent one of the most important tools in the elucidation of reaction mechanisms. Structural effects are particularly useful in deducing the composition, stereochemistry, and electron distribution in the transition state.

To discuss the role and application of structural effects in the elucidation of mechanisms, it is possible to consider the reaction of a substance R–A–X with a reagent B–Y. The structure of both R–A–X and B–Y can be varied and the effect of a particular change on reaction rate followed. In this symbolism, R is considered to be the reaction center, which is the atom, functional group, or part of the reactant molecule where bonding is changed in the course of reaction. The molecular frame (A) is the system of linked carbon atoms which carries the reactive center and possibly the substituent (X), which remains unchanged in the course of the reaction. A substituent (X) is an atom or a group introduced on the molecular frame which does not alter the molecular frame and is not part of the reactive center. In the reagent, Y is the reactive center and B the remainder of the species.

Information from structural studies can be divided into two general classes: the scope of a reaction, and the effects on rate constants. Information on the scope of a reaction is obtained by detecting those structural changes that lead to a change in the fundamental reaction type or the reaction rate equation. If, and only if, the reaction rate equation does not change for a series of compounds, the relative values of the rate constants can be compared in such series and a more detailed interpretation of the reaction mechanism can be attempted. The scope of a reaction is discussed first, followed by discussions of the effect on rate constants due to a change in the reactive center, a change in molecular frame, and substituent effects.

3.9.1. Scope of a Reaction

In the search for the scope of a reaction, the role of substantial changes in the structure both of the reactant R–A–X and the reagent B–Y is followed. At this stage the investigation is aimed at finding out how big the structural change can be, first in one, then in the other participating species, without altering the type of the reaction mechanism.

To establish the similarity of compared reactions, it must be verified whether all compared reactions yield identical (or analogous) products and intermediates. It must also be verified that all compared reactions follow the same reaction rate equation. Furthermore, assurance must be obtained that the obtained reaction rate constant depends in an analogous way on composition of the reaction medium (e.g., pH) for all compounds compared. Such a broad investigation makes it possible to find out which structural changes are too extensive and must be excluded, if only reactions following analogous mechanisms are to be compared.

Let us assume, for example, that a new reaction of reagent B–Y with aromatic aldehydes was observed. It is first checked whether aliphatic and alicyclic aldehydes react in the same way. Next, the study is extended to diaryl, aryl alkyl, and dialkyl ketones, cyclanones and benzocyclanones. Next, the behavior of aldehydes and ketones with a formyl or aroyl group attached to a heterocyclic ring can be tested. Further, it is possible to test whether acids, esters, amides or acyl chlorides undergo the same reaction. Alternatively, the examination can be extended to compounds with azomethine bonds (oximes, hydrazones, semicarbazones, ketimines, etc.) or compounds with a thiono group or azo group. After the question of the limits of analogous reactivity of the compound R–A–X has been settled, effect of variation in the structure of the reagent B–Y is dealt with in a similar way.

In the establishment of the scope of a reaction, product identification and verification of reaction rate equation is more convincing than comparison of yields. A proven change of the reaction rate equation (e.g., from first- to second-order kinetics) resulting from a variation in structure can often be an important contribution to mechanism elucidation. Even minor changes in structure, such as replacement of a substituent, can lead to a change in the reaction rate equation, for example, by altering the rate-determining step in a given mechanism. Similarly, a switch from one preferred reaction pathway to another can be induced by substituents. These effects are often detected by a systematic comparison of observed rate constants for a substituted series of compounds, as described in Section 3.9.4.

Establishment of the scope of a reaction is of importance, particularly from the synthetic point of view, offering information regarding which compounds related to the parent compound initially investigated can still be prepared by the studied reaction. It also plays an important role in kinetic studies. Failure to establish the scope of reaction can lead to uneconomical research where reactions are studied following mechanisms different from that of the parent compound, as well as to wrong conclusions, when structural effects on the values of rate constants are compared for compounds following different mechanisms.

3.9.2. Variations at the Reaction Center and Reagent

Structural variations can be divided into three categories, involving the reaction center, molecular framework, or substituents, but some variations can be classified as belonging to more than one category and the borderlines between them are not well defined. This is, nevertheless, not of major importance, since the principles on which the use of structural variations are based are similar in all three categories. The methodological approach in all studies of the effects of structural variations on rate constants is the same: Behavior of compounds which differ by a change in structure or by a series of structural changes is compared. Based on the postulated mechanism it is possible to predict the direction of the change in the value of the rate constant and possibly a sequence in which the values of rate constants will vary in a series of structurally related compounds. If the observed effect agrees with the prediction, the observation is considered to support (but does not prove) the assumed mechanism. Sometimes structural studies are performed to test the scope of a mechanism or to define more precisely the steric and electronic requirements for the reaction when the mechanism is already well established.

A reaction center can be defined as that part of the reactant molecule where bonding is changed in the course of a reaction, but such a definition could include the entire molecule in a quantum mechanical interpretation of bonding as involving molecular orbitals which have contributions from orbitals on every atom. In a more practical sense, the reaction center can include a large part of a molecule when the reaction occurs with a conjugated system, where delocalized bonding must be considered. Thus, in practice, the reaction center and structural variations that qualify as exchanges of a reaction center must be defined for each reaction.

It is proposed to call "reagents" those starting materials which participate in the conversion of the substrate into product. They usually have smaller molecular weight than the substrate and often are inorganic species. The investigations of the role of the structure of the reagent on reaction rate has been less frequently investigated in a systematic manner than the effect of the variation in the structure of the substrate. One of the reasons for this is that a change in the structure of the reagent often results in a change in mechanism, thus preventing a simple comparison of the values of reaction rate constants.

A group of reactions that serves particularly well for the demonstration of both the role of the variation of the reaction center (leaving group) and the dependence of the reaction rate on the nature of the reagent (nucleophile) are nucleophilic substitutions. Studies of the nature of the leaving group can involve comparison of halides among themselves, or with amines, mercaptans, hydroxide ion, alcohols, cyanide ion, carboxylates, sulfonates, and so forth. Alternatively, reactivity of water, p-toluenesulfonate ion, fluoride, acetate and chloride ion, pyridine, bromide, hydroxide ion, thiocyanates, iodides, cyanide, mercaptide, and SH^- ions as nucleophilic reagents can be compared.

Structural variations show a strong dependence of the rate constant on leaving group if loss of the leaving group is involved in the rate-determining step and similar dependence on the structure of the nucleophile if displacement by the nucleophile is part of the rate-determining step. Comparison of reactions with a given nucleophile indicates that the more stable the leaving group, the faster the reaction. Hence, a weaker base should be a better leaving group. Conversely, stronger bases should be better nucleophiles. However, nucleophilic reactivity can be affected by no less than 17 factors, including solvation [3-200]. It is thus not surprising that attempts to establish a single scale of nucleophilicity, which would be independent of the structure of the substrate, have been unsuccessful. Only for groups of closely related substrates is it possible to derive nucleophilicity constants which adequately describe the effect of the structure of the nucleophile on the role in a given type of reaction.

Variations in reactivity of nucleophiles in various nucleophilic attacks on carbon can be attributed [3-201] to varying contributions of an electrostatic and a covalent term. When the covalent term is dominant ("orbital controlled order of relative nucleophilicities"), the nucleophilicities decrease in the sequence:

$$SH^- > I^- > CN^- > Br^- > Cl^- > OH^- > F^-$$

When the electrostatic term predominates ("charge controlled order") the sequence is

$$OH^- > CN^- > SH^- > F^- > Cl^- > Br^- > I^-$$

as are reactivities towards proton. When both terms are comparable, the sequence follows

$$SH^- > CN^- > I^- > OH^- > Br^- > Cl^- > F^-$$

which is the sequence often observed for nucleophilic attacks on an sp^3 carbon.

A similar but empirical concept, based on the assumption that hard acids react preferably with hard bases and soft acids with soft bases [3-202], does not explain why soft bases like RS^-, CN^-, and so forth add readily to carbonyl compounds which are hard acids, and is thus of doubtful value.

Since nucleophiles are also bases, attempts have been made to compare their pK_B-values of reactions (3.9-1).

$$Nu^- + HOH \underset{pK_B}{\rightleftharpoons} NuH + OH^- \qquad (3.9\text{-}1)$$

in water

with pK_C values of their carbon basicities:

$$Nu^- + ROH \underset{pK_C}{\rightleftharpoons} NuR + OH^- \qquad (3.9\text{-}2)$$

in water

For $R = COCH_3$, values of pK_B and pK_C show similar trends [3-203], but no similarity is shown for $R = C_6H_5$ and the sequence of pK_B and pK_C is completely different if $R = CH_3$.

When the rates of the reaction of nucleophiles with substituted phenylacetates are compared with the pK_B-values of the reagent, a nonlinear relationship between the logarithm of k and the pK_B-values was found [3-204]. This has been interpreted to be caused by a two-step mechanism (3.9-3):

$$(3.9\text{-}3)$$

For strongly basic nucleophiles, where $pK_B > pK_{ArOH}$, the addition step is rate determining, for the less basic compounds the first equilibrium is rapidly established and the second step becomes rate determining.

When the comparison is restricted to groups of closely related nucleophiles involving the same attacking atom in similar surroundings (e.g., a series of ring substituted phenolates) a linear correlation can be found between pK_B and the logarithm of the rate constants of reactions following an S_N2 mechanism [3-205]. When the reaction rate is additionally affected by steric effects, the situation is more complicated. For example, triethylamine is 26 times more reactive than pyridine in the replacement of iodide ion in methyl iodide, but in the displacement of the same iodide ion in isopropyl iodide pyridine is 6 times more reactive than triethylamine. For the reaction with methyl iodide, greater basicity of triethylamine when compared with pyridine plays a decisive role, but for reaction with isopropyl iodide, which has larger steric requirements, steric interactions with the nucleophile in the transition state become important. The bulky triethylamine thus reacts more slowly than pyridine with isopropyl iodide.

Somewhat wider variation in the structure of the nucleophile can be treated by the Swain–Scott [3-206] equation (3.9-4):

$$\log k = s \cdot n + \log k_0 \tag{3.9-4}$$

where s is the reaction constant dependent on the nature of the substrate and on the medium used (defined by $s = 1.0$ for reactions of methylbromide in aqueous solutions) and n the nucleophilicity constant (defined by $n = 0.0$ for water).

For reactions of stable cations with nucleophiles, Ritchie [3-207] proposed (3.9-5):

$$\ln k = aN_+ + \text{const} \tag{3.9-5}$$

The nucleophilicity constant N_+ was chosen equal to 0.0 for water. The value of a is equal to 1.0 for reactions of phenyldiazonium and tropylium ions, reactions of phenylacetates bearing electronegative groups in the aromatic ring, and for nucleophilic substitution on 2,4-dinitrohalobenzenes [3-208]. The value of the reaction constant a is smaller than 1.0 for reactions of some triarylcarbonium ions [3-209] and oxiranes [3-210] and larger than 1.0 for those of fluorene-9-dinitro-methylene [3-211].

This similarity of the value of a for unrelated types of reactions, together with the absence of correlation between values of N_+ on one side and those of n, pK_B, or pK_C on the other, indicate that values of constants N_+ reflect a process that is common to all nucleophiles. Since values of N_+ depend on the solvent used, it is assumed that they correspond to the energy needed to detach molecule(s) of solvent from the solvent shell surrounding the nucleophile and to make space available for the electrophile. It is thus assumed that in these reactions $\Delta G^{\ddagger}$ is governed only by the destruction of the solvation shell of the nucleophile, whereas the solvation of the electrophile in the initial and transition states remains practically the same. Nevertheless, it has been proposed [3-212] that the small variation in the value of a results in contributions from two opposing factors: strong solvation of the elec-trophile resulting in an increased susceptibility to structural effects, and formation of a reactant-similar transition state, which decreases such susceptibility. Thus, reactive and strongly solvated electrophiles lose little solvent in the formation of the transition state. Alternatively, weakly solvated electrophiles are more desolvated in the transition state.

Another group of compounds, in which reactivities can be compared, are con-jugated systems. Since the entire π-electron system can be considered as one reactive center, comparisons of the effects of extent of conjugation can be thus classified as variations in the reactive center. Before comparing the effect of the extent of conjugation on relative reactivities, it is particularly important to prove that the same site of reaction is involved in all compared reactions. Structural variations often alter the relative reactivities at different positions and result in an alteration of the site of reaction. For example, when comparing interaction of a nucleophilic reagent with saturated carbonyl compounds to reactions of the same reagent with α,β-unsaturated carbonyl compounds, it is first necessary to show that the addition occurs on the carbonyl group and not on the ethylenic bond. For instance, changing the position of phenyl substitution on the double bond alters the site of attack of a Grignard reagent in reactions (3.9-6 and 3.9-7):

$$\underset{\underset{Ph}{|}}{Ph-C}=CH-\underset{\underset{O}{\|}}{C}-Ph \xrightarrow[\text{2. H}^\bullet]{\text{1. PhMgBr}} \underset{\underset{Ph}{|}}{Ph-C}=CH-\underset{\overset{Ph}{|}}{\underset{\underset{OH}{|}}{C}}-Ph \qquad (3.9\text{-}6)$$

$$Ph-CH=\underset{\overset{Ph}{|}}{\underset{\underset{O}{\|}}{C}}-C-Ph \xrightarrow[\text{2. H}^\bullet]{\text{1. PhMgBr}} Ph-CH-\underset{\overset{Ph}{|}}{\underset{\underset{Ph}{|}}{CH}}-\underset{\underset{O}{\|}}{C}-Ph \qquad (3.9\text{-}7)$$

Another example of the role of varying the reactive center can be demonstrated by comparison of reactions of various aromatic systems. These systems also have several possible sites of reaction. Meaningful comparisons of reaction rates, for example, for electrophilic substitution reactions, must be among rates determined for specific positions, not composite rates for the entire system. Many reactions of aromatic systems have been followed only by comparison of yields of products, in competitive reactions of two aromatic systems or within a single system. This methodological approach offers useful information, provided that the reactions involved are irreversible. For reactions involving equilibria, information obtained from yields can be completely misleading and kinetic studies are indispensable.

Reactivity of individual positions in polynuclear aromatic systems can be characterized by reactivity indices, N_t [3-213]. These constants, which are inversely related to the reactivity in a given position, have been calculated from the C–C resonance integral and nonbonding molecular orbital coefficients. Coefficients N_t were found to characterize the reactivity of individual sites in protonations, Diels–Alder reactions with maleic anhydride [3-214], electrophilic substitutions [3-215], and so forth.

3.9.3. Variations of Molecular Frame

The possibilities for variation in molecular frame are very wide. Assuming that a reaction occurs at a side chain or a functional group where the bond between the molecular frame and the reactive group is not severed during the reaction, it may be informative to compare the reactivity of such groups bound to an sp^3 carbon of an alicyclic molecular frame and to an sp^2 carbon of an aromatic ring. The aromatic ring may profoundly affect reactivity, particularly in those cases where charge delocalization by the aromatic ring is important in the reactant or transition state. It is also possible to compare the effects of the kind of aliphatic chain, the size of an alicyclic ring, and the nature of the aromatic ring. For example, aromatic rings differ in their capability to increase or decrease electron density in the side chain. Reactivity in such cases also depends on the position of attachment to the aromatic ring and on the size of the aromatic system (e.g., benzene, naphthalene, anthracene). As with attacks on the aromatic system, comparisons should be restricted to equivalent systems—the reactivity of benzene derivatives bearing a group R can be compared with derivatives bearing the same group in position 1 on a naphthalene or anthracene ring; compounds bearing group R in position 2 in a naphthalene ring should be compared with compounds bearing group R in position 2 of anthracene and so on. Another alternative is to compare reactivity of a group in position 2 on a naphthalene ring with the reactivity of compounds bearing the same group on a benzene ring with anelled heterocyclic aromatic ring. For heterocyclic compounds bearing the reactive center in the side chain, the position of this reactive center is essential and the effects of, for example, 2-pyridine, 3-pyridine, and 4-pyridine groups, should be discussed separately.

For alicyclic systems it is possible to compare the effects of ring size not only on reactions in the side chain, but also on reactions which involve one or more carbons of the ring. For example, addition or elimination reactions which lead to formation or disappearance of an endocyclic or exocyclic double bond can be studied, such as addition reactions of cyclanones or the elimination in cyclic 1,2-dibromides. The effects of conformationally rigid frameworks can be examined through bicyclic systems, for example, norbornane and bicyclo[2.2.2]octane.

The stereochemical requirements for reactions are often established through product studies utilizing stereoisomeric compounds as reactants. Both cyclic and acyclic reactants are often suitable. For instance, the stereospecific *anti* elimination of HBr from bromoethanes was first demonstrated with the stereoisomeric 1,2-dibromo-1,2-diphenylethanes.

This type of study, especially when coupled with determination of relative rates of reaction, reveals a great deal about the role of conformation and steric factors in a reaction. The dependence of reaction rates on known spatial arrangements of molecules can be compared, both in direction and magnitude, with the changes expected for the proposed transition state. An unexpected sequence of reactivity often leads to a revised and improved mechanism.

3.9.4. Substituent Effects

Substituents influence the reactivity of molecules by steric effects, when the site of substitution is close to the reaction center, and electronic effects, wherein the electron distribution is altered by the substituent. In theory, electronic or polar effects can be separated into field effects, which are due to electrostatic interactions through space; inductive effects, which result from polarization of the σ-bond system; and resonance effects, which operate through the π-bond system. Many substituent effect studies have been carefully designed to evaluate the various contributions to the overall substituent effect. Rather than discuss the theory of substituent effects in detail, the following discussion will focus on the usefulness of the substituent electronic effect as a probe of reaction mechanisms.

Substituent effects can often be treated quantitatively by use of a linear free energy relationship. The most commonly used variant of such an empirical treatment involves the use of the Hammett equation [3-96–3-103]. For substituent effects on reaction rates this equation can be written as (3.9-8):

$$\log \frac{k_X}{k_0} = \rho \sigma_X \tag{3.9-8}$$

where k_X is the rate constant for a compound substituted by group X, k_0 is the rate constant for the unsubstituted compound, σ_X is a substituent constant, and ρ is the proportionality factor called reaction constant which remains unchanged for a given reaction under a given set of conditions. The ρ value is obtained from the slope of the line in a $\log(k_X/k_0)$ vs. σ_X plot for a reaction series that consists of several substituted compounds. Data for literally hundreds of reactions have been successfully correlated by means of this empirical treatment.

Equilibrium constants can be treated in a similar way and even physical properties such as chemical shifts in NMR, frequencies in IR-spectra, and polarographic half-wave potentials have been successfully correlated using the Hammett equation.

The observation of a linear correlation between data for a reaction series and substituent constants σ_X establishes an analogy between the reaction series studied and the reaction series used to define the constants σ_X. Such comparisons allow deductions to be made regarding the electronic character of transition states. The linear correlation in itself indicates the likelihood that a single mechanism is operating throughout the reaction series. Treatments of the type using the Hammett equation provide three sources of information that can be used in further mechanistic interpretations: (1) the type of substituent constants (σ_X), (2) the sign and magnitude of the reaction constant (ρ), (3) deviations from linear correlations.

3.9.4.1. Choice of Substituent Constant

The Hammett equation (3.9-8) has been derived [3-96] to describe effects of a substituent (X) on a benzene ring in either the meta- or para-position relative to the reaction center. The effect of each substituent differs according to its position and hence two values, σ_{m-X} and σ_{p-X} must be used. As a reference reaction series the thermodynamic values of dissociation constants of substituted benzoic acids (K_X) in water at 25°C was chosen (Table 3-10), for which, by definition, ρ equated to 1.00. The unsubstituted compound (X = H) was chosen as internal standard ($\sigma_H = 0.00$). The substituent constants are thus defined as $\sigma_X \equiv \log (K_X/K_H)$. Electron-withdrawing substituents, which enhance the dissociation, have positive σ_X-values, electron-donating substituents, which decrease dissociation of benzoic acids, have negative σ_X-values (Table 3-10).

The standard values of substituent constants σ_X have been successfully used for characterization of substituent effects in numerous reactions. These σ_X-values represent a particular blend of resonance and polar (inductive and field) effects, which for meta-substituents is adequate for characterization of their role in all reactions.

TABLE 3-10
Selected Values of Substituent Constants

	meta	para		
Substituent	σ_m	σ_p	σ_p^+	σ_p^-
H	0	0	0	0
CH_3	-0.06 ± 0.03	-0.14 ± 0.03	-0.31	—
C_2H_5	-0.08 ± 0.12	-0.13 ± 0.09	-0.30	—
$t\text{-}C_4H_9$	-0.09 ± 0.11	-0.15 ± 0.07	-0.26	—
Ph	0.05 ± 0.1	0.02 ± 0.1	-0.18	0.08
CH_2Ph	-0.05 ± 0.18	-0.06 ± 0.13	-0.23	—
CF_3	0.46 ± 0.09	0.53 ± 0.11	—	0.7
CH_2Cl	0.11 ± 0.1	0.12 ± 0.1	-0.01	—
CHO	0.41 ± 0.13	0.47 ± 0.18	—	1.04
$COCH_3$	0.36 ± 0.07	0.47 ± 0.10	—	0.84
$CONH_2$	0.28 ± 0.1	0.31 ± 0.02	—	0.62
COOH	0.35 ± 0.18	0.44 ± 0.18	—	0.73
COO^-	0.02	0.11	—	0.34
COOR	0.35 ± 0.10	0.44 ± 0.09	—	0.74
CN	0.62 ± 0.05	0.70 ± 0.02	—	1.0
NH_2	-0.09 ± 0.05	-0.30 ± 0.11	-1.3	—
NMe_2	-0.10 ± 0.09	-0.32 ± 0.12	-1.7	—
NO_2	0.71 ± 0.04	0.81 ± 0.05	—	1.25
OH	0.02 ± 0.08	-0.22 ± 0.12	-0.92	—
OMe	0.10 ± 0.03	-0.28 ± 0.05	-0.78	—
SH	0.25 ± 0.1	0.15 ± 0.1	—	—
SMe	0.14 ± 0.18	0.06 ± 0.18	-0.60	0.04
SO_2Me	0.64 ± 0.1	0.73 ± 0.09	—	1.05
SO_3^-	0.3	0.35	—	0.5
F	0.34 ± 0.05	0.15 ± 0.06	-0.07	—
Cl	0.37 ± 0.03	0.24 ± 0.03	0.11	—
Br	0.37 ± 0.04	0.26 ± 0.04	0.15	—
I	0.34 ± 0.04	0.28 ± 0.04	0.13	—

Note: Data mostly from O. Exner, A Critical Compilation of Substituent Constants, in *Correlation Analysis in Chemistry* (N. B. Chapman and J. Shorter, Eds.), Plenum, New York, 1978, p. 439 ff.

Thus only one set of values of substituent constants σ_{m-X} is needed (Table 3-10). Nevertheless, in some reactions of para-substituted benzenoid compounds, standard values of σ_{p-X} do not sufficiently describe the resonance interaction between the reactive center and the substituent. When the *change* in resonance interactions between the initial and final (or transition) states differs significantly from the change in resonance interactions that occurs in the dissociation of benzoic acids, then a new set of values for σ_X must be used. For reactions in which either the initial or the final state has a reaction center which acts as a strong electron acceptor, the substituent effects of electron-donating substituents (such as OR, NR_2, or CH_3)

involve resonance contribution larger than reflected in values of σ_{p-X}. A special set of constants σ_{p-X}^{+} (Table 3-10) must be used, which were defined based on the rate constants for the S_N1 solvolysis of substituted phenyl dimethyl carbinyl chlorides [3-216, 3-217] using a value of ρ obtained for substituents for which direct resonance interaction with the reaction center seemed implausible.

Similarly, when a reaction center in either the initial or the final state can act as a strong electron-donor, then, in the presence of electron-withdrawing substituents (like NO_2, CHO, CH_3CO, CN) in para-position, a strong resonance interaction between the substituent and the reaction center necessitates the use of a new set of substituent constants, denoted as σ_{p-X}^{-} (Table 3-10). The tabulated values of constants are based on dissociation constants of substituted phenols or anilinium ions. In some reactions the resonance contributions may lay between those described by σ_{p-X} and σ_{p-X}^{+} or σ_{p-X} and σ_{p-X}^{-}, respectively.

Provided that a sufficiently broad range of substituents was investigated, the type of value of σ_X that provide the best fit with the data can be used for diagnosis of the type of mechanism involved in the reaction. The correlation with a particular type of constant σ_X indicates the nature of the resonance interactions in the studied reaction, which can lead to deductions regarding the structure of the transition state, and hence to a conclusion about the type of reaction mechanism. For instance, substituent effects in several electrophilic aromatic substitution reactions correlate well with values σ_{p-X}^{+}. This is an indication that these electrophilic substitutions proceed through a mechanism (3.9-9) where the electrophile attacks the benzene ring to form a benzenium ion (or σ-complex) in the rate-determining step:

$$(3.9\text{-}9)$$

A transition state resembling the benzenium ion would be expected to have the same enhanced π-resonance interactions for electron-donating substituents in para-position as are present in the model reaction, the solvolysis of phenyldimethylcarbinyl chlorides.

An example of another reaction, where rate constants can be correlated with constants σ_{p-X}^{+}, is the addition of bromine to substituted styrenes. This indicates that intermediates in this reaction are resonance stabilized carbonium ions rather than bromonium ions [3-218, but compare 3-219]. On the contrary, the rate of solvomercuration of substituted styrenes correlates well with σ_{p-X} constants, indicating the importance of formation of mercurinium ions [3-220].

3.9.4.2. Signs and Values of Reaction Constants

The value of the reaction constant ρ in the Hammett equation (3.9-8) is a measure of the susceptibility of a reaction to substituent electronic effects. When the ρ value is positive, the reaction is accelerated or the equilibrium constant is increased by electron-withdrawing groups. When the ρ value is negative, electron-donating groups aid the reaction. In diagnosing mechanism types, the sign of ρ thus indicates the direction of the change in charge at the reaction center. A positive ρ indicates the development of negative charge, as in nucleophilic substitution, and a negative ρ indicates the development of positive charge, as in electrophilic aromatic substitution.

Carbenes and carbenoids usually have an electrophilic character, but the addition of cycloheptatrienylidene to substituted styrenes shows a positive value of ρ [3-221]. This indicates a nucleophilic character of the carbene that is interpreted by the formation of a dipolar transition state (3.9-10):

$$(3.9\text{-}10)$$

The exceptional behavior of this carbene is attributed to the possibility of incorporating the empty 2p orbital into a stable tropylium system.

The absolute magnitude of ρ is a measure of the amount of charge development at the reaction center and the extent to which the substituents are able to interact with the developing charge. For example, the dissociation of benzoic acids is defined to have $\rho = 1.00$, but under the same conditions, the plot of logarithms of dissociation constants of phenylacetic acids against σ_X gives $\rho = 0.49$. Clearly, the same amount of charge is developed in these equilibrium reactions, but the CH_2 group intervening between the reaction center and the ring decreases the influence of the substituents. On the other hand, differences in the magnitude of ρ in various electrophilic aromatic substitutions have often been interpreted as reflecting differences in the amount of charge developed in the ring in the transition state and thus, differences in the extent of bond formation to the ring in the transition state. The ρ values in reactions where the ring is directly attacked are often large, for example, $\rho = -10.0$ for chlorination by Cl_2 in acetic acid at 25°C. In reactions with little polar character in which the change in the charge is small, such as those with concerted or cyclic mechanisms, the ρ values are typically quite small.

Values of reaction constants ρ can be particularly informative when substituent effects are compared involving reactions in which the nature of the interaction between the reactive center and substituent can be assumed to be similar. Thus it is possible to assume that substituent effects on benzoic acid and benzaldehydes

will be similar and hence values of ρ_R similar. Also for reactions of anions

similar values of reaction constants ρ can be expected. Alternatively, reaction constants for reactions of the uncharged molecules of the first pair can be considerably different from reaction constants for reaction of the anions of the second pair. Thus comparison of the value of the reaction constant of the studied reaction series with corresponding values for model reaction series can contribute to understanding of the charge and composition of the reactive species.

The nature of reactions compared can sometimes differ considerably and yet show a similar value of reaction constant ρ. For example, similar values of ρ were observed for electrochemical oxidation of benzaldehydes [3-222] and oxidation of substituted benzaldehydes by permanganate in alkaline media [3-223]. It has been concluded that this is due to the close resemblance of reactive intermediates whose similarity indicates that the corresponding transition states will also be similar. However, it is wise to be cautious in interpreting small differences in ρ values, as they can be quite sensitive to solvent and temperature effects.

3.9.4.3. Nonlinear log k_X vs. σ_X Plots

To be able to use a linear dependence between values of log k_X and σ_X for drawing conclusions about the structure of the transition state, it is required that the rate data correspond to a single-step reaction or that the reaction rate is governed by a single step in a mechanism. If the observed rate constants correspond to a reaction that includes one or more steps prior to the rate-determining step, the ρ value will be a composite of the ρ values for each step including the rate-determining step. For example for a reaction (3.9-11) with an equilibrium preceding the rate-determining step:

$$A + B \xrightleftharpoons{K_1} AB \xrightarrow{k_2} P \qquad (3.9\text{-}11)$$

the observed rate constant will be $k_{obs} = K_1 k_2$. In comparing rate constants for a substituted and an unsubstituted reactant:

$$\log \frac{(k_X)_{obs}}{(k_0)_{obs}} = \log \frac{(K_X)_1}{(K_0)_1} + \log \frac{(k_X)_2}{(k_0)_2} \qquad (3.9\text{-}12)$$

If both steps follow the Hammett equation,

$$\log \frac{(K_X)_1}{(K_0)_1} = \rho_1 \sigma_X \qquad (3.9\text{-}13)$$

and

$$\log \frac{(k_X)_2}{(k_0)_2} = \rho_2 \sigma_X \qquad (3.9\text{-}14)$$

and therefore

$$\log \frac{(k_X)_{obs}}{(k_0)_{obs}} = (\rho_1 + \rho_2)\sigma_X \qquad (3.9\text{-}15)$$

Thus the observed value $\rho = \rho_1 + \rho_2$ not only reflects the susceptibility of the slow step in reaction (3.9-11) but is a composite value, expressing the contribution of the susceptibility of the equilibrium to substituent effects.

When the rate determining step in reaction (3.9-11) varies with changes in the substituent X, a convex shape of the dependence of log k_X on σ_X is observed (Fig. 3-64a). For electron-donating substituents the rate of conversion of AB is rate

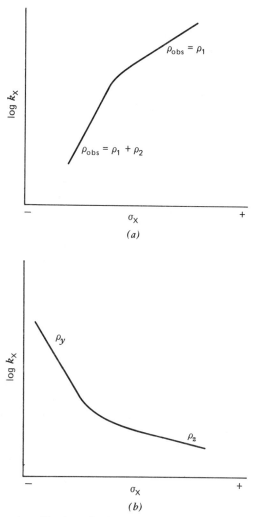

FIG. 3-64. Nonlinear plots of $\log k_X = f(\sigma_X)$. (*a*) Convex plot corresponding to reaction (3.9-11) in which the rate determining step changes with the change in substituent; (*b*) concave plot corresponding to two competitive reactions (3.9-16) and (3.9-17), for $\rho_Y > \rho_Z$.

determining and the slope of the plot is equal to ($\rho_1 + \rho_2$). For electron-withdrawing substituents the rate of the establishment of the equilibrium becomes dominant and the slope is equal to ρ_1. Since $\rho_1 < (\rho_1 + \rho_2)$, a convex curve results. An example of this type is the reaction of substituted benzaldehydes with butylamine [3-224], where the reactants are in equilibrium with carbinolamine (AB), which undergoes subsequent dehydration.

The concave shape of the $\log k_X$–σ_X plot (Fig. 3-64*b*) indicates the presence of two competing mechanisms, whose contributions change gradually with variation in the substituent X. If a substrate (S) can be converted into the same product (P) by two pathways with rate constants k^Y and k^Z and both reactions (3.9-16 and 3.9-17) follow the Hammett equations (3.9-18 and 3.9-19), a concave shape in the plot (Fig. 3-64b) results.

$$S \xrightarrow{k^Y} P \tag{3.9-16}$$

$$S \xrightarrow{k^Z} P \tag{3.9-17}$$

$$\log \frac{(k^Y)_X}{(k^Y)_0} = \rho_Y \sigma_X \tag{3.9-18}$$

$$\log \frac{(k^Z)_X}{(k^Z)_0} = \rho_Z \sigma_X \tag{3.9-19}$$

For electron-donating substituents, reaction with rate constant k^Y predominates, for which $\rho_Y > \rho_Z$; for electron-withdrawing substituents, reaction with rate constant k^Z predominates. For example, in acetolysis of threo-3-aryl-2-butyltosylates [3-225] derivatives with strongly electron-withdrawing substituents (such as NO_2 and CF_3) follow predominantly a solvent assisted solvolysis, whereas those with a more electron-donating ones (H, CH_3, OCH_3) follow also a competitive solvolysis via phenonium ions, which shows a much larger value of ρ_Y. The latter process occurs with retention of the threo-configuration, whereas the former yields inverted products. Comparison of yields of both epimers confirmed the presence of two independent reaction paths. Similar changes of the predominating reaction path with the nature of the substituent have been observed for the hydrolyses of syn-7-aryl-anti-norbornenyl-4-nitrobenzoate [3-226].

Change not only of the value but of the sign of the constant ρ are observed in those 1,3-dipolar additions [3-227] where the dipole and dipolarophile have approximately equal separation of highest occupied (HOMO) and lowest unoccupied (LUMO) molecular orbitals. Such behavior is interpreted as due to a dual mechanism participation, where one type of interaction (e.g., $HOMO_{dipole}$-$LUMO_{dipolarophile}$) in the transition state predominates in one part of the reaction series, whereas another one ($HOMO_{dipolarophile}$-$LUMO_{dipole}$) predominates in the other part.

Great caution is recommended in interpreting nonlinear $\log k_X$-σ_X plots. It is essential to distinguish this nonlinearity from normal scatter due to experimental error. In addition to the type of mechanistic changes described above, systematic curvature may also result from incorrect values of rate constants due to failure in detection of reactive or catalytic impurities, side reactions, or isolation of a single reaction. Further, this type of curvature must be distinguished from nonlinearity due to an improper choice of the type of constants σ_X, or inadequacy of the constants σ_X to properly describe the balance of resonance and polar effects. When correlation of $\log k_X$ with σ_X yields a nonlinear plot, testing correlations with σ_X^+ and σ_X^- is essential.

3.9.4.4. Isolated Points Deviating from Linear log k_X–σ_X Plot

Deviation of a value of log k_T for a substituent T from a linear log k_X–σ_X plot can be caused by an experimental error, mistaken compound, or a mistake in data handling. Alternatively, the deviation can be caused by an extraordinary property of the substituent T, not reflected in the value of σ_T. This may be caused by protonation of a substituent at low pH, formation of a complex with a Lewis acid, an interaction of the substituent with another, adjacent substituent (which can result, e.g., in steric hindrance of coplanarity), or a strong interaction of the substituent with the solvent used (often observed for substituents bearing unit charge). In all such cases the same mechanism applies for the compound R–A–T with substituent T as for other compounds belonging to the same reaction series R–A–X.

When the above possibilities are excluded, a deviation of the value of log k_T from the linear plot can be due to a change in mechanism or in a given mechanism due to a change in the rate determining step. Hence for a deviating value of log k_T, for which there is no reason to expect an experimental error and where the nature of the substituent T does not indicate the possibility of an exceptional substituent effect, the validity of the reaction rate equation, confirmed for other members of the reaction series, should be verified, as well as the nature of the products and intermediates. Deviation from the linear log k_X–σ_X plot can, in this way, provide more detailed information about the mechanism involved and ranges of its validity. If there is any question dealing with the interpretation of substituent effects, the widest range of substituents available should be investigated.

3.9.4.5. Substituents in Other Systems

Substituent constants σ_X have been successfully applied to correlations with rate and equilibrium constants in polycyclic aromatic systems, five-membered heterocycles and ethylenic compounds. For some other heterocyclic systems special sets of substituent constants were derived following the same principles used in derivation of the Hammett equation.

For treatment of substituent effects in aliphatic and alicyclic compounds, inductive substituent constants (σ_X^I) were defined by means of substituent effects on dissociation constants of 4-substituted bicyclo[2.2.2]octane-1-carboxylic acids, and polar substituent constants (σ_X^*) based on rate constants of acid and base catalyzed ester hydrolysis [3-97–3-103, 3-191]. An example of the application of the latter is the Taft equation (3.9-20)

$$\log \frac{k_X}{k_0} = \rho^* \sum \sigma_X^* \tag{3.9-20}$$

successfully used in a mechanistic study for the treatment of solvolysis of secondary tosylates. For the relatively slow solvolysis of tosylates derived from isopropyl

(3.9-21), it was assumed [3-228] that $k_{-1} \gg k_2$,

$$H-\underset{\underset{CH_3}{|}}{\overset{\overset{CH_3}{|}}{C}}-Tos \underset{k_{-1}}{\overset{k_1}{\rightleftharpoons}} H-\underset{\underset{CH_3}{|}}{\overset{\overset{CH_3}{|}}{C^+}} \; Tos^- \xrightarrow{k_2} Products \qquad (3.9\text{-}21)$$

whereas for rapidly solvolyzed pinacolyl derivatives (3.9-22) a rapid methyl migration resulting in $k_2 \gg k_{-1}$ was assumed.

$$H_3C-\underset{\underset{CH_3}{|}}{\overset{\overset{CH_3}{|}}{C}}-\underset{\underset{CH_3}{|}}{\overset{\overset{H}{|}}{C}}-Tos \underset{k_{-1}}{\overset{k_1}{\rightleftharpoons}} CH_3\underset{\underset{CH_3}{|}}{\overset{\overset{CH_3}{|}}{C}}-\underset{\underset{CH_3}{|}}{\overset{\overset{H}{|}}{C^+}} \; Tos^- \xrightarrow{k_2}$$

$$CH_3\underset{\underset{CH_3}{|}}{\overset{\overset{CH_3}{|}}{\underset{+}{C}}}-\underset{\overset{H}{|}}{C}-CH_3 \longrightarrow Products \quad (3.9\text{-}22)$$

Using eq. (3.9-20), the linear plot of $\log k_X$ vs. $\Sigma\sigma_X^*$ indicated that the increase in the rate of solvolysis for the pinacolyl derivative is due solely to the polar (inductive) effects of alkyl groups and that all reactions follow the same mechanism (3.9-21).

Nevertheless, because of additional assumptions, the application of linear free energy relationships to aliphatic and alicyclic compounds should be carried out with greater reservations than the treatment of aromatic systems.

Finally, linear free energy relationships play an important role in the planning of a mechanistic study, enabling us to choose substituents which differ widely in their effects and cover the broadest possible variations in reactivity.

3.9.4.6. Isotope Effects

Isotope effects represent a special case of either substituent effects or of an exchange of the reaction center. They involve special precautions, methodology, and assumptions in data treatment. Therefore they will not be discussed here and the interested reader should consult specialized treatments [3-191, 3-192, 3-229–3-331].

REFERENCES

[3-1]   R. M. Krupka, H. Kaplan, and K. J. Laidler, *Trans. Faraday Soc.* **62**, 2754 (1966).

[3-2]   T. M. Barbara and P. L. Corio, *J. Chem. Ed.* **57**, 243 (1980).

[3-3]   J. F. Corbett, *J. Chem. Ed.* **49**, 663 (1972).

[3-4]   C. W. Pyun and I. Lipschitz, *J. Chem. Ed.* **53**, 293 (1976).

[3-5]   R. E. Powel, quoted in J. W. Moore and R. G. Pearson, *Kinetics and Mechanism*, 3rd ed., Wiley, New York, 1981, p. 20.

[3-6]   K. J. Hall, T. I. Quickenden, and D. W. Watts, *J. Chem. Ed.* **53**, 493 (1976).

[3-7]   J. H. Flynn, *J. Phys. Chem.* **60**, 1332 (1956).

[3-8]   P. Čársky, P. Zuman, and V. Horák, *Collect. Czechoslov. Chem. Commun.* **30**, 4316 (1965).

[3-9]   P. Zuman, G. Fodor, and V. Horák, quoted in P. Zuman, Polarography and Reaction Kinetics, *Adv. Phys. Org. Chem.* (V. Gold, Ed.) **5**, 1 (1967).

[3-10] J. W. Moore and R. G. Pearson, *Kinetics and Mechanism*, 3rd ed., Wiley, New York, 1981, p. 291.

[3-11] R. K. Boyd, *J. Chem. Ed.* **55**, 85 (1978).

[3-12] O. K. Rice, *J. Phys. Chem.* **64**, 1851 (1960).

[3-13] J. C. Giddings and H. K. Shin, *Trans. Faraday Soc.* **57**, 468 (1961).

[3-14] L. Volk, W. Richardson, K. H. Lau, M. Hall, and S. H. Lin, *J. Chem. Ed.* **54**, 95 (1977).

[3-15] A. E. R. Westman and D. B. DeLury, *Canad. J. Chem.* **34**, 1134 (1956).

[3-16] A. A. Frost and W. C. Schwemer, *J. Am. Chem. Soc.* **73**, 4541 (1951); **74**, 1268 (1952).

[3-17] W. G. McMillan, *J. Am. Chem. Soc.* **79**, 4838 (1957).

[3-18] H. G. Higgins and E. J. Williams, *Austr. J. Sci. Res.* **5A**, 572 (1952).

[3-19] N. V. Riggs, *Austr. J. Chem.* **11**, 86 (1958).

[3-20] W. Y. Wen, *J. Phys. Chem.* **76**, 704 (1972).

[3-21] W. J. Svirbely and H. E. Weisberg, *J. Am. Chem. Soc.* **81**, 257 (1959).

[3-22] W. J. Svirbely and J. A. Blauer, *J. Am. Chem. Soc.* **83**, 4115, 4118 (1961).

[3-23] M. Kubín and L. Zikmund, *Collect. Czechoslov. Chem. Commun.* **34**, 1254 (1969).

[3-24] M. Kubín, S. Ševčík, J. Štamberg, and P. Špaček, *Collect. Czechoslov. Chem. Commun.* **39**, 2591 (1974).

[3-25] P. Zuman and J. Krupička, *Collect. Czechoslov. Chem. Commun.* **23**, 598 (1958).

[3-26] R. D. Brown and B. A. W. Coller, *Austr. J. Chem.* **11**, 90 (1958).

[3-27] Ch. T. Chen, *J. Phys. Chem.* **62**, 639 (1958).

[3-28] H. G. Higgins and E. J. Williams, *Austr. J. Chem.* **6**, 195 (1953).

[3-29] H. G. Higgins, *Austr. J. Chem.* **10**, 99 (1957).

[3-30] T. M. Lowry and W. T. John, *J. Chem. Soc.* **97**, 2634 (1910).

[3-31] E. McLauglin and R. W. Rozett, *J. Chem. Ed.* **49**, 482 (1972).

[3-32] D. McDaniel and C. R. Smoot, *J. Phys. Chem.* **60**, 966 (1956).

[3-33] B. G. Gowenlock and K. A. Redish, *Z. Phys. Chem.*, N.F. **31**, 169 (1962).

[3-34] G. J. Buist and C. A. Bunton, *J. Chem. Soc.* **1954**, 1406.

[3-35] P. Zuman, J. Sicher, J. Krupička, and M. Svoboda, *Collect. Czechoslov. Chem.* **23**, 1237 (1958).

[3-36] G. J. Buist, C. A. Bunton, and J. Lomas, *J. Chem. Soc.* **B 1966**, 1094, 1099; G. J. Buist and J. D. Lewis, *J. Chem. Soc.* **B 1968**, 90; G. J. Buist and C. A. Bunton, *J. Chem. Soc.* **B 1971**, 2117; G. J. Buist, C. A. Bunton, and W. C. P. Hipperson, *J. Chem. Soc.* **B 1971**, 2128.

[3-37] J. I. Seeman and W. A. Farone, *J. Org. Chem.* **43**, 1854 (1978).

[3-38] J. H. Flynn, *J. Phys. Chem.* **61**, 110 (1956).

[3-39] J. P. Birk, *J. Chem. Ed.* **53**, 704 (1976).

[3-40] G. M. Machwart and R. E. Quilici, *Ind. Eng. Chem.* **48**, 1194 (1956).

[3-41] E. A. Guggenheim, *Phil. Mag.* **2**, 538 (1926).

[3-42] F. J. Kezdy, J. Jaz, and A. Bruylants, *Bull. Soc. Chim. Belg.* **67**, 687 (1958).

[3-43] P. C. Mangelsdorf, Jr., *J. Appl. Phys.* **30**, 442 (1959).

[3-44] E. S. Swinbourne, *J. Chem. Soc.* **1960**, 2371.

[3-45] L. M. Schwartz and R. I. Gelb, *Anal. Chem.* **50**, 1592 (1978).

[3-46] W. E. Wentworth, *J. Chem. Ed.* **42**, 96, 162 (1965).

[3-47] P. Moore, *J. Chem. Soc. Faraday Trans.* **68**, 1890 (1972).

[3-48] M. J. J. Holt and A. C. Norris, *J. Chem. Ed.* **54**, 426 (1977).

[3-49] L. M. Schwartz, *Anal. Chem.* **53**, 206 (1981).

[3-50] W. E. Roseveare, *J. Am. Chem. Soc.* **53**, 1651 (1931).

[3-51] S. W. Tobey, *J. Chem. Ed.* **39**, 473 (1962).

[3-52] J. H. Espenson, *J. Chem. Ed.* **57**, 160 (1980).

[3-53] L. M. Schwartz, *J. Chem. Ed.* **58**, 588 (1981).

[3-54] Computer Series, J. W. Moore (Ed.), *J. Chem. Ed.,* e.g., **59**, 409 (1982).

[3-55] G. M. Fleck, *Chemical Reaction Mechanisms,* Holt, Rinehart & Winston, New York, 1971.

[3-56] J. Casanova, Jr. and E. R. Weaver, *J. Chem. Ed.* **42**, 137 (1965).

[3-57] D. O. Jones, M. D. Scamuffa, L. S. Portnoff, and S. Perone, *J. Chem. Ed.* **49**, 717 (1972).

[3-58] J. L. Hogg, *J. Chem. Ed.* **51**, 109 (1974).

[3-59] R. H. Schuler, *J. Chem. Ed.* **52**, 166 (1975).

[3-60] D. Wolf and R. D. Williams, *J. Chem. Ed.* **51**, 319 (1974).

[3-61] J. M. Anderson, *J. Chem. Ed.* **53**, 561 (1976).

[3-62] D. M. Shindell, C. Magagnosc, and D. L. Purich, *J. Chem. Ed.* **55**, 708 (1978).

[3-63] D. L. Langhus and G. S. Wilson, *Anal. Chem.* **51**, 1134 (1979).

[3-64] J. Janata, Analog Computer Simulation of Kinetic Models, in *Spectroscopy and Kinetics* (J. S. Mattson, H. B. Mark, Jr., and H. C. MacDonald, Jr., Eds.), Dekker, New York, 1973.

[3-65] E. Hamori, *J. Chem. Ed.* **49**, 39 (1972).

[3-66] P. R. Nott and B. K. Selinger, *J. Chem. Ed.* **49**, 618 (1972).

[3-67] C. K. Chang, *Appl. Spectroscop.* **30**, 364 (1976).

[3-68] G. G. Giachino, *J. Chem. Ed.* **55**, 201 (1978).

[3-69] D. A. Davenport, *J. Chem. Ed.* **52**, 379 (1975).

[3-70] J. P. Birk and S. K. Gunter, *J. Chem. Ed.* **54**, 557 (1977).

[3-71] R. Zahradník and P. Zuman, *Collect. Czechoslov. Chem. Commun.* **24**, 1132 (1959).

[3-72] R. Zahradník, *Collect. Czechoslov. Chem. Commun.* **23**, 1529 (1958).

[3-73] P. Čársky, P. Zuman, and V. Horák, *Collect. Czechoslov. Chem. Commun.* **29**, 3044 (1964).

[3-74] F. Hibbert and D. P. N. Satchel, *Chem. Commun.* **1966**, 516.

[3-75] T. J. Taylor, B. Soldano, and G. A. Hall, *J. Am. Chem. Soc.* **77**, 2656 (1955).

[3-76] P. Zuman and O. Manoušek, *Collect. Czechoslov. Chem. Commun.* **26**, 2134 (1961).

[3-77] P. Zuman, *Collect. Czechoslov. Chem. Commun.* **32**, 1610 (1967).

[3-78] P. D. Bartlett and J. D. McCollum, *J. Am. Chem. Soc.* **78**, 1441 (1956).

[3-79] R. E. Weston, Jr., S. Ehrenson, and K. Heinzinger, *J. Am. Chem. Soc.* **89**, 482 (1967).

[3-80] P. Jandík, L. Meites, and P. Zuman, *J. Phys. Chem.* **87**, 238 (1983).

[3-81] J. Roček and J. Krupička, *Collect. Czechoslov. Chem. Commun.* **23**, 2068 (1958).

[3-82] O. Manoušek and P. Zuman, *Experientia* **20**, 301 (1964).

[3-83] W. P. Jencks, *J. Am. Chem. Soc.* **81**, 475 (1959).

[3-84] W. P. Jencks, *Catalysis in Chemistry and Enzymology,* McGraw Hill, New York, 1969, p. 467ff.

[3-85]   R. P. Bell, *The Proton in Chemistry,* Cornell University Press, Ithaca, New York, 2nd ed., 1973, p. 172.

[3-86]   M. D. Waddington and J. E. Meany, *J. Chem. Ed.* **55,** 60 (1978).

[3-87]   A. C. Knipe, *J. Chem. Ed.* **53,** 618 (1976).

[3-88]   K. Bowden, M. J. Hanson, and G. R. Taylor, *J. Chem. Soc.* **B 1968,** 174.

[3-89]   A. Ledwith and H. J. Woods, *J. Chem. Soc.* **B 1966,** 753.

[3-90]   W. Ostwald, *Chemische Betrachtungen,* Die Aula (1895), No. 1.

[3-91]   R. P. Bell, *Acid–Base Catalysis,* Oxford Univ. Press, Oxford, 1941.

[3-92]   J. N. Brönsted and K. J. Pedersen, *Z. phys. Chem.* **108,** 185 (1924).

[3-93]   J. N. Brönsted, *Chem. Rev.* **5,** 322 (1928).

[3-94]   Ref. 3-85, p. 194ff.

[3-95]   Ref. 3-84, p. 170ff.

[3-96]   L. P. Hammett, *Chem. Rev.* **17,** 125 (1935).

[3-97]   R. W. Taft, Jr., Separation of Polar, Steric and Resonance Effects in Reactivity, in *Steric Effects in Organic Chemistry* (M. S. Newman, Ed.), Wiley, New York, 1956, p. 556ff.

[3-98]   J. E. Leffler and E. Grunwald, *Rates and Equilibria of Organic Reactions,* Wiley, New York, 1963.

[3-99]   P. R. Wells, *Linear Free Energy Relationships,* Academic, London, 1968.

[3-100] N. B. Chapman and J. Shorter (Eds.), *Advances in Linear Free Energy Relationships,* Plenum, New York, 1972.

[3-101] J. Shorter, *Correlation Analysis in Organic Chemistry,* Clarendon Press, Oxford, 1973.

[3-102] J. Hine, *Structural Effects on Equilibria in Organic Chemistry,* Wiley, New York, 1975, p. 55ff.

[3-103] N. B. Chapman and J. Shorter, *Correlation Analysis in Chemistry,* Plenum, New York, 1978.

[3-104] Ref. 3-85, p. 160ff.

[3-104a] R. A. More O'Ferrall, *J. Chem. Soc.* **B 1970,** 274; W. P. Jencks, *Chem. Rev.* **72,** 705 (1972); D. A. Jencks and W. P. Jencks, *J. Am. Chem. Soc.* **99,** 7948 (1977).

[3-105] Ref. 3-84, p. 199ff.

[3-106] Ref. 3-85, p. 188–189.

[3-107] Ref. 3-84, p. 516–517.

[3-108] Ref. 3-85, p. 168ff.

[3-109] M. Fedoroňko and P. Zuman, *Collect. Czechoslov. Chem. Commun.* **29,** 2115 (1964).

[3-110] S. Ewing, *J. Chem. Ed.* **59,** 606 (1982).

[3-111] R. S. Bulmer, E. Senogles, and R. A. Thomas, *J. Chem. Ed.* **58,** 738 (1981).

[3-112] A. Lombardo, *J. Chem. Ed.* **59,** 887 (1982).

[3-113] M. Eigen, *Angew. Chem.* **75,** 489 (1963); *Internat. Ed.* **3,** 1 (1964).

[3-114] W. P. Jencks, *Acc. Chem. Res.* **13,** 161 (1980).

[3-115] S. R. Logan, *J. Chem. Ed.* **59,** 279 (1982).

[3-116] C. Reichardt, *Solvent Effects in Organic Chemistry,* Verlag Chemie, Weinheim (W. Germany), 1979.

[3-117] M. R. J. Dack (Ed.), *Solutions and Solubilities,* Vol VIII, Parts I and II, *Techniques in Chemistry* (A. Weissberger, Ed.) Wiley, New York, 1975 and 1976.

[3-118] C. W. Davies, Salt Effects in Solution Kinetics, in *Progress in Reaction Kinetics* (G. Porter, Ed.), Vol I, Pergamon Press, Oxford, 1961, p. 161ff.

[3-119] J. N. Brønsted, *Z. phys. Chem.* **102,** 169 (1922).

[3-120] N. Bjerrum, *Z. phys. Chem.* **108**, 82 (1924).

[3-121] G. Scatchard, *Chem. Revs.* **10**, 229 (1932).

[3-122] K. J. Laidler and H. Erving, *Ann. N.Y. Acad. Sci.* **39**, 303 (1940).

[3-123] G. Corsaro, *J. Chem. Ed.* **54**, 483 (1977).

[3-124] E. S. Amis and V. K. LaMer, *J. Am. Chem. Soc.* **61**, 905 (1939).

[3-125] K. J. Laidler and P. A. Landskroener, *Trans. Faraday Soc.* **52**, 200 (1956).

[3-126] E. S. Amis, *J. Chem. Ed.* **29**, 337 (1952); **30**, 351 (1953); *Anal. Chem.* **27**, 1672 (1955).

[3-127] J. E. Quinlan and E. S. Amis, *J. Am. Chem. Soc.* **77**, 4187 (1955).

[3-128] E. S. Amis, *Solvent Effects on Reaction Rates and Mechanisms.* Academic, New York, 1966.

[3-129] J. G. Kirkwood, *J. Chem. Phys.* **2**, 351 (1934).

[3-130] S. Winstein and A. H. Fainberg, *J. Am. Chem. Soc.* **78**, 2770 (1956); **79**, 5937 (1957).

[3-131] Ref. 3-116, p. 134ff.

[3-132] M. R. J. Dack, *J. Chem. Ed.* **51**, 231 (1974); *Chem. Soc. Rev.* **4**, 211 (1975); *Austr. J. Chem.* **28**, 1643 (1975).

[3-133] V. Gutmann, *Coordination Chem. Rev.* **2**, 239 (1967); **18**, 225 (1976); *Chimia* **23**, 285 (1969); *Electrochimica Acta* **21**, 661 (1976).

[3-134] V. Gutmann, *The Donor–Acceptor Approach to Molecular Interactions.* Plenum, New York, 1978.

[3-135] V. Gutmann, *Pure Appl. Chem.* **27**, 73 (1971); *Fortschr. Chem. Forschg.* **27**, 59 (1972); *Structure and Bonding* **15**, 141 (1973); *Chimia* **31**, 1 (1977).

[3-136] K. H. Meyer, *Ber.* **45**, 2843 (1912); **47**, 826 (1914).

[3-137] P. Zuman, paper presented at IVth Internat. Conf. on Non-Aqueous Solutions, Vienna, 1974.

[3-138] J. Segretario, Ph.D. Thesis, Clarkson College of Technology, Potsdam, New York, 1982.

[3-139] E. L. Eliel and O. Hofer, *J. Am. Chem. Soc.* **95**, 8041 (1973).

[3-139a] A. J. Parker, *Chem. Rev.* **69**, 1 (1969); O. Popovych, *Crit. Rev. Anal. Chem.* **1**, 73 (1970).

[3-139b] M. H. Abraham, *Progr. Phys. Org. Chem.* **11**, 1 (1974); B. G. Cox and A. J. Parker, *J. Am. Chem. Soc.* **95**, 402 (1973).

[3-139c] A. J. Parker, V. Mayer, R. Schmid, and V. Gutmann, *J. Org. Chem.* **43**, 1843 (1978).

[3-140] E. Grunwald and S. Winstein, *J. Am. Chem. Soc.* **70**, 846 (1948).

[3-141] S. Winstein et al., *J. Am. Chem. Soc.* **73**, 2700 (1951); **78**, 2770 (1956); **79**, 1597, 1602, 1608, 4146, 5937 (1957); **83**, 618 (1961).

[3-142] Ref. 3-116, p. 233.

[3-143] S. Winstein, E. Grunwald, and H. W. Jones, *J. Am. Chem. Soc.* **73**, 2700 (1951).

[3-144] T. W. Bentley, F. L. Schadt, and P. R. Schleyer, *J. Am. Chem. Soc.* **94**, 992 (1972); **98**, 7667 (1976).

[3-145] S. G. Smith, A. H. Fainberg, and S. Winstein, *J. Am. Chem. Soc.* **83**, 618 (1961).

[3-146] Y. Drougard and D. Decroocq, *Bull. Soc. Chim. Fr.* **1969**, 2972.

[3-147] M. Gielen and J. Nasielski, *J. Organomet. Chem.* **1**, 173 (1963); **7**, 273 (1967).

[3-148] N. S. Bayliss and E. G. McRae, *J. phys. Chem.* **58**, 1002, 1006 (1954).

[3-149] P. Scheibe, S. Schneider, F. Dorr, and E. Daltrozzo, *Ber. Bunsenges. Phys. Chem.* **80**, 130 (1976).

[3-150] R. Radeglia, *Z. Chem.* **15**, 355 (1975).

[3-151] M. Wähnert and S. Dähne, *J. Prakt. Chem.* **318**, 321 (1976).

[3-152] E. M. Kosower, *An Introduction to Physical Organic Chemistry.* Wiley, New York, 1968, p. 293ff.

[3-152a] K. Dimroth, C. Reichardt, T. Siepmann, and F. Bohlmann, *Liebigs Ann. Chem.* **661**, 1 (1963).

[3-153] E. M. Kosower and M. Mohammad, *J. Am. Chem. Soc.* **90**, 3271 (1968); **93**, 2713 (1971); *J. Phys. Chem.* **74**, 1153 (1970).

[3-154] S. Dähne, F. Shob, K. D. Nolte, and R. Radeglia, *Ukr. Khim. Zh.* **41**, 1170 (1975); *C. A.* **84**, 43086j (1976).

[3-155] L. G. S. Brooker, A. C. Craig, D. W. Heseltine, P. W. Jenkins, and L. L. Lincoln, *J. Am. Chem. Soc.* **87**, 2443 (1965).

[3-156] W. Walter and D. H. Bauer, *Liebigs, Ann. Chem.* **1977**, 421.

[3-157] S. Brownstein, *Canad. J. Chem.* **38**, 1590 (1960).

[3-158] J.-E. Dubois and A. Bienvenüe, *J. Chim. Phys.* **65**, 1259 (1968).

[3-159] I. A. Zhmyreva, V. V. Zelinskii, V. P. Kolobkov, and N. D. Krasnitskaya, *Dokl. Akad. Nauk USSR, Ser. Khim.* **129**, 1089 (1959).

[3-160] A. Allerhand and P. v. R. Schleyer, *J. Am. Chem. Soc.* **85**, 371 (1963); **86**, 5709 (1964).

[3-161] I. A. Koppel and V. A. Palm in ref. 3-100, p. 203ff.

[3-162] A. G. Burden, G. Collier, and J. Shorter, *J. Chem. Soc., Perkin Trans. II*, **1976**, 1627.

[3-163] I. A. Koppel and A. I. Paju, *Org. Reactivity* (*Russ.*) **11**, 121 (1974).

[3-164] B. R. Knauer and J. J. Napier, *J. Am. Chem. Soc.* **98**, 4395 (1976).

[3-165] R. T. C. Brownlee, S. K. Dayal, J. L. Lyle, and R. W. Taft, *J. Am. Chem. Soc.* **94**, 7208 (1972).

[3-166] R. E. Uschold and R. W. Taft, *Org. Mag. Resonance* **1**, 375 (1969).

[3-167] M. J. Kamlet, J. L. M. Abboud, and R. W. Taft, *Progr. Phys. Org. Chem.* **13**, 485 (1980) and references therein.

[3-168] M. J. Kamlet, C. Dickinson, T. Gramstad, and R. W. Taft, *J. Org. Chem.* **47**, 4971 (1982).

[3-169] K. Kalyanasundaram and J. K. Thomas, *J. Am. Chem. Soc.* **99**, 2039 (1977).

[3-170] B. Chawla, S. K. Pollack, C. B. Lebrilla, M. J. Kamlet, and R. W. Taft, *J. Am. Chem. Soc.* **103**, 6924 (1981).

[3-171] R. W. Taft, T. Gramstad, and M. J. Kamlet, *J. Org. Chem.* **47**, 4557 (1982).

[3-172] R. W. Taft, N. J. Pienta, M. J. Kamlet, and E. M. Arnett, *J. Org. Chem.* **46**, 661 (1981).

[3-173] R. W. Taft, J. L. M. Abboud, and M. J. Kamlet, *J. Am. Chem. Soc.* **103**, 1080 (1981).

[3-174] D. J. Raber, R. C. Bingham, J. M. Harris, J. L. Fry, and P. v. R. Schleyer, *J. Am. Chem. Soc.* **92**, 5977 (1970).

[3-175] D. N. Kevill, K. C. Kolwyck, and F. L. Weitl, *J. Am. Chem. Soc.* **92**, 7300 (1970).

[3-176] T. W. Bentley and P. v. R. Schleyer, *J. Am. Chem. Soc.* **98**, 7658 (1976).

[3-177] C. K. Ingold, *Structure and Mechanism in Organic Chemistry*, 2nd ed., Cornell Univ. Press, Ithaca, New York, 1969.

[3-178] E. D. Hughes and C. K. Ingold, *J. Chem. Soc.* **1935**, 244; *Trans. Faraday Soc.* **37**, 603, 657 (1941).

[3-179] M. H. Abraham, *J. Chem. Soc., Perkin Trans. II*, **1972**, 1343.

[3-180] H. v. Halban, *Z. Physikal. Chem.* **67**, 129 (1909).

[3-181] E. R. Swart and L. J. LeRoux, *J. Chem. Soc.* **1956**, 2110; **1957**, 406.

[3-182] E. D. Hughes and D. J. Whittingham, *J. Chem. Soc.* **1960**, 806.

[3-183] J. L. Gleave, E. D. Hughes, and C. K. Ingold, *J. Chem. Soc.* **1935**, 236.

[3-184] G. Steiner and R. Huisgen, *J. Am. Chem. Soc.* **95**, 5054, 5055, 5056 (1973); *Tetrahedron Lett.* **1973**, 3763, 3769.

[3-185] J. E. Dubois and F. Garnier, *Bull. Soc. Chim. Fr.* **1968**, 3797.

[3-186] M.-F. Ruasse and J. E. Dubois, *J. Am. Chem. Soc.* **97**, 1977 (1975).

[3-187] C. J. O'Connor, E. J. Fendler, and J. H. Fendler, *J. Chem. Soc., Perkin Trans. II*, **1973**, 1900.

[3-188] L. P. Hammett: *Physical Organic Chemistry*, McGraw Hill, New York, 1940; 2nd ed., 1970.

[3-189] E. S. Gould, *Mechanism and Structure in Organic Chemistry*, H. Holt and Co., New York, 1959.

[3-190] J. Hine, *Structural Effects on Equilibria in Organic Chemistry*, Wiley, New York, 1975.

[3-191] G. W. Klumpp, *Reactivity in Organic Chemistry*, Wiley, New York, 1982.

[3-192] K. Schwetlick, *Kinetische Methoden zur Untersuchung von Reaktionsmechanismen*. VEB Verlag der Wissenschaften, Berlin, 1971.

[3-193] J. A. Hirsch, *Concepts in Theoretical Organic Chemistry*, Allyn & Bacon, Boston, 1974.

[3-194] L. N. Ferguson, *Organic Molecular Structure*, Willard Grant Press, Boston, 1975.

[3-195] J. M. Harris and C. C. Wamser, *Fundamentals of Organic Reaction Mechanisms*. Wiley, New York, 1976.

[3-196] T. H. Lowry and K. S. Richardson, *Mechanisms and Theory in Organic Chemistry*, 2nd ed., Harper & Row, New York, 1981.

[3-197] K. S. Connors, *Reaction Mechanisms in Organic Analytical Chemistry*, Wiley, New York, 1973.

[3-198] J. Hine, *Physical Organic Chemistry*, McGraw-Hill, New York, 1956.

[3-199] J. March, *Advanced Organic Chemistry*, McGraw-Hill, New York, 1968.

[3-200] J. F. Bunnett, Nucleophilic Reactivity, *Ann. Rev. Phys. Chem.* **14,** 271 (1963).

[3-201] G. Klopman, *Chemical Reactivity and Reaction Paths*, Wiley, New York, 1974, p. 55.

[3-202] Ref. 3-191, p. 153ff.

[3-203] J. Hine and R. D. Weimar, *J. Am. Chem. Soc.* **87,** 3387 (1970).

[3-204] W. P. Jencks and M. Gilchrist, *J. Am. Chem. Soc.* **90,** 2622 (1968).

[3-205] R. F. Hudson and G. Loveday, *J. Chem. Soc.* **1962,** 1068.

[3-206] C. G. Swain and C. B. Scott, *J. Am. Chem. Soc.* **75,** 141 (1953).

[3-207] C. D. Ritchie and P. O. I. Virtanen, *J. Am. Chem. Soc.* **95,** 1882 (1973).

[3-208] C. D. Ritchie and M. Sawada, *J. Am. Chem. Soc.* **99,** 3754 (1977).

[3-209] K. Hillier, J. M. W. Scott, D. J. Barnes, and F. J. P. Steele, *Canad. J. Chem.* **54,** 3312 (1976).

[3-210] P. O. I. Virtanen and R. Korhonen, *Acta Chem. Scand.* **27,** 2650 (1973).

[3-211] S. Hoz and D. Speizman, *Tetrahedron Lett.* **1978,** 1775.

[3-212] A. Pross, *J. Am. Chem. Soc.* **98,** 776 (1976).

[3-213] M. J. S. Dewar, *The Molecular Orbital Theory of Organic Chemistry*, McGraw-Hill, New York, 1969, p. 214.

[3-214] A. Streitwieser, *Molecular Orbital Theory for Organic Chemists*, Wiley, New York, 1969, p. 433.

[3-215] E. Berliner, *Progr. Phys. Org. Chem.* **2,** 253 (1964).

[3-216] H. C. Brown and Y. Okamoto, *J. Am. Chem. Soc.* **80,** 4979 (1958).

[3-217] A. J. Hoefnagel and B. M. Webster, *J. Am. Chem. Soc.* **95,** 5357 (1973).

[3-218] J. A. Pincock and K. Yates, *Canad. J. Chem.* **48,** 2944 (1970).

[3-219] J. E. Dubois, M. F. Ruasse, and A. Argile, *Tetrahedron Lett.* **1978,** 177.

[3-220] G. Müller-Hagen and W. Pritzkow, *J. Prakt. Chem.* **311,** 874 (1969).

[3-221] L. W. Christensen, E. E. Waali, and W. M. Jones, *J. Am. Chem. Soc.* **94,** 2118 (1972).

[3-222] W. J. Bover and P. Zuman, *J. Electrochem. Soc.* **122,** 368 (1975).

[3-223] K. B. Wiberg and R. Stewart, *J. Am. Chem. Soc.* **77,** 1786 (1955).

[3-224] J. O. Schreck, *J. Chem. Ed.* **48,** 103 (1971).

[3-225]   H. C. Brown, C. J. Kim, C. J. Lancelot, and P. v. R. Schleyer, *J. Am. Chem. Soc.* **92,** 5244 (1970).

[3-226]   P. G. Gasman and A. F. Fentiman, Jr., *J. Am. Chem. Soc.* **92,** 2549 (1970).

[3-227]   R. Sustmann, *Tetrahedron Lett.* **1974,** 963; *Pure Appl. Chem.* **40,** 576 (1974).

[3-228]   V. J. Shiner, Jr., R. D. Fischer, and W. Dowd, *J. Am. Chem. Soc.* **91,** 7748 (1969).

[3-229]   R. P. Bell, *The Proton in Chemistry,* 2nd ed., Cornell Univ. Press, Ithaca, New York, 1973, p. 226ff.

[3-330]   C. J. Collins and N. S. Bowman (Eds.), *Isotope Effects in Chemical Reactions,* Van Nostrand Reinhold, New York, 1970.

[3-331]   E. Caldin and V. Gold (Eds.), *Proton-Transfer Reactions,* Chapman and Hall, London, 1975.

BIBLIOGRAPHY

(a)   Handling of Kinetic Data

S. L. Friess, E. S. Lewis, and A. Weissberger (Eds.), *Investigation of Rates and Mechanisms of Reactions,* Parts I, II, 2nd ed., *Techniques of Organic Chemistry,* Vol. VIII, Interscience, New York, 1961.

E. S. Lewis (Ed.), *Investigation of Rates and Mechanism,* Parts I, II, 3rd ed., *Techniques of Chemistry,* Vol. VI, Wiley, New York, 1974.

C. H. Bamford and C. F. H. Tipper (Eds.), *Comprehensive Chemical Kinetics,* Vols. 1–22, Elsevier, Amsterdam, 1969–1980.

J. C. Jungers, J. C. Balaceanu, F. Coussemant, F. Eschard, A. Giraud, M. Hellin, P. LePrince, and G. E. Limido, *Cinétique Chimique Appliquée,* Soc. Edit. Technip, Paris, 1958.

S. W. Benson, *The Foundations of Chemical Kinetics,* McGraw-Hill, New York, 1960.

G. M. Panchenkov and B. P. Lebedev, *Chemical Kinetics and Catalysis* (in Russian), Izdat. Moskov. Univ., Moscow, 1961.

G. Pannetier and P. Souchay, *Chemical Kinetics,* Elsevier, Amsterdam, 1967.

R. Schaal, *Chemical Kinetics of Homogeneous Systems,* D. Reidel Publ. Co., Dordrecht, Nederland, 1974.

W. C. Gardiner, Jr., *Rates and Mechanisms of Chemical Reactions,* W. A. Benjamin, Menlo Park, California, 1969.

C. Capellos and B. H. J. Bielski, *Kinetic Systems,* Wiley, New York, 1972.

N. M. Emanuel and D. G. Knorre, *Chemical Kinetics-Homogeneous Reactions,* Wiley, New York, 1973.

G. B. Skinner, *Introduction to Chemical Kinetics,* Academic, New York, 1974.

C. D. Ritchie, *Physical Organic Chemistry,* M. Dekker, New York, 1975.

E. N. Yeremin, *The Foundations of Chemical Kinetics,* Mir Publishers, Moscow, 1979.

W. Drenth and H. Kwart, *Kinetics Applied to Organic Reactions,* Dekker, New York, 1980.

J. H. Espenson, *Chemical Kinetics and Reaction Mechanisms,* McGraw-Hill, New York, 1981.

R. Schmid and V. N. Sapunov, *Non-Formal Kinetics,* Verlag Chemie, Weinheim, 1982.

See also refs. [3-10] and [3-55].

(b)   Effects of Composition of Reaction Mixture

J. E. Gordon, *The Organic Chemistry of Electrolyte Solutions,* Wiley, New York, 1975.

C. D. Ritchie, *Physical Organic Chemistry,* Dekker, New York, 1975.

E. F. Caldin and V. Gold (Eds.), *Proton-Transfer Reactions.* Chapman and Hall, London, 1975.

S. G. Entelis and R. P. Tiger, *Reaction Kinetics in the Liquid Phase,* Wiley, New York, 1976.

O. Popovych and R. P. T. Tomkins, *Nonaqueous Solution Chemistry,* Wiley, New York, 1981.

See also refs. [3-84], [3-116], [3-117], [3-128], [3-134], and [3-152].

(c)   Intermediates and Structural Effects

J. E. Leffler, *The Reactive Intermediates of Organic Chemistry,* Interscience, New York, 1956.

C. D. Ritchie, *Physical Organic Chemistry,* Dekker, New York, 1975.

See also refs. [3-97]–[3-103], [3-177], [3-188]–[3-199], [3-201], [3-213], [3-214], [3-330], [3-331].

CHAPTER FOUR

STUDY OF FAST REACTIONS

Common techniques of reaction kinetics discussed so far can usually be applied to the study of reactions with half-times greater than about 10 s. For the study of faster reactions it is necessary to develop special methods that enable measurement of time changes during very short time intervals. Techniques developed for such purposes can be divided into several groups: (1) modified classical methods (2) techniques based on rapid preparation of the reaction mixture, (3) competitive methods, (4) methods based on the line width in NMR and ESR, (5) relaxation (perturbation methods). These methods can be used for the studies of reactions with the range of the lifetime indicated in Fig. 4-1.

In classical methods it is sometimes possible to modify reaction conditions or conditions of measurement so they enable following even the fastest reaction. In another approach, specially designed reaction chambers allow fast mixing of the reaction mixture. The composition of the reaction mixture or a measured quantity that is a function of this composition is then determined at constant flow of the reaction mixture at a chosen distance from the mixing chamber.

The latter two types of techniques, competition and perturbation (relaxation), are predominately used for equilibrium reactions. In competition methods one component of the reaction mixture participates in a competitive reaction, for example, electrolysis or photolysis. The establishment of a steady state in the presence of the competitive reaction is followed.

In relaxation methods the system is displaced by one or more pulses of an external parameter by means of a sudden change in temperature, pressure, electric field, and so forth from equilibrium [4-1]. The approach of the system to the new equilibrium position is then followed. In these cases the course of the reaction is followed under near equilibrium conditions.

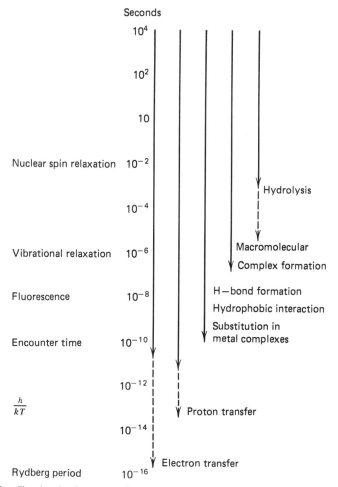

FIG. 4-1.   Showing the time range of commonly encountered chemical reactions and events.

4.1.   CLASSICAL METHODS

4.1.1.   Use of Low Concentrations

When the reaction rate follows equations for reactions of second and higher order, it decreases with decreasing concentration of reactants. If sufficiently sensitive analytical methods are available, large rate constants can be determined at sufficient dilution.

For example, in the bromination of N,N-dimethyl-m-toluidine in aqueous solution a rate constant of 4.3×10^9 liter mol^{-1} s^{-1} was found [4-2]. This was achieved by using electrometric methods for bromine determination, which enable a determination of 10^{-8} M bromine concentration by measurement of the potential of a smooth platinum electrode or by measurement of the voltammetric limiting current of bromine by means of a rotating platinum electrode. A coulometric arrangement also allows the generation of small amounts of bromine electrolytically by a known, low, constant current.

The bromination takes place only on the unprotonated, free base form of the toluidine derivative. Hence the rate of the reaction can be decreased and brought into an accessible range by buffering, for example, by addition of a strong acid that can decrease the concentration of the unprotonated free amine form to the order of 10^{-8} M. The same principles can be applied to the study of bromination of a number of amines or phenols.

Also, spectrophotometric methods, if the extinction coefficient is larger than 1×10^4 liter mol^{-1} cm^{-1}, are sufficiently sensitive for such analyses. Low concentrations can be used also in connection with other classical, flow, or relaxation methods to widen the applicable range of the determined rate constant.

4.1.2. Measurement of the Equilibrium Constant and the Rate Constant of the Reverse Reaction

For the equilibrium reaction

$$A + B \underset{k_{-1}}{\overset{k_1}{\rightleftharpoons}} C + D \tag{4.1-1}$$

it is possible to evaluate the equilibrium constant $K = k_1/k_{-1} = [C][D]/[A][B]$. If the value of k_{-1} is accessible to measurement, whereas the rate of the forward reaction is too high, it is possible to calculate the value of k_1 by means of the expression $k_1 = k_{-1}K$.

To be able to apply this approach it is necessary to be sure that the reaction follows the simple scheme shown above. Even when equilibrium measurements indicate simple stoichiometry and when the decrease in concentration of C and D with time (when the reverse reaction is followed) follows simple second-order kinetics, the reaction can still be more complex. For example,

$$A + B \underset{k'_{-1}}{\overset{k'_1}{\rightleftharpoons}} E \underset{k'_{-2}}{\overset{k'_2}{\rightleftharpoons}} C + D \tag{4.1-2}$$

The experimental evidence as shown above does not exclude the possibility that whereas the equilibrium constant corresponds to $K = [C][D]/[A][B]$, the measured rate constant corresponds to k_{-2}. Without the possibility of determining independently either the equilibrium concentration of species E or the value of the rate constants k_1 or k_2, it is impossible to decide which of the forward reactions is the slower step and to compute the value of its rate constant.

Existence of a more complex scheme can be revealed by a search for an intermediate E and by a detailed study of the medium effects on the measured rate constant. Finally, a more complex scheme is clearly indicated if the calculated rate constant for the forward reaction, assumed to be k_1, has a value which is larger than the rate constant for a diffusion-controlled process (about 3×10^{10} liter mol^{-1} s^{-1}, according to Onsager)[4-3].

An example of a system in which the simple calculation of the second rate constant, using equilibrium constant, is virtually impossible is the condensation of amines with carbonyl compounds, which instead of the simple process

$$>\!C\!=\!O + H_2NR \underset{k_{-1}}{\overset{k_1}{\rightleftharpoons}} >\!C\!=\!NR + H_2O \tag{4.1-3}$$

was proved to be more complex [4-4]:

$$>\!C\!=\!O + H_2NR \underset{k'_{-1}}{\overset{k'_1}{\rightleftharpoons}} >\!C\!\!\begin{array}{c} O^- \\ \diagdown \\ NH_2R \\ + \end{array} \tag{4.1-4}$$

$$>\!C\!\!\begin{array}{c} O^- \\ \diagdown \\ NH_2R \\ + \end{array} \underset{k'_{-2}}{\overset{k'_2}{\rightleftharpoons}} >\!C\!=\!NR + H_2O \tag{4.1-5}$$

The detailed mechanism is even more complex, and includes a rapidly established acid–base equilibrium involving the hydrated intermediate.

Similarly, when the base catalyzed hydration of some carbonyl compounds (e.g., aliphatic aldehydes) was assumed to follow the reaction path

$$>\!C\!=\!O + H_2O \underset{k'_{-1}}{\overset{k'_1}{\rightleftharpoons}} >\!C\!\!\begin{array}{c} OH \\ \diagdown \\ OH \end{array} \tag{4.1-6}$$

and k_{-1} was determined [4-5]. Calculation of k_1 from k_{-1} and $K = [C(OH)_2]/[CO]$ gave a value of k_1 which had no physical meaning, since in the presence of base the reaction follows the scheme [4-6]:

$$>\!C\!=\!O + OH^- \underset{k'_{-1}}{\overset{k'_1}{\rightleftharpoons}} >\!C\!\!\begin{array}{c} O^- \\ \diagdown \\ OH \end{array} \tag{4.1-7}$$

$$>\!C\!\!\begin{array}{c} O^- \\ \diagdown \\ OH \end{array} + H_2O \underset{k'_{-2}}{\overset{k'_2}{\rightleftharpoons}} >\!C\!\!\begin{array}{c} OH \\ \diagdown \\ OH \end{array} + OH^- \tag{4.1-8}$$

On the other hand, applications of this approach to simple systems such as recombination of C- acids [4-7], were successful; for example,

$$H_3O^+ + CH(CN)_2^- \underset{k_{-1}}{\overset{k_1}{\rightleftharpoons}} CH_2(CN)_2 + H_2O$$

$$k_1 = 2.3 \times 10^9 \text{ liter mol}^{-1} \text{ s}^{-1} \quad (4.1-9)$$

or

$$H_3O^+ + CH_2COCH_2^- \underset{k_{-1}}{\overset{k_1}{\rightleftharpoons}} CH_3COCH_3 + H_2O$$

$$k_1 = 5 \times 10^{10} \text{ liter mol}^{-1} \text{ s}^{-1} \quad (4.1-10)$$

4.1.3. Competitive Reactions

If values of the rate constants k_1 and k_{-1} in the system

$$A^- + H_3O^+ \underset{k_{-1}}{\overset{k_1}{\rightleftharpoons}} HA + H_2O \quad (4.1-11)$$

are known, it is possible to determine the value of the rate constant k_2 of the reaction

$$A^- + M^{2+} \xrightarrow{k_2} MA^+ \quad (4.1-12)$$

by determining the ratio of $[MA^+]$:$[IIA]$ in the product. On the other hand, when $k_1/k_{-1} = K$ is known and k_2 can be determined from an independent experiment (at sufficiently high pH-values), it is possible to calculate the value of k_1 (and consequently of k_{-1}).

This type of approach is most frequently used in complexation kinetics [4-8]. In organic reactions another type of competitive reaction has been reported: in the study of the rate of dehydration, use was made of the fact that the rate of semicarbazone formation is much higher than the rate of dehydration [4-9]. Addition of semicarbazide to aqueous solutions of aliphatic aldehydes results in formation of a semicarbazone. The observed rate of semicarbazone formation is equal (or at least is assumed to be) to the rate of dehydration with constant k_{-1}:

$$>C=O + H_2O \underset{k_{-1}}{\overset{k_1}{\rightleftharpoons}} >C\begin{matrix} OH \\ OH \end{matrix} \quad (4.1-13)$$

$$>C=O + NH_2NHCONH_2 \underset{k_{-n}}{\overset{k_n}{\rightleftharpoons}} >C=NNHCONH_2 + H_2O \quad (4.1-14)$$

The actual interpretation of the value of the measured constant is more complex due to the stepwise mechanism of the dehydration reaction mentioned above.

Competitive reactions are also used in conjunction with flow techniques (see Section 4.2).

4.1.4. Low Temperatures

The reaction rate usually decreases considerably with decreasing temperature, for example, reaction with E_a = 10 kcal mol^{-1} shows a decrease in the rate constant of approximately five orders of magnitude when the temperature of the reaction mixture is decreased by 100°C. The lowest temperature experimentally accessible for a given system is dependent on the melting point of the solvent, on the solubility of solutes (which usually decreases with decreasing temperature), and sometimes even on increased viscosity, which prevents a thorough mixing of the reaction mixture.

To achieve the low temperatures, baths prepared from solid CO_2 and acetone or ethanol (-78°C), or from methanol with liquid nitrogen (-100°C), are used. Temperatures up to -40°C can be attained by means of commercially available refrigeration units. Also, use can be made of the low boiling ponts of NH_3 or SO_2. Special thermostats (cryostats) for low temperatures have been developed. The lowest temperature reported for a kinetic study is about 4°K[4-10].

The lowering of temperature can be used in connection with both classical and special methods. From values of rate constants determined at several low temperatures it is possible to calculate the activation energy and the value of the rate constant at a temperature of 20° or 25°C. To minimize errors resulting from extrapolations it is recommended to include measurements at as high a temperature as possible.

4.2. RAPID PREPARATION OF A REACTION MIXTURE

When slower reactions are followed, usually one component is added by a pipette or syringe to the reaction mixture. In such instances a local increase in the concentration of the added component in the reaction mixture necessarily occurs over a short period of time. The interval between the beginning of the addition and the homogenization of the solution (so that the concentration of all components is equal in the whole volume) can take a few seconds or a fraction of a minute according to the intensity of stirring, the kind of stirring, the volume of the reaction mixture, and the shape of the reactor. For reactions with half-times over 30 s, such ill-defined conditions at the beginning of the reaction are of little importance, but play a role for reactions with half-times of seconds.

A somewhat faster achievement of homogeneity of the solution is possible when a glass bulb containing the added component is broken below the surface of the reaction mixture. Considerable improvement is achieved in the method of tearing off a barrier. In this method the reaction mixture is prepared in two comparable volumes of solutions which are separated by a barrier. By tearing off of the barrier and intensive stirring it is possible to achieve "complete" mixing within 10–15 milliseconds. Significantly shorter mixing times can be achieved by using specially

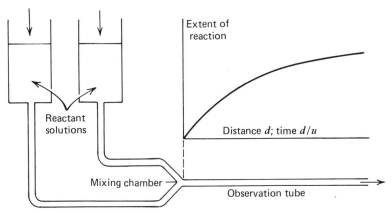

FIG. 4-2. Arrangement and principle of measurement using the continuous-flow method (according to E. F. Caldin, *Fast Reactions in Solution*, Wiley, New York, 1964, p. 30).

constructed mixing chambers that were introduced in 1923 by Hartridge and Roughton [4-11] in their flow method, which has found renaissance over the last decade. The principle of their method is as follows: the components of the reaction mixture are placed in two reservoirs (Fig. 4-2) and are introduced under pressure into a special mixing chamber (Fig. 4-3). After mixing, the reaction mixture passes at the rate of several meters per second through a tube in which, at several distances from the mixing chamber, the composition of the solution is determined by optical, electrical, chemical, and other methods which enable a sufficiently rapid analysis of the rapidly moving liquid.

The conditions for application of continuous flow methods are (1) the time needed

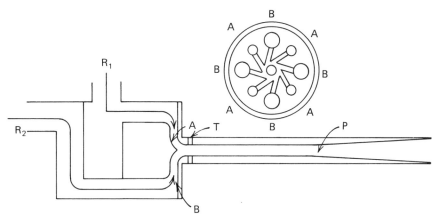

FIG. 4-3. Scheme of an early type of mixing chamber. Alternate jets A, B, A, B . . . deliver two reagents tangentially into a central cavity shown in cross section (top). Longitudinal section shows the entry of reagent R_1 and R_2, mixing occurs close to T, P is the observation tube. [According to F. J. W. Roughton, in *Techniques of Organic Chemistry*, 2nd ed., Vol. VIII, Part 2 (S. L. Friess, E. S. Lewis, and A. Weissberger, Eds.), Wiley, New York, 1963, p. 709.]

for a thorough mixing of reaction components must be short when compared with the half-time of the studied reaction and (2) the flow of the reaction mixture through the observation tube must be free from turbulence [4-12]. A pneumatic, or motor-driven device forces the individual reactants contained in two syringes through the mixing chamber. Grease must be carefully removed from both syringes and the rate of outflow from the reservoirs must be kept constant up to 1%. Usually, the two syringes are activated by means of a pushing block. Air bubbles in the syringe should be avoided.

The two solutions are introduced into the mixing chamber by a larger number (usually four or eight) of jets. These jets must be oriented in such a way as to achieve a tangential rather than a radial streaming of the liquid. The position of the jets must ensure sufficiently rapid mixing of both components of the reaction mixture and a proper kind of flow during and after passage through the mixing chamber. Under these conditions 97% complete mixing in intervals ranging from 0.003 to 0.004 s can be achieved. Consequently, reactions with half-times from 0.003 s (or 0.004 s respectively) can be studied. To establish completeness of mixing, measurement of refractive index (Schlieren method) is used in connection with a reaction resulting in a rapid color change, such as the reaction of iodine with thiosulfate or of hydrogen ions with a suitable color indicator. Flow methods are not necessarily restricted to the use of mixing chambers. Initiation of the reaction can be carried out also by other methods, for example, photochemically or electrochemically. For photochemical initiation, the reaction mixture flowing through a tube is irradiated by light of appropriate wavelength at some distance from the position at which the measurement of concentration is carried out. For electrochemical initiation the investigated solution is allowed to flow through an electrolytic cell in which a controlled-potential or constant-current electrolysis is carried out.

Efficient mixing requires turbulent rather than laminar flow of the two liquids. From the Reynolds equation in fluid mechanics it follows that to achieve a turbulent flow, the rate of flow must exceed a critical value (for water at 20°C this value is equal to $200/d$ cm s^{-1} where d is the inner diameter of the observation tube). Flow velocities cannot, however, be arbitrarily increased in order to achieve better mixing times. At very high flow velocities the appearance of tiny bubbles (cavitation) in the mixed solution leads to disturbances in the detection of concentration changes. This condition is a decisive factor in limiting the application of flow measurements to times shorter than about 0.0001 s. The shortest times reported using stopped-flow measurements are about 0.0002 s, by means of a special spherical ball-jet mixing chamber [4-13]. Comparable performance has been reported recently using continuous flow methods [4-14]. It should be realized that the condition for efficient mixing requires turbulence, whereas the subsequent observation process requires a stable solution.

The concentration of one or more components of the reaction mixture is measured in the observation tube at certain distances from the mixing chamber. The interval between the initiation of the reaction and the monitoring of concentration is given by the time needed for the reaction mixture to reach the observation point. This time is directly proportional to the distance of the observation point from the mixing chamber and indirectly proportional to the rate of flow of the liquid. The rate of flow is chosen based on deliberations mentioned in the preceding paragraph and is kept constant. Then the time interval from the beginning of reaction depends only on the distance of the observation point from the mixing chamber. Measurements carried out at several observation points correspond to measurements done after several time intervals. From the known flow velocity and the distance of the observation point from the mixing chamber it is possible to calculate the time interval t which at the given point elapsed from the moment of mixing. Measurements of concentrations at various positions along the tube thus make it possible to obtain the usual concentration–time plot.

The rate of flow of the liquid can be determined by measuring the inner diameter of the tube and the volume of the liquid that has flowed from the tube during a measured time interval. It is possible to maintain outflow velocities of the order of tens of meters/second. In such cases the distance of 1 cm from the mixing chamber corresponds to 10^{-3} s and it is possible to follow reactions with half-times of about 0.001 s.

Mixing chambers are fabricated from translucent plastics (e.g., Perspex or Lucite) or Kel F, whereas a quartz tube whose inner diameter is typically 0.5–15 mm is used as the observation tube. The length of the tube can vary between 1 and 100 cm. If the conditions for efficient mixing and detection sensitivity are fulfilled, the accuracy of the determination of first- and second-order rate constants is about 2–3%. For a first-order reaction the errors resulting from nonimmediate mixing are negligible if $k_1 < 1.5 \times 10^3$ s^{-1} and $\tau_{1/2} > 0.0005$ s. In second-order reactions the errors due to noninstantaneous mixing are negligible for reactions with half-times larger than 0.005 s.

The methods used for determination of concentration change along the observation tube do not have to be very fast; usually it is sufficient if the measurement can be carried out in several seconds. The volumes of the reagents needed for a single flow experiment depend on the diameter of the observation tube. If this diameter is 2 mm and the concentration is determined spectrophotometrically, or the temperature is measured by a thermocouple or thermistors, some 100–500 ml of the reaction mixture is needed. The use of 1-mm tubing and detection by a photocell reduces the volume to about 20–30 ml. Finally, when the syringe activation is synchronized with recording, the volume can be reduced to 10 ml and even less [4-15].

The most frequently used technique for the determination of the concentration is spectrophotometry (Fig. 4-4) followed by thermometry (Fig. 4-5). Recently developed diode-array, rapid-scan spectrophotometry (DARSS) enables the recording of complete spectra during the stopped flow time scale [4-16].

For the former, it is of importance that temperature changes caused by thermal conductivity, frictional heat released by flowing liquids, or eventually by thermoelastic effects do not exceed 0.001°C. For reactions where the total change in temperature is 0.05°C, such effects can be neglected, but it is necessary to consider them for reactions where the overall change in temperature is between 0.05°C and 0.003°C [4-17]. Thermometric measurements have been applied to the measurement

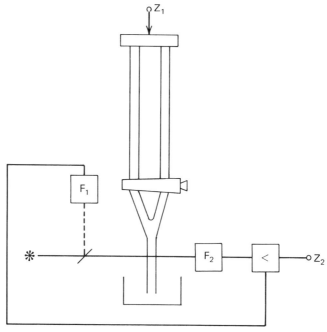

FIG. 4-4. Scheme of an apparatus for spectrophotometric recording of the rates of fast reactions. F_1, F_2—photocells, *—source of monochromatic light, Z_1—registration of the rate of the shift of the piston; Z_2—recording of variations in absorbance.

of the rate of reaction of some acids with bases, in particular with proteins, or the reaction

$$CO_2 + 2NH_3 \longrightarrow NH_2COONH_4 \qquad (4.2\text{-}1)$$

The thermometric measurements reported were based on the use of thermoelements and the concentration of solutions investigated was at least $5 \times 10^{-3} M$, in some instances $5 \times 10^{-4} M$. If the temperature changes related to liquid movement do not become predominant, the lower limit of concentration can probably be decreased by the use of thermistors. Optical absorption enables in some instances the use of considerably lower concentrations (down to $1 \times 10^{-6} M$). Spectrophotometric measurements thus allow, in general cases, a determination of second-order rate constants in the order of 10^7 or 10^8 liter $mol^{-1} s^{-1}$, thermometric measurements with thermocouples up to 10^6 or 10^5 liter $mol^{-1} s^{-1}$. For first-order rate constants where the half-times are independent of concentrations, both types of measurements make it possible to measure constants in the order of $10^3 s^{-1}$.

Thermometric measurements can be used even for such reactions where none of the components of the reaction mixture gives an absorption spectrum in the wavelength range accessible with commercial spectrophotometers. On the other hand, spectrophotometry, by measurement of absorbance at suitable wavelengths,

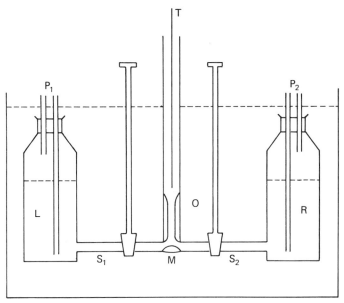

FIG. 4-5. Scheme of an apparatus for thermometric studies of fast reactions. L, R are two reagents under gas pressure; P_1, P_2, S_1, S_2 are two stopcocks; M is the mixing chamber; O is the observation tube, T is a thermocouple or thermistor. [According to F. J. W. Roughton, in *Techniques of Organic Chemistry*, 2nd ed., Vol. VIII, Part 2 (S. L. Friess, E. S. Lewis, and A. Weissberger, Eds.), Wiley, New York, 1963, p. 760.]

enables us to follow separately concentration changes of individual components of the reaction mixture and is thus more advantageous for the study of more complex reactions. For the application of thermometric measurements it is necessary that the path of the reaction is well understood. In those instances where the application of thermometric measurements is possible, such measurements are often more accurate than spectrophotometric measurements. However, they can be used in instances when the amount of the substance needed for the kinetic study is not a limiting factor.

Other methods that have been used for the determination of concentration along the observation tube are conductometry, potentiometry (measurement of pH and of oxidation reduction potentials) [4-18], and voltammetry [4-19], usually as measurement of the current flowing at solid electrodes at a selected potential. For electrochemical methods electrodes are placed at varying distances along the observation tube. Platinum or silver electrodes can be directly fused into the walls of the observation tube. For the study of reactions in which radicals participate, ESR measurements can be used. The use of gasometric and chemical methods have also been described.

In the application of chemical methods the principle of consecutive quenching or mixing is used. For mixing chambers connected in series (Fig. 4-6) it is possible to quench the reaction by addition of the substance Z and the resulting solution. By changing the distance between the two mixing chambers M_1 and M_2 or by

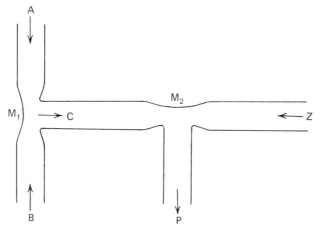

FIG. 4-6. Scheme of arrangement of two mixing chambers (M_1, M_2). A and B, reagents; Z, trapping reagent; P, products.

varying the relative rates of flow of substances C and Z it is possible to alter the moment of mixing C and Z or the quenching of the reaction, respectively. The condition which has to be fulfilled for such an application is that the time needed for the quenching of the reaction (i.e., for the reaction of C with Z) is short when compared with the time needed for reaction of A with B.

The choice of the quenching agent Z is critical and therefore only relatively few examples of this method have been reported (e.g., the reaction of carbon monoxide with hemoglobin involving the use of oxygen as the quenching agent). In systems where the use of this method is suitable, the results are accurate.

In addition to the possibility of carrying out concentration measurements at several positions along the observation tube, it is possible to measure the concentration at one fixed position and to change the rate of flow of the reaction mixture. By varying the flow rate, the time elapsed between mixing and measurement can be changed. The rate of flow can be changed in steps and individual points for the concentration–time curve determined. The other alternative is to change the flow-rate continuously. This latter technique of accelerated flow, introduced by Chance [4-20], offers the great advantage of consumption of a small volume of the reaction mixture (even as little as 0.1 ml can be sufficient), but requires the use of a rapid measuring technique, spectrophotometry for example. To record the concentration change with time the measured signal is transformed into an electrical one and recorded by means of an oscilloscope or analog to digital (A/D) converter and storage device. With a suitable arrangement, reactions with half-times down to 0.00004 s can be measured [4-14].

Another modification of the flow-technique is the widely used stopped-flow method, in which the flow of the reaction mixture is suddenly stopped. Immediately the measurement or recording of the time change of concentration is started at a point close to the mixing chamber. Most frequently optical absorbance methods are used for the concentration measurement.

The stopped-flow method offers the following advantages when compared with the continuous flow method: (1) It is independent of the rate and mode of the flow of the reaction mixture through the observation tube; (2) It makes it possible to obtain a continuous recording corresponding to concentration–time dependence starting at a millisecond after mixing and continuing up to practically infinite time; (3) Only a very small volume of the reaction mixture is needed.

The position at which the measurement is carried out along the observation tube must be chosen to be several millimeters from the mixing chamber to ensure (with an effective mixing chamber) complete mixing. Nevertheless, the distance of the observation point from the mixing chamber should be as short as possible to ensure the possibility of starting the measurement in the shortest possible time after mixing. With the stopped flow method, the observation tube can be more effectively replaced by means of an observation chamber equipped with appropriate focusing lenses to achieve higher signal to noise (S/N) ratios in optical detection methods (absorption, fluorescence, etc.).

The stopping of the flow must be carried out in such a way that the moving liquid is brought to a standstill in the shortest possible time.

In the instrument developed originally by Gibson [4-15] (Fig. 4-7), the observation window is placed about 8 mm from the mixing chamber and at a flow velocity of 5 m/s it is possible to start the observation 0.003 s after mixing. For each experiment about 0.2 ml of the solution is needed and it is possible to follow reactions with half-times of 0.01 s and longer. Recording of the time change of the absorbance is carried out by an oscilloscope or transient recorder.

With stopped-flow methods it is possible to use (in addition to the measurement of absorbance) measurements of conductivity, changes in pH by means of a glass electrode, and voltammetric and polarographic limiting currents. The dropping mercury electrode can be used in connection with this method, because the solution in the moment of measurement is stationary [4-21].

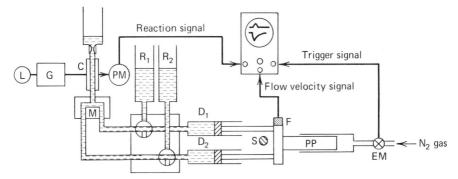

FIG. 4-7. Block diagram of pneumatically driven stopped-flow apparatus in which the flow of fluid is stopped by abrupt halt of the *drive* syringes. L, light source (a halogen lamp or deuterium lamp can be used); G, monochromator; C, observation cell (2 mm and 10 mm optical paths are available); M, mixer; PM, photomultiplier; D$_1$ and D$_2$, driving syringes; F, flow velocity detector; PP, gas pressure driving piston; EM, electromagnetic valve; S, pin for stopping the flow. (According to K. Hiromi, *Kinetics of Fast Enzyme Reactions*, Halsted Press, Division of Wiley, New York, 1979, p. 76.)

Constant flow methods must be used for systems where the sensor indicates concentration changes with some small time-lag. They are preferred for reactions where sufficient amount of reagents and solvents are available and are best suited for reactions with half-times between 0.001 and 0.1 s. The apparatus is simple and does not require synchronization of the beginning of reaction, measurement, and recording. With sufficiently sensitive methods (e.g., fluorimetry capable of following concentration changes down to 10^{-9} M) it is possible to measure second-order rate constants greater than 10^{10} liter mol^{-1} s^{-1} reproducible to $\pm$ 5%.

Methods of accelerated and stopped-flow require a detector with very rapid response. These methods are recommended for the study of systems where the amount of reagents available is limited and/or solvents expensive. Their optimum range of half-times is longer than about 0.001 s. Some stopped flow instruments are available commercially (Durrum, Aminco). The reproducibility of rate constants is in the order of $\pm$ 2%.

Over the past two decades, mainly as a result of the interest in problems in complexation chemistry, interest in flow techniques increased considerably.

4.3. COMPETITIVE METHODS

In competitive methods the concentration of one of the components taking part in the reaction is altered by a competing process. In electrochemical methods such reactions competing with chemical reactions can be electrolysis (which is a rapid process when limiting currents are compared) and transport (diffusion, migration). In optical methods processes which compete with the chemical reaction can be photochemical generation of reagents or photochemical excitation shown by changes in the fluorescence spectrum.

4.3.1. Electrochemical Methods [4-22–4-31]

In electrochemical methods either a chosen voltage is applied to a system of two electrodes and the current flowing through the circuit measured or a certain current is imposed on the electrodes and their potential measured. Current is measured in polarographic and other potentiostatic methods, method of rotating disk, and of the stationary field. Potential of the electrode as a function of the imposed current and time is measured in galvanostatic methods and chronopotentiometry.

4.3.1.1. Polarographic Methods [4-22–4-25, 4-27, 4-29–4-31]

In polarography, as discussed in Section 2.4.3.6., current–voltage curves are recorded during constant potential electrolysis with a dropping mercury electrode. Under conditions generally used, in the presence of excess of the so-called supporting electrolyte (buffer, solution of acid, base, or a neutral salt), the organic compound which undergoes reduction or oxidation at the electrode surface is transported from the bulk or the solution towards the surface mainly by diffusion.

When we restrict ourselves to the potential region where the wave reaches its limiting value, the electrode process can be considered fast and sufficiently rapid to transform all species transported to the surface of the mercury drop. If the species which takes part in the electrode process undergoes a chemical transformation, the chemical reaction generating or deactivating an electroactive species is in competition with diffusion.

The chemical reaction can take place before the electron transfer (preceding electrolysis), or after the electron transfer (subsequent or consecutive reaction), or concurrently. Chemical reactions can also be interposed between two electron transfers.

To demonstrate how chemical reactions affect polarographic curves the case of preceding reactions will be discussed in some detail, because this type of reaction is most frequently encountered in practice.

A simple example of a preceding reaction is an equilibrium in which the electroactive species C is formed in the reaction of electroinactive compounds A and B:

$$A + B \underset{k_{-1}}{\overset{k_1}{\rightleftharpoons}} C \tag{4.3-1}$$

$$C + ne \xrightarrow[E_C]{} \text{Products} \tag{4.3-2}$$

This reaction is frequently studied under conditions when reagent B is present in the reaction mixture in excess or the solution is buffered in B. In both cases the concentration of B does not change in the course of reaction. Under such conditions when C is removed by electrolysis, only electroinactive substance A is transported from the bulk of the solution to the electrode surface by diffusion and the system is simplified to:

$$A \underset{k_{-1}}{\overset{k'_1}{\rightleftharpoons}} C \tag{4.3-3}$$

$$C + ne \xrightarrow[E_C]{} \text{Products} \tag{4.3-4}$$

where the formal rate constant k'_1 is given by $k'_1 = k_1[B]$.

It has been stressed that only species C is electroactive at the investigated potential. This does not exclude the possibility that species A is electroactive at more negative potentials (E_A)

$$A + ne \xrightarrow[E_A]{} \text{Products} \tag{4.3-5}$$

The possibility of application of polarography for the determination of rate constants of fast reactions depends on the relative rate of establishment of the equilibrium between A and C and the rate of diffusion. Equilibria established in 10 s or more can be considered as slowly established, those in less than 3 s as rapidly established (at a normal drop time of 3 s).

The shape of polarographic curves for slowly established equilibria will be discussed first. When the equilibrium is slowly established, the polarographic curves show two waves: That of compound $C(i_c)$ at potential E_c and that of compound A (i_A) at more negative potential E_A. The height of both waves is governed only by diffusion, and the height of i_c is proportional to the concentration of substance C in the bulk of the solution whereas the height of i_A is proportional to the concentration of A. According to the concentration of reagent B (present in excess) three possible cases can occur:

(a)   At high excess of compound B, the equilibrium is shifted in favor of substance C. On polarographic curves only wave i_C at potential E_C is observed (Fig. 4-8a).

(b)   At very low concentration of compound B and a suitable value of the constant $K = k_1/k_{-1}$, the equilibrium will be shifted in favor of the substance A and this is manifested by the appearance of only wave i_A at potential E_A (Fig. 4-8b).

(c)   Finally, at such concentration of compound B when the value of $k_1[B]$ is comparable to the value of k_{-1}, the concentrations of substances A and C are comparable. On the polarographic curves two waves are observed: The height of wave i_C at potential E_C is proportional to the equilibrium concentration of C in the

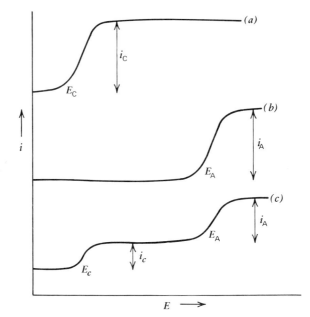

FIG. 4-8.   Polarographic curves corresponding to reduction of species A and C which are in equilibrium [equations (4.3-3)–(4.3-5)]. (a) Form C predominates, (b) form A predominates, (c) concentrations of A and C are comparable. E_A, E_C are half-wave potentials of species A and C, i_A and i_C their limiting currents. [According to P. Zuman, in *Advances in Physical Organic Chemistry* (V. Gold, Ed.), Vol. 5, Academic, New York, 1967, p. 31.]

bulk of the solution, and that of wave i_A at potential E_A to equilibrium concentration of A (Fig. 4-8c). From the ratio of $i_A:i_C$ the value of the equilibrium constant K can be calculated, but for a slowly established equilibrium no information on the rate of the equilibrium establishment can be obtained. Next the same system can be discussed under conditions when the rate of establishment of the equilibrium between A and C is not relatively slow (i.e., the establishment takes less than some 3 s), but also not extremely fast. In such systems, at potentials where the reduction of substance C occurs, the equilibrium between A and C is disturbed as the surface concentration of substance C is diminished by electrolysis. To re-establish the equilibrium, substance C is depleted in the vicinity of the electrode by a reaction between A and B with rate constant k_1 (or from A with reaction rate constant k_1'). By this chemical transformation of the electroinactive A (at potential E_C) into electroactive C, more of the species C is present in the vicinity of the electrode than in the bulk of solution.

On polarographic curves we observe an increase in the height of wave i_C in comparison with conditions, when the equilibrium is slowly established (Fig. 4-9). The magnitude of this increase depends (under given experimental conditions) on the value of the formal rate constant k_1', that is, on the value of the constant k_1 and on the concentration of substance B present in excess. In the region of concentration of B where the equilibrium is shifted in such a way that the equilibrium concentration of C is so small that it does not give a measurable wave on the polarographic current–voltage curves, then with a decrease in the concentration of B we observe changes in the polarographic waves shown in Fig. 4-10.

1. At sufficiently large concentrations of the compound B, the rate of formation of compound C is so high that this reaction can transform all of the substance A

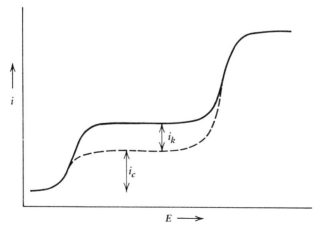

FIG. 4-9. Effect of fast chemical reactions on polarographic waves, i_C is the diffusion-controlled limiting current of substance C in equilibrium (4.3-3) with A; i_k is the kinetic current of substance C, governed by the rate of the chemical reaction by which species C is formed from species A in the vicinity of the electrode. [According to P. Zuman, in *Advances in Physical Organic Chemistry* (V. Gold, Ed.), Vol. 5, Academic, New York, 1967, p. 32.]

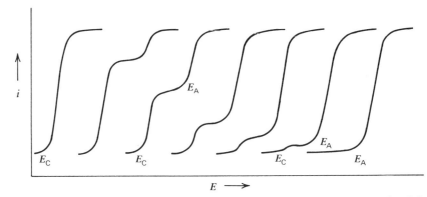

FIG. 4-10. Change of polarographic waves of the species C [reduced according to equation (4.3-4)] and of species A [reduced according to equation (4.3-5)]which are in equilibrium (4.3-1) with increasing concentration of substance B present in excess. (According to P. Zuman, in *Advances in Physical Organic Chemistry* (V. Gold, Ed.), Vol. 5, Academic, New York, 1967, p. 33).

present in the vicinity of the mercury drop surface (where it was brought by diffusion) into the electroactive form C. Under such conditions, diffusion is slower than the reaction A $\rightarrow$ C and even when the species C, which undergoes electrolysis, is almost completely formed by a chemical reaction, the height of the polarographic wave is governed only by the rate of diffusion of A.

2. At sufficiently low concentration of the compound B, the rate of the reaction A $\rightarrow$ C is too slow to produce a measurable amount of species C. Compound A, which is present in excess in the bulk of the solution, is transported towards the electrode surface by diffusion and, since it is *not* converted into C, undergoes reduction in wave i_A at potential E_A (more negative than E_C). On polarographic curves at low concentration of compound B only one single wave, i_A, at negative potentials is observed.

3. Finally, in a suitable range of concentration of B, the rate of formation of species C from A is such that it can transform part of the species A present in the solution into form C which then undergoes electrolysis. Under such conditions two waves, i_C and i_A are observed on polarographic current–voltage curves. The ratio of heights of the two waves i_A and i_C changes with the concentration of compound B.

If the ratio $i_C/(i_C + i_A)$ is then plotted as a function of log [B], the plot shows the shape of a dissociation curve, called a "polarographic dissociation curve." (Fig. 4-11).

The polarographic dissociation curve is next compared with ordinary dissociation curves, that is, with $[C]_e/([C]_e + [A]_e) = f(\log [B])$ plots obtained under conditions when the equilibrium between A and C is established and not affected by removal of C by electrolysis. Such curves can be obtained, for example, by spectrophotometry or potentiometry. The comparison shows that polarographic and ordinary dissociation curves have similar shape, but the polarographic dissociation curve is shifted along the concentration axis towards larger values of log [B].

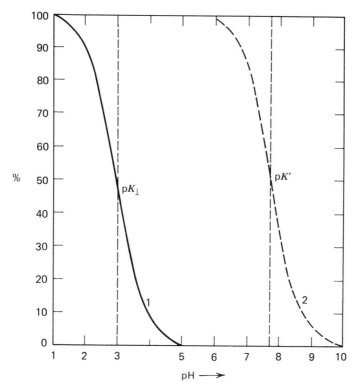

FIG. 4-11. Comparison of polarographic (dotted) and potentiometric (or spectrophotometric—full) dissociation curves. Polarographic dissociation curve corresponds to the dependence of ratio of wave-heights i_k/i_d on pH. The shift of polarographic pK'-value when compared with potentiometric pK_1 results from formation of C in the vicinity of the electrode from A and B. The shift increases with increasing rate of formation of C in reaction (4.3-1). [According to P. Zuman, in *Advances in Physical Organic Chemistry* (V. Gold, Ed.), Vol. 5, Academic, New York, 1967, p. 34.]

Just as in an ordinary dissociation curve the value of $-\log [B]$ at which $[C]_e = [A]_e$ is denoted pK, for a polarographic dissociation curve we denote the value of $-\log [B]$ at which $i_C = i_A$ as pK' and call K' the polarographic dissociation constant of reaction $A + B \rightleftharpoons C$. For systems where during the drop-time a steady state is established between the removal of C by electrolysis and its formation by chemical reaction $A \rightleftharpoons C$, the numerical value of pK' is larger than the numerical value of pK corresponding to equilibrium conditions.

From the difference between the polarographic pK' and the equilibrium pK it is possible to calculate the value of the constant k_1 in the reactions (4.3-6) and (4.3-7):

$$A + B \underset{k_{-1}}{\overset{k_1}{\rightleftharpoons}} C \tag{4.3-6}$$

$$C + ne \xrightarrow[E_C]{} \text{Products} \tag{4.3-7}$$

by means of the equation [4-32]:

$$\log k_1 = 2pK' - pK - 2 \log 0.886 - \log t_1 \qquad (4.3\text{-}8)$$

where 0.886 is a numerical factor and t_1 the drop-time.

Conditions for the application of equation (4.3-8) can be summarized as follows:

1. At least substance C must be polarographically active in the accessible potential range.
2. The height of the polarographic wave in the region where $i_C < [0.1(i_A + i_C)]$ is governed by the rate of chemical reaction rather than by diffusion. The current i_C is then a kinetic current rather than a diffusion current.
3. The reaction must take place as a homogeneous reaction in solution in close vicinity of the electrode rather than as a heterogeneous reaction at the electrode surface. The reaction must be thus a volume reaction rather than a surface reaction.
4. The value of the equilibrium pK must be available from independent measurements.

Condition 1 clearly indicates the limits of applicability, but the limitation is not too severe; about 70% of known organic compounds other than saturated hydrocarbons, alcohols, and amines give polarographic waves. When the reduction of A occurs at such negative potentials that its wave is superimposed upon that of the supporting electrolyte current, special techniques have to be used to obtain the value of $i_C + i_A$.

Experimental verification for the kinetic character of current i_C [4-22–4-25] when $i_C < [0.1(i_A + i_C)]$ is described in monographs on electroanalytical chemistry [4-27]. Similarly, verification of the volume character of the wave i_C in question is beyond the scope of this treatment and can be found in special treatises [4-24]–[4-25].

To calculate the rate constant from polarographic data it is thus necessary to record polarographic waves i_A and i_C at several concentrations of [B], measure the wave-heights, and find log [B] when $i_C = i_A$ [or $i_C = (i_C + i_A)/2$]. After ensuring that the measured current i_C is kinetic and that the reaction involved is a volume reaction, the drop-time t_1 is measured at the potential corresponding to the limiting current and the value of the equilibrium dissociation constant is found in the literature or determined experimentally. The rate constant can then be calculated using the equation for log k_1 (4.3-8).

The application of this procedure will be demonstrated using some examples:

Rather frequently investigated preceding reactions involve acid–base equilibria, where [B] = [H$^+$] and $-$log [B] = pH. As the conjugate acid HA is invariably reduced at more positive potentials than the base, it corresponds to species C and the couple of chemical and electrochemical processes become:

$$\mathrm{A^- + H^+} \underset{k_{-1}}{\overset{k_1}{\rightleftharpoons}} \mathrm{HA} \qquad (4.3\text{-}9)$$

$$HA + ne \xrightarrow[E_{HA}]{} Products \qquad (4.3\text{-}10)$$

With such systems, pH-changes in the height of waves i_{HA} and i_A—in the shape of a dissociation curve—are often observed.

The only real problem with preceding acid–base equilibria is that they are frequently surface rather than volume reactions.

An example of an acid–base system which is claimed to be free from surface phenomena, is the reduction of the monoanion of maleic acid [4-33], which follows the reduction path

$$
\begin{array}{l}
CH\ COO^- \\
\parallel \qquad\qquad + H^+ \\
CH\ COO^-
\end{array}
\xrightleftharpoons[k_{-1}]{k_1}
\begin{array}{l}
CH\ COO^- \\
\parallel \\
CH\ COOH
\end{array}
\qquad (4.3\text{-}11)
$$

$$
\begin{array}{l}
CH\ COO^- \\
\parallel \qquad\qquad +2e + 2H^+ \longrightarrow \\
CH\ COOH
\end{array}
\begin{array}{l}
CH_2\ COO^- \\
\mid \\
CH_2\ COOH
\end{array}
\qquad (4.3\text{-}12)
$$

For the rate constant k_1 the value 4×10^{11} liter $mol^{-1}\ s^{-1}$ was found, indicating that practically every encounter of the dianion with the proton is effective.

Another example where it has been demonstrated that the preceding acid–base reaction takes place predominantly as a volume reaction, is tropylium ion solvolysis [4-34]:

$$(4.3\text{-}13)$$

$$C(HA) \qquad\qquad A(A^-)$$

$$\boxed{\ } + e \longrightarrow Dimer \qquad (4.3\text{-}14)$$

Here the rate constant for the formation of tropylium ion $k_1 = 2 \times 10^6$ liter $mol^{-1}\ s^{-1}$ and that of the reverse reaction $k_{-1} = 50\ s^{-1}$ [4-34] are in reasonable agreement with data found for the same system by relaxation techniques [4-35].

Acid–base equilibria can be more complex as is the case with 1,3-diketones, β-ketoacids, and related compounds [4-36], where keto–enol equilibria are to be considered as well. In the system (where X is an electronegative group):

$$RCOCH_2X + B \xrightleftharpoons[k_1]{k_{-1}} RCO\bar{C}HX + BH^+ \qquad (4.3\text{-}15)$$

$$\updownarrow$$

$$
\begin{array}{l}
RC{=}CHX + B \xrightleftharpoons[k_2]{k_{-2}} RC{=}CHX + BH^+ \\
\ \ \mid \qquad\qquad\qquad\qquad\quad \mid \\
\ \ OH \qquad\qquad\qquad\qquad\quad\ O^-
\end{array}
\qquad (4.3\text{-}16)
$$

$$
\begin{array}{l}
RCOCH_2X + 2e + 2H^+ \longrightarrow RCHCH_2X \\
\qquad\qquad\qquad\qquad\qquad\qquad\quad \mid \\
\qquad\qquad\qquad\qquad\qquad\qquad\quad OH
\end{array}
\qquad (4.3\text{-}17)
$$

only the keto form $RCOCH_2X$ is electroactive in the positive potential range available in aqueous solutions (most investigations have been carried out with $R = C_6H_5$). Equation (4.3-8) for the rate constant k_1 corresponding to the rate of formation of the keto form from the electroinactive carbanion-enolate can be used when pK_1' is measured from polarographic curves and pK_1 (for $K_1 = [RCO\bar{C}HX][H^+]/[RCOCH_2X]$) is evaluated from spectroscopic measurements and keto–enol equilibria.

Experimentally, from the change in wave-height with pH, the value of pK_1' can be obtained. To obtain the value for pK_1, spectra of the compound $RCOCH_2X$ are recorded as a function of pH. The pH-dependence of the absorption band corresponding to [Keto] versus $-pH$, [Enol] versus $-pH$, or [Enolate] versus $-pH$ shows dissociation curves with the same inflexion point. The pH-value corresponding to this inflexion point, pK_{exp} is related to acid dissociation constants of the keto and enol form by the relation $K_{exp} = K_1 K_2/(K_1 + K_2)$ where $K_2 = [Enolate][H^+]/[Enol]$. To separate the values of K_1 and K_2, the keto–enol ratio must be determined in an independent experiment. It is then possible to make use of the fact that $[Enol]/[Keto] = K_1/K_2$ and by transformation

$$K_1 = \frac{K_{exp}}{1 + [Enol]/[Keto]} \qquad (4.3\text{-}18)$$

Thus to calculate the rate constant k_1 it is necessary to record the pH-dependence of polarographic waves and determine pK', record the pH-dependence of the spectra

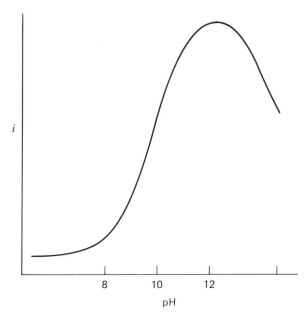

FIG. 4-12. Dependence of polarographic limiting current (i) on pH for reduction of aliphatic aldehydes. Dependence is due to variations of rate of formation of the free aldehydic form with pH according to the reaction scheme (4.3-23) to (4.3-25).

of the 1,3-diketone or related compounds and determine K_{exp}, determine the [Enol]/[Keto] ratio, calculate K_1, and insert pK_1 into the equation for log k_1.

Values found for k_1 are as follows:

$$R = C_6H_5 \quad X = COCH_3 \qquad 1 \times 10^{10} \text{ liter mol}^{-1} \text{ s}^{-1}$$
$$CN \qquad 2 \times 10^8$$
$$COOC_2H_5 \qquad <3 \times 10^9$$

Polarographic measurements are susceptible to the same source of errors as classical methods when the rate constant is calculated using a wrong equilibrium constant. This happens when the equilibrium is more complex than originally assumed. An example of this type is the study of the hydration of aldehydes.

The polarographic wave of aldehydes, which exist in solution predominantly in the hydrated form as geminal diols, are in numerous cases reduced in a wave which increases with pH to a maximum. With further increase in pH, a decrease in current is observed (Fig. 4-12).

Originally, this behavior was ascribed to the fact that the hydrated geminal diol form is electroinactive and not reduced and only the unhydrated free carbonyl form undergoes reduction [4-5]. The following equilibrium was considered:

$$>C\!\!\begin{array}{c} OH \\ \\ OH \end{array} \underset{k_A}{\overset{k}{\rightleftharpoons}} >C{=}O + H_2O \qquad (4.3\text{-}19)$$

$$>C{=}O + 2e + 2H^+ \longrightarrow >CHOH \qquad (4.3\text{-}20)$$

The rate constant k was assumed to be general base catalyzed. The increase in the current was interpreted as due to the increase in the rate constant because of base catalysis. The decrease in current was ascribed to a competitive reaction

$$\begin{array}{c} \diagdown \\ C \\ \diagup \end{array}\!\!\begin{array}{c} OH \\ \\ OH \end{array} \rightleftharpoons C\!\!\begin{array}{c} O^- \\ \\ OH \end{array} + H^+ \qquad (4.3\text{-}21)$$

where the anion of the hydrated form was assumed to be a nonreactive side product.

The calculation of the rate constant was based on measurement of the current i (as a fraction of the total diffusion current i_d) and on the overall hydration–dehydration equilibrium constant $K = [>C{=}O]/[C(OH)_2]$. Numerical values calculated in this way have no physical meaning.

The proposed scheme must express the effect of base catalysis, explain the decrease of current at higher pH-values, and be in accordance with other reactions of the carbonyl group. Moreover, for some aldehydes, the pH where the decrease was observed was found to parallel the pK of the dissociation

$$>C\!\!\begin{array}{c} OH \\ \\ OH \end{array} \rightleftharpoons >C\!\!\begin{array}{c} O^- \\ \\ OH \end{array} + H^+ \qquad (4.3\text{-}22)$$

All these factors were reflected in the reaction scheme [4-6]:

$$>C\underset{OH}{\overset{OH}{<}} + B \underset{k_{-1}}{\overset{k_1}{\rightleftharpoons}} >C\underset{OH}{\overset{O^-}{<}} + BH^+ \tag{4.3-23}$$

$$>C\underset{OH}{\overset{O^-}{<}} \underset{k_{-2}}{\overset{k_2}{\rightleftharpoons}} >C{=}O + OH^- \tag{4.3-24}$$

$$>C{=}O + 2e + 2H^+ \longrightarrow >CHOH \tag{4.3-25}$$

The rate of the dehydration reaction increases with increasing pH because the first acid–base equilibrium (4.3-23) is shifted to the right-hand side. The rate of the elimination reaction is given by $k_2[>C(OH)O^-]$ and thus an increase in anion concentration leads to an increase in the formation of the free carbonyl form $>C{=}O$ and hence to an increase in current with increasing pH.

At the same time, with increasing pH not only does the rate of the reaction with the constant k_2 increase, but also that of the reverse reaction with the constant k_{-2}. The rate of this nucleophilic addition is linearly proportional to the concentration of $[OH^-]$. At sufficiently high pH, $k_{-2}[>C{=}O][OH^-]$ becomes larger than $k_2[>C(OH)O^-]$ and with increasing pH the steady state concentration of the free carbonyl form that can undergo reduction is decreased. Consequently the current at sufficiently high pH-value decreases, because the second equilibrium (4.3-24) is shifted to the left, in favor of the electroinactive anion $>C(OH)O^-$.

To calculate the rate constant k_2 from the increasing portion of the pH-dependence, it is necessary to know the value of the equilibrium constant K_2 which is not directly accessible to measurement. To obtain this value, it is possible to measure the overall dehydration equilibrium constant $K_d = [>C{=}O]/[>C(OH)_2]$ and the acid dissociation constant of the hydrated form: $K_1 = [>C(OH)O^-][H^+]/[>C(OH)_2]$. Both of these measurements can be done spectrophotometrically; the value of K_d can also be determined electrochemically and by NMR. The value of K_2 needed for the evaluation of the rate constant is then $K_2 = K_d K_w/K_1$.

The main limitation in the calculation of k_2 is the uncertainty in the determination of K_1, as with aliphatic aldehydes the pK_1 is in a region where competitive reactions, in particular aldolization, occur. The most successful application seems to be the study of hydration of aldehydes and ketones with adjacent trihalogeno groups such as CCl_3CHO [4-37] and $C_6H_5COCF_3$ [4-38], the hydrated forms of which are more acidic (pK about 11).

Polarography proved to be a useful technique for the study of fast reactions with second-order rate constants in the range from 10^4 to 10^{10} liter mol^{-1} s^{-1}. The essential limitation is the condition that the reaction must take place as a volume rather than a surface reaction, which excludes strongly adsorbable organic substances. If this condition is fulfilled, the rate constants obtained are in good agreement with the data obtained with other techniques.

Experimentally the technique is simple; the whole study can be carried out in a short time, and for numerous types of reactions the mathematical apparatus for the calculation of rate constants is available. Qualitative and partly quantitative treatment has been applied to numerous organic systems and information on more systems is available from polarography than from all other electrochemical methods combined.

On the other hand the treatment of polarographic data needs experience. A sizable portion of the values of rate constants reported in the literature is doubtful, because either the reaction studied was a surface reaction or the actual reaction scheme was more complex than anticipated. But basically, the technique is sound. It should not be forgotten that it was the first technique which enabled measurement of reactions too fast to be studied by flow methods. This occurred more than a decade before the advent of relaxation techniques, which became more widely known due to the pioneering work of Eigen, who was awarded the Nobel Prize.

Nevertheless, there are numerous intriguing problems in the application of polarography to rapid reactions that are still open to investigation.

4.3.1.2. Rotating Disk Electrode [4-28, 4-39–4-42]

The rotating disk electrode can also be used for the recording of current–voltage curves. The electrode consists of a platinum wire (diameter about 1 mm) fused into a glass or teflon tube (Fig. 4.13). The tubing and the wire are ground and the active electrode surface is given by the disk of platinum. Such an electrode is then rotated around a vertical axis with angular velocity of more than 100 c/s.

Under such conditions a diffusion layer is formed at the electrode surface that is carried together with the electrode. To the outer surface of this layer, species are transported from the bulk of the solution by convection–stirring and streaming. Into this layer the electroactive species are nevertheless transported only by diffusion,

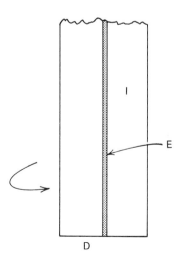

FIG. 4-13. Principle of the rotating disk electrode (D). E, electrode material; l, insulator; arrow indicates direction of rotation.

and for the limiting current it is possible to derive [4-43] the equation (4.3-26):

$$i_{\lim} = 0.62 \, n \, F \, A \, D^{2/3} \, v^{-1/6} \, \omega^{1/2} \, C \qquad (4.3\text{-}26)$$

where n is the number of electrons transferred per molecule; F is the Faraday charge; A, electrode surface; D, diffusion coefficient of the electroactive substance; v, kinematic viscosity; ω, angular velocity of the disk; and C, concentration of the electroactive species. For a given electrode and a given substance the equation simplifies to

$$i_{\lim} = K \, C \, \frac{\omega^{1/2}}{v^{1/6}} \qquad (4.3\text{-}27)$$

where $v = \eta/d$. To obtain a linear relationship between concentration and limiting current, the rotation velocity must be very constant and solutions of the same kinematic viscosity (η) and density (d) must be compared.

For a reaction of the type

$$A \underset{k_{-1}}{\overset{k_1}{\rightleftharpoons}} C \qquad C \overset{\text{el}}{\longrightarrow} P \qquad (4.3\text{-}28)$$

the following equation is approximately valid:

$$\frac{i}{\omega^{1/2}} = \frac{i_d}{\omega^{1/2}} - \frac{D^{1/6} i_k}{\text{const } v^{1/6} (k_1 + k_{-1})^{1/2}} \qquad (4.3\text{-}29)$$

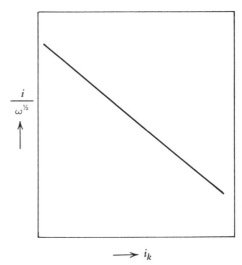

FIG. 4-14. Dependence of the limiting kinetic current (i_k) obtained with a disk electrode on the angular velocity of the disk (ω). Linear plot of $i/\omega^{1/2}$ on i_k indicates process (4.3-28).

When $i/\omega^{1/2}$ is plotted as a function of i_k (Fig. 4-14) a linear plot is obtained. From the slope of the plot it is possible to calculate values of k_1 and k_{-1} provided that $K = k_1/k_{-1}$ is known.

Even when the evaluation is relatively simple, the rotating disk electrode voltammetry has been applied only to a few systems.

Among organic substances the rate of recombination and dissociation of acetic and some other aliphatic carboxylic acids has been investigated [4-44]. The system involved was

$$HA + H_2O \underset{k_{-1}}{\overset{k_1}{\rightleftharpoons}} A^- + H_3O^+ \tag{4.3-30}$$

$$H_3O^+ + e \longrightarrow 1/2\ H_2 + H_2O \tag{4.3-31}$$

and the electrolytic process was thus the reduction of hydrogen ions. The value of k_{-1} determined by the above method in a differential measurement against a solution of strong acid was of the same order of magnitude as the value obtained by relaxation method [4-45]. For pivalic (trimethylacetic) acid a smaller value of k_{-1} was found and this was interpreted as due to steric factors.

The rotating ring-disk electrode is a modification of the rotating electrode, which is particularly suitable for studies of chemical reactions consecutive to the primary electrode process. In the construction of these electrodes the central disk is surrounded by a ring of an insulating material around which the circular ring electrode is concentrically placed (Fig. 4.15). As with the simple disk electrode, the ring-disk electrode is again rotated along the vertical axis. It is then possible to apply a constant potential to the disk electrode and to use the ring electrode for identification and determination of the electrolytically generated product. In such cases the ring electrode acts as an indicator electrode in recording of a current–voltage curve. Alternatively, it is possible to generate the product on the ring and detect it on the disk electrode. In practice it is preferable to keep the width of both the insulation and the electrode rings small. The insulation ring is usually made of a resin or teflon and ensures also the mechanical stability of the electrode. The concentric arrangement is important. The characteristic parameter with this type of

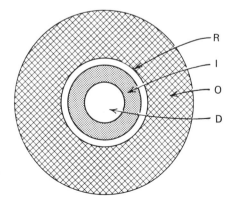

FIG. 4-15. Frontal view of the ring-disk electrode. D, disk electrode; I, insulator ring; R, ring electrode; O, outer insulating shell.

electrode is N, the collecting efficiency of the ring disk electrode. It is defined as

$$N = i_r/i_d \qquad (4.3\text{-}32)$$

where i_r is the current on the ring electrode and i_d is the current on the disk electrode. The following system is considered:

$$A + ne \xrightarrow[\ E_1\ ]{\text{disk el}} B \qquad (4.3\text{-}33)$$

$$B \xrightarrow[\ k\ ]{\text{chem reaction}} C \qquad (4.3\text{-}34)$$

$$C + me \xrightarrow[\ E_2\ ]{\text{ring el}} D \qquad (4.3\text{-}35)$$

where D can be identical with A. Reactant B is generated at the disk electrode where the rate of its generation can be governed. In the vicinity of the electrode a flow of the reactant is established, but contrary to flow methods the flow is laminar. The rates of this flow can be described at each point, because the convectional velocities are known functions of the distance of the moving particle from the center of the electrode in radial and normal directions. The species B which is formed at the disk electrode is transported by diffusion to the bulk of the solution and is carried away in a radial direction by convection towards the outer diameter. The ring electrode then monitors the resulting flow.

The collecting efficiency is measured under conditions when the chemical reaction does not take place (N) and when the chemical reaction with rate constant k transforms the primary electrolysis product (N_k). Their ratio is then given by the expression [4-41]:

$$\frac{N}{N_k} = 1 + 1.28 \left(\frac{\nu}{D}\right)^{1/3} \frac{k}{\omega} \qquad (4.3\text{-}36)$$

The validity of this expression is limited to conditions when the value of N_k is neither very small when compared with N nor has a value comparable to N, which can be achieved by a proper choice of the rotation speed. If this condition is fulfilled, the ratio of the thickness of the convection layer to the thickness of the reaction layer lies between 0.5 and 3.5.

To determine the value of k from the last equation the value of $1/N_k$ is plotted as a function of $1/\omega$ (where the velocity of rotation (ω) is measured in revolutions per second). From the slope of the linear plot the first-order rate constant k can be determined and can be further handled.

The method was used to study the hydrolysis of the electrolysis product of p-aminophenol [4.46].

$$+ 2e + 2H^+ \qquad (4.3\text{-}37)$$

$$\text{(structure)} + H_2O \xrightarrow[k]{\text{chem. r.}} \text{(structure)} + NH_3 \qquad (4.3\text{-}38)$$

$$\text{(structure)} + 2e + 2H^+ \xrightarrow[E_2]{\text{ring el}} \text{(structure)} \qquad (4.3\text{-}39)$$

A carbon paste electrode was used for both ring and disk electrodes.

The main advantage of the rotating disk electrode is the possibility of application of rigorously derived hydrodynamic equations. The electrode can also be relatively easily manufactured. It is essential that the rotation velocity be kept constant.

The advantage of the rotating ring-disk electrode, apart from the possibility of studying reactions of products and intermediates directly, is that the chemical reaction studied takes place in the solution rather than at the electrode surface and thus the surface effects, including the double-layer effects and those of the electric field, are negligible. The method makes it possible [4-41], [4-42] to determine first-order rate constants in the range from 0.04 to 1000 s^{-1}.

4.3.1.3. Potentiostatic Method (Chronoamperometry) [4-26]

In potentiostatic methods the potential of an indicator electrode is kept constant and the change of current with time is followed from the moment of application of the potential. The indicator is usually a mercury pool or hanging mercury drop electrode, and the potential is kept constant by means of a potentiostat. The current change resulting in the processes in the vicinity of the electrode is recorded by an oscilloscope.

Evaluation of i–t curves (for $t \ll 1$ s) enables a determination of the rate of a chemical reaction accompanying the electrode process. The recombination rate of monochloroacetic acid was determined in this way [4-47]. The technique can also be used for the study of consecutive reactions. The range of values of first-order rate constants which can be measured (0.04 s^{-1} to 1000 s^{-1}) is the same as for ring-disk methods.

For slower reactions the changes in the bulk concentrations of electroactive species, resulting in electrolysis of stirred solutions can be followed [4-47a].

Numerous special electrochemical techniques have been developed, which involve applied potential either constant or varying with time, such as double potential step chronoamperometry. Other methods, particularly useful for the study of chemical reactions of electrochemically generated species, include linear sweep voltammetry and cyclic voltammetry. For the study of chemical reactions of species originally present in the solution application of such methods is less convenient than that of dc polarography. The same applies for methods where sinusoidal or pulsed voltage is superimposed over an increasing voltage ramp. For applications of such methods and their limitations the reader is referred to the original literature [4-22, 4-26–4-28].

4.3.1.4. Method of a Stationary Field [4-48]

If a solution containing only a few ionizable species is placed between two electrodes placed close (e.g., 0.01 cm apart) and if a high voltage is applied to those two electrodes, the strong electric field transports all the ions towards the electrodes and the saturated current which results is governed only by the rate of formation of new ions. The method can be used only for solutions with extremely low concentrations of ions ($C < 10^{-7} M$) and is thus suitable for following ionic reactions in nonaqueous media.

4.3.1.5 Chronopotentiometric (Galvanostatic) Methods [4-26, 4-28]

In all electrochemical techniques discussed so far the applied voltage (or potential) was kept constant and current measured. In chronopotentiometry the situation is reversed: On a system consisting of indicator and another working electrode, current is imposed and by means of a third, reference electrode the potential of the indicator electrode is measured as a function of time. Mercury (pool or hanging drop) or platinum electrodes are used as indicator, the reference electrode is connected with a capillary (Luggin capillary), the tip of which is placed about 1 mm from the surface of the indicator electrode (Fig. 4-16). If a mercury pool electrode is used, the surface of the glass vessel is treated with silicone or the vessel is manufactured from teflon. Hydrofobization of the surface prevents creeping of the solution under the mercury layer. For high current densities and short transition times τ [i.e., the time from the beginning of electrolysis and the instant when the potential begins to change strongly with time (Fig. 4-17)] the use of a hanging mercury drop electrode proves to be advantageous. Undesirable effects of convection to which this type of electrode is more susceptible do not play an important role during short time intervals.

A simple galvanostat which imposes a time-independent current on the electrodes, consists of a voltage source (100–200 V), adjustable resistance R_X and electrolytic cell in series (Fig. 4-18). More than 99% of the voltage drop occurs on the resistance R_X so that changes in the vessel in the course of electrolysis are practically without effect on polarizing current. Before the beginning of the experiment the current flows through an auxiliary resistance R_c approximately equal to the resistance of the electrolytic cell. This enables us to adjust the resistance R_X in such a way so as to achieve a chosen current density, measured by an ammeter, A. To record E–t curves recorders are used for $\tau \geq 10$ s, for faster reactions (shorter transition times) an oscilloscope is used.

For a diffusion controlled system in an unstirred solution, when conditions for

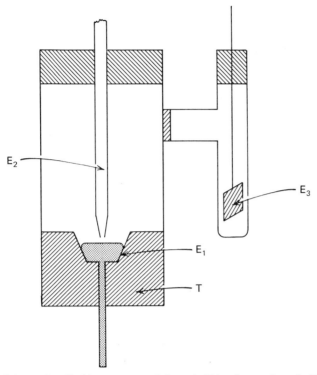

FIG. 4-16. Scheme of a cell with a mercury pool electrode (E_1), reference electrode (E_2), and second working, for example, platinum, electrode (E_3), T insulator, Teflon for example.

linear diffusion are valid (i.e., for pool or plate electrodes) the following equation can be derived

$$i\tau^{1/2} = \sqrt{\frac{\pi}{2}} \, nF \, A \, D^{1/2} \, C \qquad\qquad (4.3\text{-}40)$$

where A is the electrode surface and all other symbols have been previously defined.

For a given electrode and a given electroactive substance the first five factors on the right-hand side remain constant and hence

$$i\tau^{1/2} = \text{const } c \qquad\qquad (4.3\text{-}41)$$

The imposed current i is chosen in such a way that on one side the convection of components of the electrolysed solution would not affect diffusion transport, on the other it would be possible to neglect the potential changes resulting from the charging of the double-layer and adsorption phenomena. In practice the periods used are between 1 millisecond and 2 min.

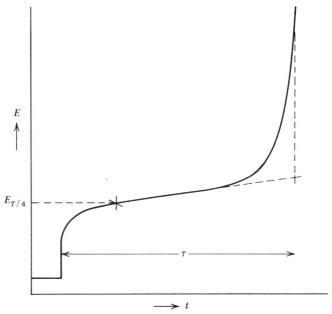

FIG. 4-17. Chronopotentiometric curve, dependence of the measured potential (E) as a function of time (t). τ, transition time; $E_\tau/4$, quarter transition time potential (corresponds to $E_{1/2}$ in polarography).

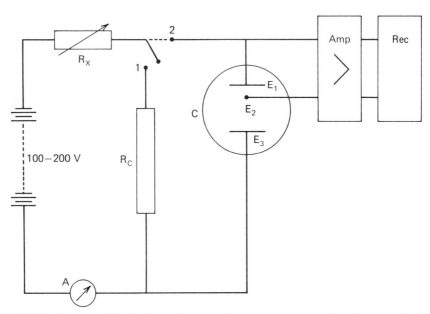

FIG. 4-18. Scheme of the circuit for chronopotentiometric measurements. C, electrolytic cell; E_1, indicator electrode; E_2, Luggin capillary; E_3, second working electrode; R_C, auxiliary resistor; R_X, variable resistor; A, ammeter; Rec, recorder or oscilloscope with an amplifier.

The rates of fast chemical reactions of the type:

$$A + B \underset{k_{-1}}{\overset{k_1}{\rightleftharpoons}} C \qquad (4.3\text{-}42)$$

$$C + ne \xrightarrow[E_C]{} Product \qquad (4.3\text{-}43)$$

can be calculated from a comparison of the transition time (τ_d) under diffusion controlled conditions when the transformation of the electroinactive species A into electroactive compound C is fast, with the transition time τ_k obtained under conditions of governing chemical reaction.

If

$$erf\,[(k_1 + k_{-1})^{1/2}\,\tau^{1/2}] \approx 1 \qquad (4.3\text{-}44)$$

the relationship between τ_k and τ_d is given by [4-50].

$$i_0\tau_k^{1/2} = i_0\tau_d^{1/2} - \frac{\pi^{1/2}}{2K\,(k_1' + k_{-1})^{1/2}}\,i_0 \qquad (4.3\text{-}45)$$

where i_0 is the current density used and $k_1' = k_1[B]$. The dependence of $i_0\tau_k^{1/2}$ on i_0 is linear (Fig. 4.19) and from the slope it is possible to calculate $K(k_1' +$

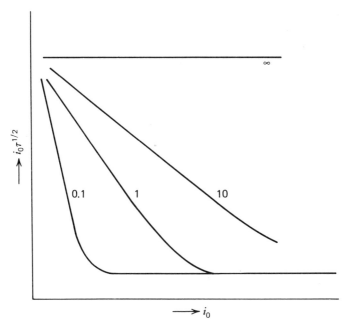

FIG. 4-19. Dependence of the product $i\tau^{1/2}$ on the value of the imposed current (i) for various values of the product $K(k_1 + k_{-1})^{1/2}$ (numerical values of the product given).

$k_{-1})^{1/2}$, and if K is independently measurable, the value of $(k_1' + k_{-1})$ can be obtained.

From the dependence of $(k_1' + k_{-1})$ on [B] it is possible to calculate k_1 and k_{-1}.

For $\tau > 0.001$ s (which is the experimental limit) it is possible to calculate $K(k_1 + k_{-1})^{1/2} < 1$ s$^{-1/2}$. Chronopotentiometric methods can thus be applied to considerably slower reactions than it is possible for classical polarographic methods which have the limit of $K(k_1 + k_{-1})^{1/2} < 5$ s$^{-1/2}$. Systems which show diffusion-controlled behavior in classical polarography can still be affected by the rate of chemical reactions in chronopotentiometry.

For high current densities the equation (4.3-45) simplifies to (4.3-46)

$$\lim i_0 \to \infty, \; i_0 \tau_k^{1/2} = \frac{i_0 \tau_d^{1/2}}{1 + 1/K} \tag{4.3-46}$$

Hence at high current density and for $K \gg 1$ the value of τ_k reaches the value of τ_d and becomes independent of the current density i_0. The magnitude of the rate constant determines whether or not it is possible to reach such current densities where the condition of independence is fulfilled. With increasing value of the product $K(k_1' + k_{-1})^{1/2}$, the range of current densities at which it is possible to reach independence of the product $i_0 \tau_k^{1/2}$ on current density is shifted towards higher current densities (Fig. 4-19).

This approach has been applied [4-51] to the oxidation of p-phenylendiamine in acidic media which is described by the scheme:

$$\tag{4.3-47}$$

$$\tag{4.3-48}$$

In addition to the simple chronopotentiometry where the current is kept constant, it is possible to modify these methods in such a way that it is possible to alter the direction of current at or before the transition time, or reverse the direction of current repeatedly (cyclic chronopotentiometry), to change the magnitude of the imposed current in steps, continuously linearly, or exponentially with time or with the square root of time [4-26–4-28].

4.3.1.6. Evaluation of Electrochemical Methods

The ranges of values of reaction rates in which the most frequently used electro-chemical techniques can be adopted are not significantly different:

Polarography	$k_1[H+] \leqq 10^5 \text{ s}^{-1}$
Chronopotentiometry	$k_1[H+] \leqq 10^6 \text{ s}^{-1}$
Rotating disk electrode	$k_1[H+] \leqq 5 \times 10^5 \text{ s}^{-1}$

Chronopotentiometry thus enables us to study reactions faster by one order of magnitude.

The decision whether polarography is suitable for the solution of a given mechanistic study is simplified by the availability of extensive, published, experimental material [4-22–4-26, 4-31]. A comparison with reported data makes it possible to decide (1) if the application of polarography is at all possible and (2) to pay attention to factors which limit its application, such as adsorption.

Chronopotentiometry has predominantly been used for the study of reactions involving products of electrolysis rather than starting materials, as has cyclic voltammetry.

It has not yet been established to which degree the chemical reactions occurring in the vicinity of an electrode are affected by the electrical fields present. Only in a few instances have the values of rate constants been obtained in electrochemical studies compared with values found by using other techniques.

4.3.2.   Optical Methods

In addition to the method of flash photolysis, discussed briefly among relaxation techniques, two other types of techniques were used for the study of fast reactions based on photochemical reactions. In both of these techniques, the reaction mixture is irradiated by light or other radiation. In the first, so-called photostationary method, radiation generates a new species (usually a free radical) whose reaction is followed. In the second technique the radiation causes creation of an excited state and reactions of the molecules in excited states (which usually differ from reactions of the ground state) are studied.

4.3.2.1.   Photostationary Methods

In photostationary methods the reaction mixture is irradiated by light of a chosen wavelength and given intensity in such a way that a steady state is established, that is, as much radical is generated by irradiation as is consumed by the chemical reaction. Although most frequently visible or UV light is used, ultrasonic waves or ionizing radiation can be applied similarly. The reaction involved can be in general

$$A + h\nu \longrightarrow B \qquad\qquad\qquad (4.3\text{-}49)$$

$$B + C \longrightarrow D \qquad\qquad\qquad (4.3\text{-}50)$$

Here only the simplest case of dimerization of radicals $B\cdot$ generated by irradiation of molecules B_2 will be discussed:

$$B_2 + h\nu \xrightarrow{\ v_1\ } 2B\cdot \qquad\qquad\qquad (4.3\text{-}51)$$

$$B_a\cdot + B_b\cdot \xrightarrow[k]{\ v_2\ } B_2 \qquad\qquad\qquad (4.3\text{-}52)$$

At a steady state, when concentration of radicals $[B\cdot]_{st}$ remains constant, $v_1 = v_2$ and the rate constant k is defined as

$$k = \frac{\phi\, q}{[B\cdot]_{st}^2}$$

where k is the second-order rate constant [liter mol^{-1} s^{-1}] for the reaction with velocity v_2 under the assumption that radicals $B_a\cdot$ and $B_b\cdot$ were formed from different molecules B_2. ϕ is the quantum yield expressing the ratio of pairs of radicals which are permanently separated, q is the rate of absorption of radiation in Einstein units liter^{-1} s^{-1}.

The treatment has been simplified, since radicals $B\cdot$ can also react with the parent compound B_2 and form a polymer-radical (which can react further), or with the solvent.

To calculate k it is necessary to know q, ϕ, and $[B\cdot]_{st}$. Rate of absorption (q) can be determined spectrophotometrically or actinometrically [4-52]. The quantum yield ϕ (always smaller than 100% because of the possibility of recombination of freshly formed, twin radicals and because of possible interaction with the solvent) can be determined by means of scavengers and by tagging. The scavenger must react rapidly with radicals $B\cdot$ but not with molecules B_2. Tagging can be done either chemically (by introducing a substituent which affects the reaction rate little or in a predictable way) or by isotopes on one part of the molecule B—$B*$ and the resulting products (e.g., B—B and $B*$—$B*$) are analyzed.

The determination of radical concentration poses a difficult problem and can be achieved by ESR or UV spectra by comparison with signals of solutions of known radical concentration.

The mean lifetime of a radical can be determined by using the technique of intermittent illumination (rotating sector) in which the sample is intermittently irradiated and kept in the dark [4-53]. After irradiation the concentration of radicals increases and after interruption of irradiation decreases (Fig. 4-20). The magnitude of the difference between the increase and decrease depends on the ratio of the period of illumination and darkness to the mean lifetime. By changing the ratio of illumination and darkness the mean lifetime can be estimated.

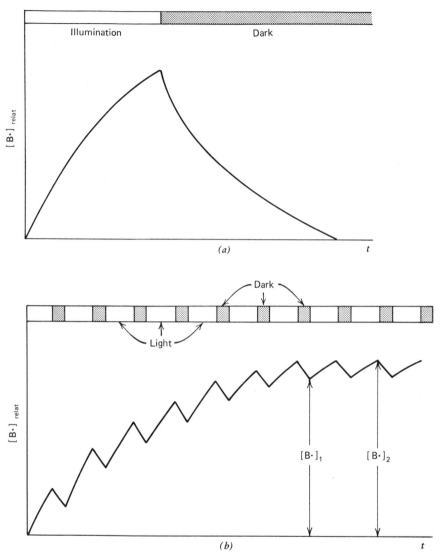

FIG. 4-20. Changes in relative radical concentration ($[B\cdot]_{relat}$) with time for (*a*) single illumination period; (*b*) for intermittent illumination [according to E. F. Caldin, *Fast Reactions in Solution*, Wiley, New York, 1964, p. 128].

4.3.2.2. Fluorescence Methods

Numerous organic substances (e.g., anthracene, β-naphthol, or quinine) after irradiation by light of the proper wavelength (usually ultraviolet) show fluorescence, which results in an emission of light.

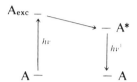

This emission is caused by the fact that some molecules which by absorption of radiation hv are brought to an excited state A_{exc} undergo nonradiative transformation to a lower excited state $A*$, and on return to the ground state A emit radiation hv [4-54]:

$$\text{Excitation} \qquad A + hv \longrightarrow A_{exc} \qquad\qquad (4.3\text{-}53)$$

$$A_{exc} \longrightarrow A*$$

$$\text{Emission} \qquad A* \xrightarrow{\ k_f\ } A + hv^1 \qquad\qquad (4.3\text{-}54)$$

The emitted light has always longer wavelength than the exciting radiation. Even when the interval between formation of A_{exc} and emission of $A*$ is very short (of the order of 10^{-7}–10^{-9} s), a number of the molecules in the excited state can be deactivated by reaction with the solvent or by "internal quenching," which is caused by interaction of excited and nonexcited molecules of the fluorescent compound A. These competing reactions can be expressed by (4.3-55)

$$A* \xrightarrow{\ k_i\ } A \qquad\qquad (4.3\text{-}55)$$

If the fluorescent substance is irradiated from a source of constant intensity, a steady state is established, in which the rate of formation of excited molecules is just equal to the deactivation of excited molecules by all the deactivating reactions. At constant intensity of the exciting radiation the intensity of fluorescence is a linear function of the concentration of molecules in the excited state.

In concentrated solutions it is possible to measure the fluorescence in the direct optical path using a normal spectrophotometer, since all of the exciting radiation is absorbed.

In dilute solutions used to eliminate effects due to internal quenching (k_i) the emission is measured in special spectrofluorimeters in a direction perpendicular to the direction of the exciting radiation [4-55].

The fluorescence spectrum [log $I = f(\lambda)$] is measured first and fluorescence intensities can then be measured at a given wavelength.

It is essential to remove all absorbing and fluorescing impurities since concentrations of fluorescent compounds as small as 10^{-12} M can be detected. The solution is purged with nitrogen or carbon dioxide to remove oxygen (which often quenches fluorescence). The cell temperature must be controlled.

The ratio of fluorescence emission for two solutions, I_a/I_b, is equal to the ratio of relative quantum yields of fluorescence, φ_a/φ_b. The latter is related to the mean lifetime of the excited molecule τ_0 by the relations:

$$\tau_0 = \frac{1}{k_f + k_i} \tag{4.3-56}$$

$$\varphi_0 = k_f \tau_0 \tag{4.3-57}$$

After addition of compound B to a solution containing compound A and its molecules in excited state A*, fast reactions between A* and B (different from the reaction between A and B) can take place

$$A* + B \xrightarrow{\;k\;} \text{Products} \tag{4.3-58}$$

This reaction with constant k is competitive with deactivation reactions characterized by k_i and k_f. The original quantum yield φ_0 will diminish to φ for which

$$\varphi = \frac{k_f}{k_f + k_i + k\,[B]} \tag{4.3-59}$$

and hence

$$\varphi = \frac{\varphi_0}{1 + k[B]\tau_0} \tag{4.3-60}$$

$$\frac{\varphi_0}{\varphi} = 1 + k[B]\tau_0 \tag{4.3-61}$$

As [B] is known, it is possible from a measurement of φ_0/φ to determine the relative rate constant $k\tau_0$.

The absolute value of k can be found, if the mean lifetime of the excited state τ_0 is determined. This can be done using excitation by short ($<10^{-9}$ s) pulses, by a radiation modulated by high frequency, or by measurement of the area under the fluorescence emission band, or from the fluorescence quantum yield [4-54]. If necessary, only the relative value ($k\tau_0$) is compared, as it is possible, for example, in the investigation of the effect of temperature, ionic strength, pH, or solvent on the reaction rate.

Fluorescence measurements enable studies of fluorescence quenching

$$A* + B \longrightarrow A + B + \text{energy} \tag{4.3-62}$$

formation of adducts

$$A* + A \longrightarrow A_2* \qquad (4.3\text{-}63)$$

$$A* + B \longrightarrow A*B \qquad (4.3\text{-}64)$$

energy transfer

$$A* + B \longrightarrow A + B* \qquad (4.3\text{-}65)$$

and acid–base reactions

$$A*H + B \longrightarrow A* + BH+ \qquad (4.3\text{-}66)$$

$$A* + BH^+ \longrightarrow A*H + B \qquad (4.3\text{-}67)$$

An example is the acid–base reaction of acridine. In nonirradiated aqueous solutions acridine behaves like a weak base with a pK_a of 5.45. Acidic solutions of acridine after irradiation with a proper wavelength give a green fluorescence and this color does not change up to pH 9. At pH > 11.8 the acridine solutions show a blue fluorescence. The green fluorescence was ascribed to the excited cation, the blue one to the excited free base. The dependence of the intensity of the emitted light at the proper wavelength on pH has a shape of a dissociation curve with $pK_{exc} = 10.65$

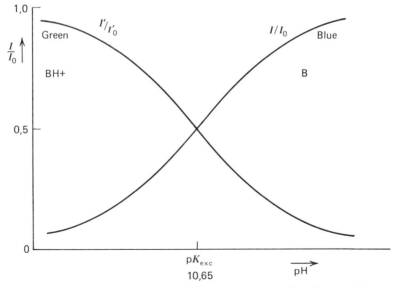

FIG. 4-21. Dependence of the relative intensity I/I_o of fluorescence of acridine on pH. Decrease of the intensity of the green fluorescence of the cation BH^+ on pH in the shape of a dissociation curve, increase in intensity of the blue fluorescence of the free base. Inflection point corresponds to the pK-value of the excited state ($pK_{exc} = 10.65$).

(hence $pK_{exc} - pK_a = 5.21$) (Fig. 4-21). After the addition of a proper proton donor, for example, NH_4^+, to a solution containing predominantly the free base of acridine, the following reactions take place:

$$Acr* + H_2O \rightleftharpoons Acr* H^+ + OH^- \qquad (4.3\text{-}68)$$

$$Acr* + NH_4^+ \underset{k_{-1}}{\overset{k_1}{\rightleftharpoons}} Acr* H^+ + NH_3 \qquad (4.3\text{-}69)$$

For a small value of the ratio $[NH_3]/[NH_4^+]$ it is possible to neglect the reaction with rate constant k_{-1}. From the increase in the intensity of the green fluorescence the value of $k_1 = 5.7 \times 10^8$ liter mol^{-1} s^{-1} has been found. Similarly the rate constants for the reactions of other proton donors have been found, for example for boric acid $k_1 = 3.8 \times 10^8$ liter mol^{-1} s^{-1}.

A condition for the use of fluorescence methods is the knowledge of the value of the rate constant of deactivation of the excited molecule in pure solvent. Only from the known value of this rate constant (k_i) it is possible to calculate the values of second-order rate constants corresponding to the reaction between the excited molecule and the reagents quenching the fluorescence. Determination of the mean lifetime τ_0 usually limits the accuracy of determination of the value of k with an error range which is usually greater than $\pm 10\%$.

4.3.3. Methods Based on Magnetic Resonance

At proper values of a strong magnetic field superimposed by periodically changed electromagnetic one a resonance between the resulting magnetic field and the nuclear (in NMR) or electron spin moment (ESR) takes place. Energy is absorbed for the transfer of the molecule from one energetic state to another which is shown by a peak in the shape of a resonance curve in NMR or ESR spectra. The position of the peak (or the energy involved in the transition between the two states) depends on the chemical nature of the groups or radicals present, the area under the peaks shows the abundance of the given group and the fine structure of peaks reflecting the spin–spin coupling reflects the effects of the environment on the energy levels involved. The width of a given band in NMR depends on the lifetime of the given nucleus in a given spin state.

The relationship between the band width in ESR ($\delta'\nu$) and the mean lifetime of the electron in one of the two possible energetic states (τ) can be in the simplest case defined by

$$\delta'\nu = \frac{1}{\pi T} \qquad (4.3\text{-}70)$$

and was derived from Heisenberg uncertainty principle. If it is possible to prove that other factors affecting the band width are small, it is possible to calculate from the band width the lifetime of the radical in one energetic state. The values of τ found in this way are of the order of 10^{-6} to 10^{-9} s.

Similarly, the width of a band in NMR is given by Bloch equation which for small amplitudes of the oscillating magnetic field can be simplified to

$$\delta \ (c/s) \ = \ \frac{1}{\pi T_2} \tag{4.3-71}$$

where T_2 is relaxation time characterizing the lifetime of the nucleus in a given energy state, considering so-called spin–spin relaxation where the energy is transferred from nucleus to nucleus by mutual exchange of spins. This relaxation time T_2 in the absence of a chemical reaction shows values of 1–10 s.

Application of broadening of ESR bands for the study of reactions of radicals will be discussed first. Addition (to a solution containing the studied radical) of a substance, which reacts with the radical, results in decrease of the lifetime of the radical and therefore to a broadening of the band. If the effect of other factors on band width can be excluded, it is possible from the widening to calculate the rate constant of the reaction between the radical and the added reagent.

For a reaction

$$A_1\cdot \ + \ A_2 \underset{k'_-}{\overset{k'_+}{\rightleftharpoons}} A_1 \ + \ A_2\cdot \tag{4.3-72}$$

the rate constants k'_+ and k'_- are equal and the rate is defined by

$$\frac{d[A\cdot]}{dt} \ = \ - \ k'[A\cdot][A] \tag{4.3-73}$$

The mean lifetime of the radical $A\cdot (\tau)$ is given by the concentration of the radicals divided by the rate of their disappearance by reaction. Thus

$$\tau \ = \ -\frac{[A\cdot]}{d[A\cdot]/dt} \ = \ \frac{1}{k'[A]} \ = \ \frac{1}{k} \tag{4.3-74}$$

where k is first-order rate constant, for which

$$k \ = \ k'[A] \ = \ \frac{1}{\tau} \ = \ \pi\delta'\nu \tag{4.3-75}$$

From increase in width ($\delta'\nu$) it is thus possible directly to calculate the value of the first-order rate constant.

An example of this type can be reaction of naphthalenide radical anion $C_{10}H_8^{\bar{\cdot}}$ with added naphthalene [4-56]. Naphthalene dissolved in tetrahydrofuran (THF) shows, after addition of metallic sodium, an ESR spectrum of the radical which has 25 main bands 27 gauss apart, accompanied by a fine structure resulting from interaction of the magnetic moment of the electronic spin with the magnetic moment of the nuclear spin of protons.

Addition of naphthalene to the solution of the naphthalenide anion results in widening of the bands due to reaction

$$C_{10}H_8^{\cdot} + C_{10}H_8 \overset{k}{\rightleftharpoons} C_{10}H_8 + C_{10}H_8^{\cdot} \qquad (4.3\text{-}76)$$

In solutions containing $1 \times 10^{-3}\, M$ sodium naphthalenide the broadening of the band was measurable in the presence of $0.1\, M$ naphthalene and higher. The measured rate constant depends on the solvent (where dimethoxyethane is denoted as DMOE) and the nature of the cation:

Cation	Solvent	k (liter mol^{-1} s^{-1})
K$^+$	DMOE	$(7.6 \pm 3) \times 10^7$
Na$^+$		3×10^8
K$^+$	THF	$(5.7 \pm 1) \times 10^7$
Na$^+$		4×10^6
Li$^+$		$(4.6 \pm 3) \times 10^8$

Higher value of k in DMOE than in THF is correlated to the fact that in DMOE sodium naphthalenide is dissociated, whereas in THF a formation of ion pairs plays a role.

The width of the NMR bands increases in the presence of reactions of the absorbing substances that have relaxation times comparable with T_2. The observed increase is given by

$$\delta'\nu = \frac{1}{\pi\tau_{HA}} \; (\text{c/s}) \qquad (4.3\text{-}77)$$

so that the overall band width is given by

$$\delta'\nu = \frac{1}{\pi T_2} + \frac{1}{\pi\tau_{HA}} \qquad (4.3\text{-}78)$$

where τ_{HA} is the mean lifetime of proton in substance HA which is responsible for the absorption band and is equal to relaxation time. The width of the band can be also expressed by means of the measured relaxation time T_2' as $1/\pi T_2'$. Then

$$\frac{1}{\tau_{HA}} = \frac{1}{T_2'} - \frac{1}{T_2} = \pi\delta'\nu \qquad (4.3\text{-}79)$$

and the value of τ_{HA} is directly proportional to the difference in relaxation times T_2 in presence and absence of a chemical reaction.

To demonstrate techniques used in determination of reaction rate, NMR spectra of equimolar mixtures of two substances HA and HB will be considered first. In the absence of interaction of the two substances the proton NMR will show two absorption bands which can differ by 100 c/s and the position of each of them will primarily depend on chemical shifts in HA and HB. In the present discussion the spin–spin interaction can be neglected.

If, on the other hand an exchange of protons between substances HA and HB can take place, the following system will be involved:

$$HA + H'B \underset{k_-}{\overset{k_+}{\rightleftharpoons}} H'A + HB \qquad (4.3\text{-}80)$$

If the rate of exchange increases with increasing temperature or concentration of hydrogen ions, it is possible by increase of temperature or pH to change the NMR spectrum (Fig. 4-22). A specific example can be reaction

$$CH_3COCH_2COCH_3 + CH_3COOH'$$

$$\underset{k_-}{\overset{k_+}{\rightleftharpoons}} CH_3COCHH'COCH_3 + CH_3COOH \qquad (4.3\text{-}81)$$

To determine the rate constant it is possible to make use of characteristic properties of the spectra, which differ in the reaction rate, in three regions.

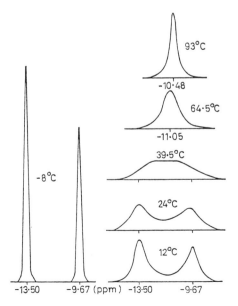

FIG. 4-22. Line broadening in proton NMR spectra. Spectra of acetylacetone and acetic acid (58:42) recorded at various temperatures. Absorption bands due to the OH groups, chemical shifts in p.p.m. relative to chemical shift of the methyl protons in the enol form of acetylacetone. (According to E. F. Caldin, *Fast Reactions in Solution*, Wiley, New York, 1964, p. 233).

(a) SLOW EXCHANGE, AT $\tau\Delta\nu_0 \gg 1$.   For such range of experimental variables (such as temperature, pH, or concentration of one reactant) where the exchange reaction is not too fast, two bands are observed in the NMR spectrum which are separated and at the same frequencies as under conditions when the reaction does not take place. The higher the reaction rate, the shorter the mean lifetime (τ) of the species HA and HB and the longer is the period in which the proton is in the position when it is uncombined either with A^- or with B^-. The shorter the τ the less determinate is the energy and the broader the band. With increasing reaction rate the widening of both bands by $\delta'\nu$ (measured at half the peak height) is observed and

$$k = \tau^{-1} = 2\,\pi\delta'\nu \qquad\qquad (4.3\text{-}82)$$

Separated bands are observed as long as $\tau > 1/\Delta\nu_0$. The observed lifetime broadening is thus analogous to broadening of ESR bands by a chemical reaction.

(b) MEDIUM RANGE OF EXCHANGE RATE ($\tau\Delta\nu_0 \approx 1$).   Under experimental conditions where the above condition is fulfilled, it is possible to observe either overlapping bands or formation of one, broad indistinct band. If $\Delta\nu$ is the difference of frequencies separating the ill-defined peaks at exchange and $\Delta\nu_0$ under conditions when the exchange reaction does not take place, then

$$k = \tau^{-1} = \pi\,\sqrt{2}\,(\Delta\nu_0^2 - \Delta\nu^2)^{1/2} \qquad\qquad (4.3\text{-}83)$$

If a single band is formed, a proton interacts with the average value of the resonance frequency and cannot be localized either at HA or at HB but somewhere in between. The energy is indeterminate and henceforth the observed band wide. At reaction conditions at which the two bands just coalesce to give one band, the following relation is valid

$$k = \tau^{-1} = \pi\,\sqrt{2}\,\Delta\nu_0 \qquad\qquad (4.3\text{-}84)$$

Because $\Delta\nu_0$ is usually of the order of 100 c/s for proton NMR, the coalescence of the bands correspond to a lifetime of the order of $\tau \approx 10^{-2}$ s.

(c) FAST EXCHANGE, AT $\tau\Delta\nu_0 \ll 1$.   Under reaction conditions when the reaction can be considered as fast (when compared with previous examples) the accidental variations of energy from the mean value are less frequent and the band has a sharper form. To find the value of the rate constant under such conditions it is first necessary to determine the lower limit of the band width, that is, the narrowest band-width corresponding to $\tau = 0$. This is achieved by gradually changing reaction conditions so that the reaction rate increases and following the band width until no further increase in rate results in decrease in band width. This band width at $\tau = \nu$ is compared with band widths measured under conditions when for a given chemical shift only one single band is observed which is wider than for $\tau = 0$. Under these conditions the reaction takes place rapidly, but not extremely rapidly. The increase in half-width of the band measured under such conditions ($\delta''\nu$ for $\tau \ll T_2$) makes it possible to calculate the rate constant by means of expression

$$k = \tau^{-1} = \frac{4\pi\Delta\nu_0^2}{\delta''\nu} \qquad (4.3\text{-}85)$$

(d) COMPARISON OF THE SHAPE OF SPECTRA. In addition to the techniques mentioned above based on broadening of two or one band, or on finding conditions under which the two bands just merge, it is also possible to determine the mean lifetime and the rate constants from the shape of the NMR spectra (Fig. 4-23). To achieve this, shape of the resonance curves are calculated by means of Bloch equations for various chosen values of τ_{HX} and such curves are compared with those obtained experimentally.

Another example of comparison of theoretical curves with experimental ones involves systems which show a simple doublet. Such curves are obtained under conditions when the number of molecules HA and HB involved in the exchange reaction is equal and so are the lifetimes of proton in molecule HA (τ_{HA}) and HB (τ_{HB}). The total lifetime of the proton τ is then given by the relation $\tau_{HA} = \tau_{HB} = 2\tau$. Comparison of curves calculated for varying $\Delta\nu_0$ (which is the difference in frequencies of both bands before the reaction) allows us to estimate the value of τ. The effect of the relaxation time T_2 can be neglected.

One example concerns exchange reactions involving alcohols. Addition of a strong acid or a strong base simplifies considerably the NMR spectra of alcohols. The triplet corresponding to the OH group becomes a singlet, quadruplet of the

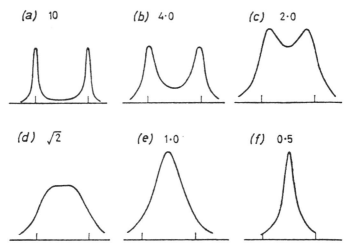

FIG. 4-23. Line broadening in proton NMR spectra. The effect of chemical exchange between acids HA and HB with rate increasing from (a) to (f). Values of $2\pi\tau\Delta\,\nu_0$ given (τ relaxation time and ν_0 initial separation of the lines), intensity scale is arbitrary. (According to E. F. Caldin, *Fast Reactions in Solution*, Wiley, New York, 1964, p. 236.)

CH_2 group with fine structure becomes a simple quadruplet. This simplification is attributed to the prevention of the spin–spin interaction and to the exchange of hydroxyl protons with other molecules. In alkaline media the exchange reaction can occur between the alcohol RCH_2OH and water, hydroxide ions, or alkoxy ions, in acidic media between the alcohol RCH_2OH and water, hydrogen ions, or $RCH_2OH_2{}^+$.

With increasing concentration of the acid or base the change in the NMR spectrum of the alcohol is gradual: First only a broadening of the peaks is observed, until at about 10^{-5} M acid or base the change in the shape of spectra (resulting in formation of singlet from triplet and decrease in fine structure) is complete. In aqueous solutions of alcohol the addition of acid or base results only in widening of the water band. To estimate the rate of the proton transfer it is possible to use the change in shape of the CH_2 band, the increase in width of the water band, and changes in the shape of the OH band.

The comparison of the shape of the CH_2 band with the theoretical curve makes it possible to determine the overall first-order rate constant k_T. This consists of the contribution corresponding to reaction of alcohol with water (k_A) and those corresponding to reactions with various forms of ethanol (k_B) and $k_T = k_A + k_B$.

The value of k_A can be determined from the broadening of the water band ($\delta'\nu$) using equation (4.3-86):

$$k_A = 2\pi \, \delta'\nu \qquad (4.3\text{-}86)$$

Combination of the above results makes it possible to determine also the value of k_B. τ varies between 10^{-2} and 1 s.

The band of OH in aqueous alcohol depends both on k_A and k_B and is thus unsuitable for their determination, but the change in the shape of the OH bands is often a useful diagnostic tool for verification of the mechanism of proton transfer in aqueous solutions of alcohols.

Next, values of k_A and k_B are determined at several concentrations of the acid or base and k_A and/or k_B is plotted as a function of $[H^+]$ or $[OH^-]$. From the shape of linear plots and from equilibrium constants $K_{RCH_2OH} = [OH^-]/[RCH_2O^-]$ or $K_{RCH_2OH_2^+} = [H_3O^+]/[RCH_2OH_2^+]$ it was possible to calculate values of second-order rate constants given in Table 4-1.

Examples of systems where the rate constant was determined from the shape of resonance curves can be found among such compounds that can exist in two forms which differ in chemical shift. This can take place for example in molecules where the rotation along a single bond is hindered where, in addition to the form with a single bond, it is possible to assume existence of a form with a double bond. If the transformation of one of these forms into another is very fast and the molecule shows free rotation, the NMR spectrum shows only one band. When the rotation is hindered and the transition of one form into another is slow, formation of two bands is observed which are attributed to two different types of protons in the vicinity of unequal groups. Separation of the two bands is observed only at lower temperatures, where the rotation is sufficiently hindered, whereas at higher temperatures only one band is observed indicating free rotation. Participating forms,

TABLE 4-1
Rate of Proton Transfers in Alcohols

Reaction	CH_3OH	C_2H_5OH	$k_2{}^b$ $i\text{-}C_3H_7OH$	$i\text{-}C_4H_9OH$
(A)	$2.6 \cdot 10^6$	$2.8 \cdot 10^6$	—	—
(B)	$7.4 \cdot 10^{8\,a}$	$1.4 \cdot 10^6$	$0.6 \cdot 10^6$	$2.7 \cdot 10^6$
(C)	<3	0.8	—	—
(D)	$\sim 10^8$	$2.8 \cdot 10^6$	—	—
(E)	$3.5 \cdot 10^{9\,a}$	$1 \cdot 10^8$	$8 \cdot 10^6$	$2 \cdot 10^7$

Source: E. F. Caldin, Fast Reactions in Solution, Wiley, New York, 1964, p. 241.
^{a}24.8°C.
^{b}In liter mol^{-1}s^{-1}, at 22 ± 2°C.

together with the temperature below which formation of two bands is observed, are given in Table 4-2.

In addition to open-chain compounds, similar treatment can also be applied to cyclic compounds including systems which involve inversion of configuration, in which axial hydrogens become equatorial and vice versa (Table 4-2). Differences in chemical shifts can be observed for such compounds if the shielding of the axial protons differs from that of equatorial. For ethylene imines the high velocity of the mutual transformation, even at -70°C, indicates why it was impossible to isolate optical isomers at room temperature.

The use of NMR techniques proved successful for studies of equilibrium reactions, in particular symmetrical exchange reactions, especially in those cases where such reactions are too fast to be studied by isotope exchange methods. Because NMR offers information about the nature and identity of the proton participating in exchange, the technique offers in some cases detailed information about the mechanism, for example about the role of the water (or other solvent) in proton transfers in alcohols and amines. In numerous cases the possibility exists to verify our conclusion, based on measurement of one band (e.g., the methylene proton) by measurement and interpretation of another band (e.g., of proton from water or a hydroxyl band). Moreover, in addition to the information obtained by measurement of the proton resonance it is possible to follow resonance of other nuclei (e.g., O^{17} or F^{19}). The possibility of simultaneous interpretation of several bands increases the probability of the proposed mechanism.

Measurement of NMR spectra enables us to determine the mean lifetime τ_X of a certain species or molecule X from which the first-order rate constant $k = \tau_X^{-1}$ can be calculated. Characteristic for the approach using NMR is the way of evaluation of τ_X. Usually, when a rate constant of a chemical reaction is to be determined under varying conditions, an attempt is made to use the same method for the determination of varying rate constants. If the studied reaction due to a change in reaction conditions becomes too slow or too fast for the method used, it is necessary to introduce the use of another method. For NMR spectra a different technique of evaluation is used according to whether the studied reaction is relatively very fast,

TABLE 4-2
Examples by Systems Investigated by NMR

Bond	Form I	Form II	Temperature Below Which 2 Pairs Can Be Distinguished	τ	k
C—N	(structure)	(structure)	$> +30°C$	~ 0.1 s	—
C—C	(structure)	(structure)	$-60°C$ at $-80°C$	—	—
N—O	(structure)	(structure)	$-40°C$ at $-60°C$	—	200 s^{-1} ($-35°C$)
N—N	(structure)	(structure)	$+180°C$	—	84 s^{-1} ($+180°C$)
	(structure) a	(structure) b	$-65°C$	—	88 s^{-1} ($-65°C$)
	(structure)	(structure)	$-10°C$	—	26 s^{-1} ($-10°C$)
	(structure)	(structure)	$-70°C$	10^{-2} s ($-70°C$)	—

Note: Data mostly from E. F. Caldin, *Fast Reactions in Solution,* Wiley, New York, 1964, p. 251 ff.

moderately fast, or relatively slow. This is in principle equivalent to s . . . on when we have three NMR methods available for the study of fast reactions, from which we select one which is most suitable for the rate of the studied reaction.

To be able to determine the value of τ it is necessary to work under conditions when the changes in the shape in particular band widths are measurable. This is possible when $\tau \Delta \nu_0$ is between 10 and 0.01. For proton NMR the value of $\Delta \nu_0$ is of the order of $10 - 100 \text{ s}^{-1}$ and consequently the NMR enables us to determine the value of τ within the range from 1 to 10^{-4} s (most frequently only to 10^{-3} s.) Hence the values of the first-order rate constant will be usually between 1 and 10^3 s^{-1} (occasionally up to 10^4 s^{-1}). For first-order reactions it is possible to bring the reaction rate into measurable range by decrease of temperature, by change in solvent or in pH. Second-order reactions taking place too rapidly can be brought into the range of measurable rates also by lowering the concentration of one reagent (in addition to lowering the temperature and variation in solvent) for the determination of τ. In this way rate constants of the order of 10^8 to 10^{10} liter $\text{mol}^{-1} \text{ s}^{-1}$ have been measured.

Shorter lifetimes τ can be measured for nuclei other than proton, because both the chemical shifts and the band widths are larger for heavier atoms. Hence for O^{17} the values of $\tau = 10^{-7}$ s and for Cu^{63} $\tau = 10^{-5}$ s have been measured.

Measurement of the broadening of the original bands or the narrowing of the one resulting band and comparison of the shape of the spectra usually offers more accurate values of the rate constant (up to $\pm 10\%$ of k) than observation of the coalescence of the band with the change in composition of the solution (e.g., pH) or with temperature change. Nevertheless, even the investigation of the coalescence of bands offers important information.

It is preferable to use relatively simple spectra for kinetic studies. For such spectra it is namely possible to carry out sufficiently accurate measurements of the shapes of individual bands. Similarly, it is recommended to restrict the study—at least in the beginning—to the study of sufficiently simple systems to allow interpretation of spectral changes and discussion of reaction mechanism.

Most of the commercially available instruments require, for obtaining sufficiently strong signals, about 1 ml of a 0.1 M (or more concentrated) solution of the sample. This indicates the limitation of investigated systems to those sufficiently soluble in the given solvent. Sometimes it is possible to follow changes in the spectrum of the solvent, resulting from addition of the studied compound.

The choice of types of compounds and reactions which can be investigated by NMR is very broad and the reaction types investigated so far undoubtedly do not cover the spectrum of possible systems. NMR seems to be very promising for future investigations.

4.3.4. Relaxation Methods

In relaxation methods a chemical reaction which is already at equilibrium is suddenly perturbed by changing an external variable such as temperature (T), pressure (P), or electric field (E). The finite time required to approach the new equilibrium position is a function of the rate constants for the elementary steps in the reaction, as well as of the equilibrium reagent concentrations and reaction medium (excess of complexing agent, pH, solvent, etc.). In general, the perturbation is relatively small, resulting in relative concentration changes of at most a few percent, so that the mathematical treatment involves first-order reaction kinetics or rate equations, even for complicated reaction schemes. This may be contrasted with techniques belonging to the same general category such as flash photolysis or pulse radiolysis, in which a major disturbance of the equilibrium takes place.

Relaxation methods can be further subdivided according to whether the displacement from equilibrium is caused by a single perturbing pulse or periodic displacement. In the single pulse the forcing function—usually temperature, pressure, or electric field—is suddenly changed in such a way that the time dependence of this function can be ideally depicted as a rectangular step. Examples of forcing function causing periodic displacement can be ultrasonic waves or changing electric field. It should be emphasized that time-dependent concentration changes are observed directly in relaxation methods, as compared with the more indirect methods that have been discussed previously. Since the perturbations can be achieved in nanoseconds (or even picoseconds by the use of lasers), even the fastest reactions in solution can be readily investigated by relaxation techniques.

4.3.4.1. A Single Displacement from Equilibrium

A suitable illustration is provided by considering the following reaction:

$$A + B \underset{k_{21}}{\overset{k_{12}}{\rightleftharpoons}} C \qquad (4.3\text{-}87)$$

The rate equation for this system can be written as (4.3-88)

$$\frac{-dC_A}{dt} = \frac{-dC_B}{dt} = \frac{dC_C}{dt} = k_{12}C_A C_B - k_{21}C_C \qquad (4.3\text{-}88)$$

For small perturbations, it is convenient to define the following concentration difference (X_i) for any species i:

$$X_i = C_i - C_i^0 = \Delta C_i \qquad (4.3\text{-}89)$$

where C_i^0 is the concentration belonging to reference external conditions, which is time independent, and C_i is the concentration at time t after application of the forcing function. When a rectangular step perturbation is applied (e.g., a temperature jump), C_i^0 may be chosen to be the time independent, initial equilibrium concentration ($\bar{C}_i^{\text{initial}}$). The final concentration differences, $\bar{X}_i$, after establishment of the new equilibrium are defined as

$$\bar{X}_i = \bar{C}_i - C_i^0 \qquad (4.3\text{-}90)$$

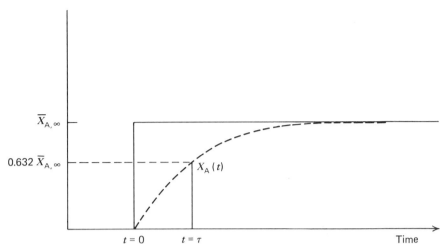

FIG. 4-24. Relaxational response to rectangular step function.

where $\bar{C}_i$ is the new final equilibrium concentration (Fig. 4-24), and can be time dependent for oscillating forcing functions such as sound waves.

For reaction (4.3-87) it follows:

$$C_A = C_A^0 + X_A \qquad \bar{C}_A = C_A^0 + \bar{X}_A \qquad (4.3\text{-}91)$$

$$C_B = C_B^0 + X_B \qquad \bar{C}_B = C_B^0 + \bar{X}_B \qquad (4.3\text{-}92)$$

$$C_C = C_C^0 + X_C \qquad \bar{C}_C = C_C^0 + \bar{X}_C \qquad (4.3\text{-}93)$$

For small perturbations, the difference between the new equilibrium concentration $\bar{C}_i$ and instantaneous concentration C_i is small and thus

$$\bar{C}_i - C_i = \bar{X}_i - X_i \ll C_i \qquad (4.3\text{-}94)$$

For a number of reactions the magnitude of the perturbation used can be, nevertheless, quite large as discussed recently [4-58]. The rate equation (4.3-88) can now be rewritten [4-57]:

$$\frac{dC_A}{dt} = \frac{dX_A}{dt} = k_{21}\,(\bar{C}_C + X_C - \bar{X}_C)$$
$$- k_{12}(\bar{C}_A + X_A - \bar{X}_A)\,(\bar{C}_B + X_B - \bar{X}_B) \qquad (4.3\text{-}95)$$

Using the mass balance condition, $X_A = X_B = -X_C, \bar{X}_A = \bar{X}_B = -\bar{X}_C$, the equilibrium condition $k_{21}\,\bar{C}_C = k_{12}\,\bar{C}_A\,\bar{C}_B$, and neglecting terms such as $X_A X_B, \bar{X}_A \bar{X}_B, X_A \bar{X}_B$, it is possible after multiplication of terms in eq. (4.3-95) to obtain the following linearized differential equation:

$$\frac{dX_A}{dt} + \frac{X_A}{\tau} = \frac{\bar{X}_A}{\tau} \qquad (4.3\text{-}96)$$

where

$$\tau = [k_{12}(\bar{C}_A + \bar{C}_B) + k_{21}]^{-1} \tag{4.3-97}$$

It should be emphasized that, for small perturbations, the same form of linearized differential equation (4.3-96) results for any type of one-step chemical reaction, only the expression for τ (the relaxation time) will be different.

It can be shown that the solution to equation (4.3-96), when a rectangular step-forcing function is applied, is given by (4.3-98):

$$X_A(t) = \bar{X}_{A,\infty}(1 - e^{-t/\tau}) \tag{4.3-98}$$

where the total amplitude $\bar{X}_{A,\infty}$ is defined by $\bar{X}_{A,\infty} = \bar{C}_A^{final} - \bar{C}_A^{initial}$, that is, the difference between the final and initial equilibrium concentration of the species A, and X_A is the difference between the instantaneous and initial equilibrium concentration of the species A, $C_A - \bar{C}_A^{initial}$. The subscript ∞ is introduced to indicate $t \to \infty$. This result follows from equations (4.3-89) and (4.3-90) if the initial equilibrium concentration of A, $\bar{C}_A^{initial}$, is substituted for C_A^0.

The meaning of the relaxation time τ can be best understood from equation (4.3-98) and Fig. 4-24. It is the characteristic time at which the concentration difference X_A [also denoted as ΔC_A; see equation (4.3-89)] is equal to 0.632 of the total amplitude $\bar{X}_{A,\infty}$. This condition is fulfilled when $X_A = \bar{X}_{A,\infty}(1 - e^{-1})$, which results when $t = \tau$.

The evaluation of the relaxation time is frequently simple. In practical measurements it is not necessary to evaluate the concentration from the usually measured physical quantity. As long as the measured quantity is a linear function of ΔC_i (as in the case of absorbance, conductance or fluorescence), values of τ can be obtained directly from the time dependence of the measured quantity. This is possible because the proportionality terms in X_A and $\bar{X}_{A,\infty}$ in equation (4.3-98) cancel.

From the complete solution of the relaxation equation (4.3-96), it can be shown that an alternate solution is given by:

$$X_A(t) = \bar{X}_{A,0} e^{-t/\tau} \tag{4.3-99}$$

if $C_A^0 = \bar{C}_A^{final}$, $X_A(t) = C_A - \bar{C}_A^{final}$, and $\bar{X}_{A,0} = \bar{C}_A^{initial} - \bar{C}_A^{final}$. The subscript 0 is introduced to indicate the beginning of the experiment when $t = 0$. This expression is useful in the analysis of experiments in which an existing constant perturbation is suddenly removed, such as in pressure jump measurements. Both equations (4.3-98) and (4.3-99) are applicable only for single perturbations.

The evaluation of the rate constants k_{12} and k_{21} can be achieved by evaluating the relaxation time for solutions with different final equilibrium concentrations $\bar{C}_A$, $\bar{C}_B$, and plotting τ^{-1} vs. $(\bar{C}_A + \bar{C}_B)$ according to equation (4.3-97). (Slope of the linear plot is equal to k_{12}, intercept to k_{21}.)

This treatment implies prior knowledge of the value of the final equilibrium constant in order to calculate $\bar{C}_A$ and $\bar{C}_B$. There are two cases when knowledge of

K is not essential. First, if $S_B (= \bar{C}_B + \bar{C}_C) \gg \bar{C}_A$, $\tau^{-1} = [k_{12}(S_B) + k_{21}]$ so that τ^{-1} depends on the total analytical concentration of B.

Second, when $S_A (= \bar{C}_A + \bar{C}_C)$ and S_B are known, and the reaction follows the simple scheme given by equation (4.3-87), then using the mass balance condition for C_A, C_B, C_C, as well as the equilibrium expression ($K = k_{12}/k_{21}$), it is possible to derive for reaction (4.3-87) the expression (4.3-100):

$$\frac{1}{\tau^2} = k_{12}^2 [(S_A - S_B)^2 + 2K^{-1}(S_A + S_B) + K^{-2}] \qquad (4.3\text{-}100)$$

If S_A is equal to S_B, equation (4.3-100) becomes (4.3-101):

$$\frac{1}{\tau^2} = 4k_{12}k_{21}(S_A) + (k_{21})^2 \qquad (4.3\text{-}101)$$

Then a plot of τ^{-2} vs. S_A results in a straight line with a slope $4k_{12}k_{21}$, and an intercept $(k_{21})^2$. Hence, it is possible in such cases to evaluate from measurement of relaxation time τ not only rate constants (k_{12} and k_{21}), but also the value of the equilibrium constant K. Amplitudes (e.g., $\bar{X}_{A,\infty}$ in Fig. 4-24) enable a very precise, independent, determination of thermodynamic parameters.

The relaxation time expressions for some common reactions derived as previously described are shown in Tables 4-3 and 4-4. It can be shown from the thermodynamics of irreversible processes [4-59] that the relaxation times are completely independent of the type of perturbation used (temperature, pressure, electric field, sound) as long as the reactions take place in dilute solutions.

TABLE 4-3
Important Types of One-Step Reaction Systems Studied by Relaxation Methods

($\bar{C}_i$ are concentration at the new equilibrium)

1. $A \rightleftharpoons B$ $1/\tau = k_{12} + k_{21}$

2. $A + B \rightleftharpoons C$ $1/\tau = k_{12}(\bar{C}_A + \bar{C}_B) + k_{21}$

3. $2A \rightleftharpoons A_2$ $1/\tau = 4k_{12}\bar{C}_A + k_{21}$

4. $A + B \rightleftharpoons C + D$ $1/\tau = k_{12}(\bar{C}_A + \bar{C}_B) + k_{21}(\bar{C}_C + \bar{C}_D)$

5. $A + C \rightleftharpoons B + C$ $1/\tau = k_{12}\bar{C}_C + k_{21}\bar{C}_C$
   (C = catalyst)

6. $A + B \rightleftharpoons C$ $1/\tau = k_{12}\bar{C}_B + k_{21}$
   (B = buffered)

7. $A + B \rightleftharpoons (AB) \rightleftharpoons C$ $1/\tau = k_{12}p(\bar{C}_A + \bar{C}_B) + k_{21}(1 - p)$
   (AB = stationary intermediate)   where $p = k_{23}/(k_{21} + k_{23})$

8. $A + B + C \rightleftharpoons D$ $1/\tau = k_{12}(\bar{C}_A\bar{C}_B + \bar{C}_A\bar{C}_C + \bar{C}_B\bar{C}_C) + k_{21}$

9. $A + B + C \rightleftharpoons (AB) + C \rightleftharpoons D$ $1/\tau = k_{12}p/\bar{C}_C(\bar{C}_A\bar{C}_B + \bar{C}_A\bar{C}_C - \bar{C}_B\bar{C}_C)$
                                $+ k_{32}(1 - p)$
                              with $p = k_{23}\bar{C}_C/(k_{21} + k_{23}\bar{C}_C)$

TABLE 4-4
Relations Between Relaxation Time and Analytical Concentrations of Participating Species[a]

2. $A + B \rightleftharpoons C$	$(1/\tau)^2 = k_{12}^2(S_A^0 - S_B^0) + 2k_{12}k_{21}(S_A^0 + S_B^0) + (k_{21})^2$
	$S_A^0 = S_A + S_C; S_B^0 = S_B + S_C$
3. $2A \rightleftharpoons A_2$	$(1/\tau)^2 = 8k_{12}k_{21}S_A^0 + (k_{21})^2;$
	$S_A^0 = S_A + 2S_{A_2}$
4. $A + B \rightleftharpoons C + D$	$(1/\tau)^2 = [k_{12}(S_C^0 - S_B^0) - k_{21}(S_C^0 - S_A^0)]^2 + 4k_{12}k_{21}S_A^0S_B^0$
	$S_A^0 = S_A + S_C; S_B^0 = S_B + S_C; S_C^0 = S_A + S_C$

[a]For Reaction Systems 2, 3, and 4, described in Table 4-3.

So far only one step reactions have been considered, whereas in practice it is necessary frequently to deal with coupled elementary reactions. An example can be given by:

$$A + B \underset{k_{21}}{\overset{k_{12}}{\rightleftharpoons}} C \underset{k_{32}}{\overset{k_{23}}{\rightleftharpoons}} D \qquad (4.3\text{-}102)$$

$$\text{I} \qquad\qquad \text{II} \qquad \text{III}$$

in which the first step (I $\rightleftharpoons$ II) is associated with the formation of an encounter complex, which is frequently much faster than the subsequent step (II $\rightleftharpoons$ III). Such a scheme has been used for example to explain the binding of drugs (such as actinomycin) to DNA [4-60]. The exact solution of eq. (4.3-102) involves solving two linearized, simultaneous differential equations. For systems involving two or more steps this is most readily accomplished by the use of matrix methods. The two relaxation times (τ_1 and τ_2) for the three state system (4.3-102) are described by

$$(\tau_1^{-1} + \tau_2^{-1}) = \Sigma k = k_{12}(\bar{C}_A + \bar{C}_B) + k_{21} + k_{23} + k_{32} \qquad (4.3\text{-}103)$$

$$(\tau_1^{-1})(\tau_2^{-1}) = k_{12}[k_{23} + k_{32}](\bar{C}_A + \bar{C}_B) + k_{21}k_{32} \qquad (4.3\text{-}104)$$

From the linear plots of $\tau_1^{-1} + \tau_2^{-1}$ and $(\tau_1^{-1})(\tau_2^{-1})$ vs. $(\bar{C}_A + \bar{C}_B)$, all four rate constants $k_{12}, k_{21}, k_{23}, k_{32}$ can be evaluated as shown in (Fig. 4-25). Various limiting cases can be readily obtained from equation (4.3-104), for example, if the first equilibrium (I $\rightleftharpoons$ II) is established very rapidly compared to the second one (II $\rightleftharpoons$ III) then:

$$\tau_1^{-1} = k_{12}(\bar{C}_A + \bar{C}_B) + k_{21} \qquad (4.3\text{-}105)$$

Substituting this result in equation (4.3-104) gives:

$$\tau_2^{-1} = \frac{k_{12}k_{23}(\bar{C}_A + \bar{C}_B)}{k_{21} + k_{12}(\bar{C}_A + \bar{C}_B)} + k_{32} \qquad (4.3\text{-}106)$$

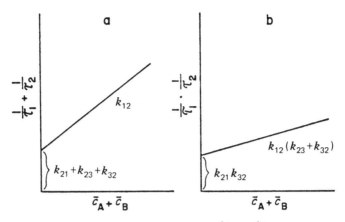

FIG. 4-25. Evaluation of kinetic parameters for $A + B \underset{k_{21}}{\overset{k_{12}}{\rightleftharpoons}} C \underset{k_{32}}{\overset{k_{23}}{\rightleftharpoons}} D$ system, when both steps reach the equilbrium at similar rates. The four rate constants are determined as follows: k_{12} = slope (a), k_{32} = (intercept (b)/k_{21}), k_{21} = intercept (a) − slope (b)/slope (a), k_{23} = intercept (a) − k_{21} − k_{32}. (According to C. F. Bernasconi, *Relaxation Kinetics*, Academic, New York, 1976, p. 28.)

A detailed discussion of this type of reaction, including practical procedures for the calculation of τ_1 and τ_2 from experimental data, can be found in ref. 4-61.

4.3.4.2. Relaxation Amplitudes

In addition to the relaxation times, a knowledge of the corresponding relaxation amplitudes ($\bar{X}_{A,\infty}$ in Fig. 4-24) is important in order to determine whether a relaxation effect will be large enough to be measured with the available material. Sufficiently large total relaxation amplitude is of particular importance as a result of the high demands on detection sensitivity using small perturbations. Furthermore, total relaxation amplitudes provide a very powerful method for the determination of thermodynamic equilibrium quantities as well as enthalpies ($\Delta H°$), volume changes ($\Delta V°$), and so forth. This possibility is illustrated by the example based on the temperature jump method using optical absorption detection. Extensions to other types of perturbation using changes in pressure or electric field and different detection (fluorescence, conductance, etc.) can be carried out easily.

A sudden change in the temperature of a thermally equilibrated solution can be accomplished by means of a capacitive discharge (Joule heating). Typically, the temperature of approximately one milliliter of the sample required for capacitive discharge is raised by a few degrees centigrade in several microseconds. This time interval poses a limit on the fastest reactions which can be studied, and can be extended to approximately 10 nanoseconds, if the energy transfer is achieved by means of a cable discharge. For reaction (4.3-87) the magnitude of the concentration changes can be calculated from the modified van't Hoff equation (4.3-107):

$$\Delta \ln K = \frac{\Delta K}{K} = \frac{\Delta H°}{RT^2} \Delta T = \frac{\Delta H°}{RT} \left(\frac{\Delta T}{T} \right) \qquad (4.3\text{-}107)$$

provided the relative change in equilibrium constant is small ($\Delta K/K \ll 1$). Thus, the relative change in K corresponds to about 5.6% for a 2°C change in temperature at 25°C, if $\Delta H° = 5$ kcal. It is important to express the resulting changes in the observable physical property explicitly as a function of K, $\bar{C}_i(S_i)$, and $\Delta H°$. Consider a general one-step reaction:

$$a\text{A} + b\text{B} + c\text{C} \cdots \rightleftharpoons x\text{X} + y\text{Y} + z\text{Z} \cdots \qquad (4.3\text{-}108)$$

It can be shown that the *total* change in the observed physical property (which is a linear function of concentration) $\Delta P°$ is given by

$$\Delta P° = \Delta\phi \,\Gamma\Delta \ln K = \Delta\phi \,\Gamma\,\frac{\Delta K}{K} \qquad (4.3\text{-}109)$$

where $\Delta P°$ may be the total change in absorbance, $\Delta\phi$ is defined as

$$\Delta\phi = \sum_i \nu_i\phi_i \qquad (4.3\text{-}110)$$

where ν_i are stoichiometric coefficients with positive values for products and negative for reactants and $\phi_i = l\epsilon_i$ where l is the optical path length and ϵ_i molar absorptivities of species i. Hence for reaction (4.3-108) $\Delta\phi$ is given by

$$\Delta\phi = l(x\epsilon_X + y\epsilon_Y + z\epsilon_Z + \cdots - a\epsilon_A - b\epsilon_B - c\epsilon_C) \qquad (4.3\text{-}111)$$

The value of Γ in equation (4.3-109) is defined as

$$\Gamma = \left(\sum_i \nu_i^2/\bar{C}_i\right)^{-1} = \left(\frac{a^2}{\bar{C}_A} + \frac{b^2}{\bar{C}_B} + \cdots \frac{x^2}{\bar{C}_X} + \frac{y^2}{\bar{C}_Y} + \cdots\right)^{-1} \qquad (4.3\text{-}112)$$

Equation (4.3-109) has general validity, regardless of the type of perturbation used or the type of detection employed, provided that appropriate expressions describing the proportionality constants are substituted for ϕ_i in equation (4.3-110). For reaction (4.3-87), application of eqs. (4.3-107) and (4.3-109) gives

$$\Delta P° = (\epsilon_{AB} - \epsilon_A - \epsilon_B)\left[\frac{1}{\bar{C}_A} + \frac{1}{\bar{C}_B} + \frac{1}{\bar{C}_{AB}}\right]^{-1}\frac{\Delta H°}{RT}\left(\frac{\Delta T}{T}\right) \qquad (4.3\text{-}113)$$

The terms containing ϵ_i and $\Delta H°$ represent linear proportionality constants, whereas the nonlinear term in brackets corresponds to Γ, and can be expressed from:

$$\frac{\Gamma}{S_A} = \frac{\alpha\,(1 - \alpha)}{1 + \dfrac{\alpha\,(1 - \alpha)}{\beta - (1 - \alpha)}} \qquad (4.3\text{-}114)$$

where

$$\alpha = \frac{\bar{C}_A}{S_A} \qquad \beta = \frac{S_B}{S_A}$$

and S_i represents the analytical concentration of species i. Several limiting forms of equation (4.3-114) may be considered

(I) for $\beta = 1$, which occurs when $(S_A = S_B)$

$$\frac{\Gamma}{S_A} = \frac{\alpha (1 - \alpha)}{(2 - \alpha)} \qquad\qquad (4.3\text{-}115)$$

In this case the plot of (Γ/S_A) vs. α results in an asymmetrical curve with a maximum at $\alpha = 0.586$, $(\Gamma/S_A) = 0.173$. The shape of this dependence indicates that possibility of evaluating K using equation (4.3-113) without having to know the molar absorptivities of $\Delta K/K$ from equation (4.3-109). The value of K is found from the shape of the curve obtained from the plot of experimentally found ΔP° vs. α. In practice, the terms ϵ and $\Delta K/K$, as well as Γ/S_A values calculated for different values of K, can be used as a single linear scaling factor until the calculated curve fits the experimental one. Alternatively, a computerized, nonlinear, least-squares procedure may be used to determine the best value of K.

(II) $K^{-1} \gg S_A$, $S_B \gg S_A$, that is, when the values of β are very large

$$\frac{\Gamma}{S_A} = \alpha (1 - \alpha) \qquad\qquad (4.3\text{-}116)$$

In this case the plot of (Γ/S_A) as a function of α shows a shallow maximum when $(\Gamma/S_A) = 0.25$, $\alpha = 0.5$ (Fig. 4-26a), and $S_B (\approx \bar{C}_B) = K^{-1}$. Under these conditions, it can be shown using equation (4.3-109) that K and ΔH° can be determined by using the expression:

$$\Delta H^\circ = \left[\frac{\Delta P^\circ}{\epsilon_{AB} - \epsilon_A - \epsilon_B} \right]_{max} \frac{RT^2}{(0.25) \, S_A(\Delta T)} \qquad\qquad (4.3\text{-}117)$$

Provided the values of molar absorptivities (ϵ_i) are accurately known (which is essential for the determination of ΔH° by this method), a single amplitude measurement can give a precise value of ΔH° using equation (4.3-117) for measurements even over a very narrow temperature range of one degree centigrade or less.

(III) A general solution to equation (4.3-114) can be obtained by using the following equation, derived from mass balance conditions for S_A, S_B, and the equilibrium constant expression for reaction (4.3-87):

$$\alpha^2 + \alpha[\beta - 1 + (KS_A)^{-1}] - (KS_A)^{-1} = 0 \qquad\qquad (4.3\text{-}118)$$

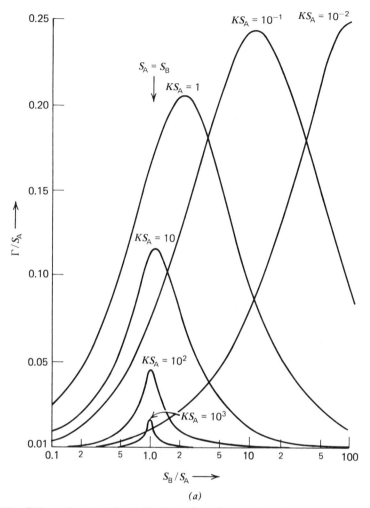

FIG. 4-26. Γ/S_A as a function of $\beta = S_B/S_A$. (According to Bernasconi, *Relaxation Kinetics*, Academic, New York, 1976, p. 95.)

The best approach is to use a computer or a programmable pocket calculator to solve for α and (Γ/S_A) using (KS_A) and β as adjustable parameters. Characteristic curves, with well-defined maxima, are obtained for varied values of Γ/S_A (Figs. 4-26a,b). When $K \ll 1/S_A$, the $(\Gamma/S_A)_{max}$ value occurs when $S_B = K^{-1}$. On the other hand, when $K \gg 1/S_A$, the $(\Gamma/S_A)_{max}$ values decrease, with a corresponding decrease in the experimentally observed total amplitudes (equation (4.3-109)). At the same time, the value $(\Gamma/S_A)_{max}$ is found when $S_A = S_B$, and the curve becomes extremely sharp (Fig. 4-26b). From the shape of the curve, the value of K can be obtained with great accuracy [4-62]. A similar behavior is found also for the function $\tau k_{12}S_A$, (Fig. 4-26b), and it can be seen that the best conditions for the experimental

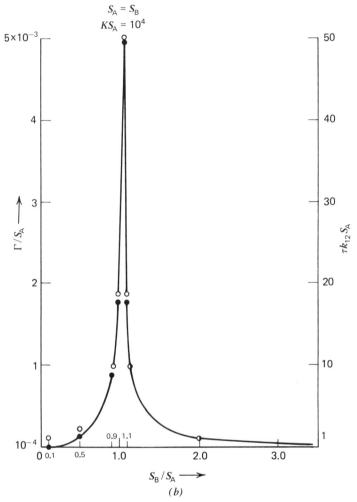

FIG. 4-26. (*Continued.*) Γ/S_A (left) and $\tau\, k_{12}\, S_A$ (right) as a function of $\beta = S_B/S_A$.

measurement of the value of k_{12} are achieved in the neighborhood of the maximum, corresponding to longer relaxation times. The types of curves shown in Fig. 4-26b are particularly important for biological or macromolecular systems, where values of K of 10^6 liter mol^{-1} or greater are often encountered. In such cases, conventional static spectrophotometric measurements combined with graphical data treatment yield little information regarding values of K.

The most significant source of errors in amplitude measurements is in the value of $\Delta P°$, since ϵ_i, T, and ΔT (equations (4.3-107) and (4.3-109)) can ordinarily be determined to $\pm$ 1%. In the more sensitive temperature jump instruments, relative changes in $\Delta P° \geq 10^{-4}$ ($\pm \approx 2\%$) can be determined, especially when one is interested solely in the amplitude, and an appropriate filter is used to virtually eliminate the noise from the phototube (in such a case, nevertheless, the information regarding τ would be lost). A typical calculation can be carried out using equation (4.3-117) and the following values:

$$\Delta P^\circ = 5 \times 10^{-3}; T = 298^\circ K; (\epsilon_{AB} - \epsilon_A - \epsilon_B) = 2 \times 10^3 \text{ liter mol}^{-1} \text{ cm}^{-1};$$

$$\Delta T = 1^\circ C; S_A = 10^{-3} M; (\Gamma/S_A)_{max} = 0.25 \text{ (Fig. 4.26a)};$$

$$K = 10; (S_B/S_A) = 100$$

gives a ΔH° value of 1.76 kcal. Thus, under appropriate experimental conditions, a ΔH° value as small as approximately 2 kcal can be readily determined over a temperature interval of only $1^\circ C$.

4.3.4.3. Temperature Jump Amplitudes for a Coupled Chemical Reaction [4-61–4-66]

In cases when the available experimental methods do not allow observation of equilibria involving an acid–base couple HA $\rightleftharpoons$ A, information regarding the value of the corresponding equilibrium constant K_B can be obtained when an indicator HI with a dissociation constant K_A is added to the solution.

The following two reactions are then considered:

$$H + I \rightleftharpoons HI \qquad K_A = \frac{\bar{C}_{HI}}{\bar{C}_H \bar{C}_I} \qquad (4.3\text{-}119)$$

$$H + A \rightleftharpoons HA \qquad K_B = \frac{\bar{C}_{HA}}{\bar{C}_H \bar{C}_A} \qquad (4.3\text{-}120)$$

Such systems are usually followed in the visible region of the spectrum where the only absorbing species is HI, I, or both. If the two equilibria (4.3-119) and (4.3-120) are perturbed in a temperature jump experiment, it is obvious that the concentration of I and HI changes, and hence the absorbance change at the chosen wavelength will be influenced by the change in the hydrogen ion concentration due to reaction of the studied acid. This is quantitatively described by the following equation:

$$\Delta P^\circ_{total} = \Delta \epsilon \frac{\bar{C}_H \bar{C}_I}{\alpha_X \bar{C}_I + \bar{C}_H + K_A^{-1}} \left[\Delta \ln K_A - (1 - \alpha_X) \Delta \ln K_B \right] \qquad (4.3\text{-}120)$$

where $\Delta \epsilon = \epsilon_{HI} - \epsilon_I$ and $\alpha_X = (\bar{C}_H + K_B^{-1})/\bar{C}_H + \bar{C}_A + K_B^{-1})$. The expression for the coupling term, α_X, contains only quantities corresponding to the studied acid and not the indicator, and $\Delta \ln K$ is directly proportional to the corresponding ΔH° according to equation (4.3-107). It can be seen from the equation for ΔP° (4.3-120) that although components of the studied acid–base system themselves do not contribute to the absorbance, an accurate determination of K_B and ΔH°_B can be made by means of the indicator reaction (4.3-119). A numerical example may illustrate this:

Let $S_{HA} = 10^{-3} M$, $S_1 = 10^{-4} M$, $K_A = K_B = 10^7 M^{-1}$, $\bar{C}_H = 10^{-7} M$ (pH $= 7 = p(K_A^{-1}) = p(K_B^{-1})$), $\Delta\epsilon = 5 \times 10^4$, $\Delta T = 5°C$, $\Delta H_A° = 2$ kcal, $\Delta H_B° = 10$ kcal, $\alpha_X = 4 \times 10^{-4}$, and hence $(1 - \alpha_X) \approx 1$. From equation (4.3-120), it follows that $\Delta P°_{total} = -0.252$, which is an enormous change for temperature jump measurements. A similar calculation can be carried out for reaction (4.3-119) alone under identical conditions using equation (4.3-109) or (4.3-113), and $\Delta P° = 2.77 \times 10^{-4}$. The absorbance increases by about three orders of magnitude due to coupling of the reaction (4.3-118) with reaction (4.3-120). This example also illustrates the possibility of studying a reaction with $\Delta H° = 0$, by coupling with a suitable reaction with a finite $\Delta H°$, since according to the expression for $\Delta P°$ (4.3-120) a finite amplitude will be determined by the expression in square brackets. It should also be emphasized that even in those cases where $\Delta\phi$ in equation (4.3-109) may not be known, this equation offers valuable information regarding the stoichiometry, from the marked dependence of Γ on the stoichiometric coefficients defined by equation (4.3-112). For experiments in which the total, unresolved amplitude is measured, a correction due to thermal expansion of the solution (dilution), as well as the temperature dependence of $\Delta\phi$, and hence ϵ, has to be included [4-61]. This would be especially important if static measurements were carried out using a sensitive spectrophotometer.

4.3.4.4. Temperature Jump Method

A sudden change in the temperature of an electrically conducting solution is brought about by discharging a high-voltage capacitor with low inductance using a triggered spark gap. The time course of the temperature change is given by:

$$\Delta T(t) = \Delta T_\infty (1 - e^{-2t/RC}) \qquad (4.3\text{-}121)$$

$$\Delta T_\infty \approx \frac{C E_0^2}{2 C_p \rho V} \qquad (4.3\text{-}122)$$

where T is temperature in °C; E_0 is the applied high voltage; ρ is the density of the sample solution; V is the volume of the solution between the electrodes; C_p is the specific heat of the solution; ΔT_∞ is the maximum temperature change, expressed assuming constancy of R and C_p in the given temperature interval; and $\tau_H = RC/2$ is the heating time constant. For a 0.04μF capacitor, with a sample solution resistance of 100 Ω, the heating time constant (τ_H) is 2μs. Sample volumes from about 40 μl in a microcell to several milliliters can be used. For example, a total temperature change (ΔT_∞) of $\sim 5°C$, eq. (4.3-122) results if the applied high voltage is 15 kV and the sample volume (V) is 0.2 ml. The rapid heating takes place under essentially constant volume conditions, resulting in a pressure shock wave whose amplitude is 25 atmospheres for a 5°C change at room temperature. The possible disturbances generated by this effect have to be minimized by careful cell design enabling throttled expansion of the sample solution (Fig. 4-27).

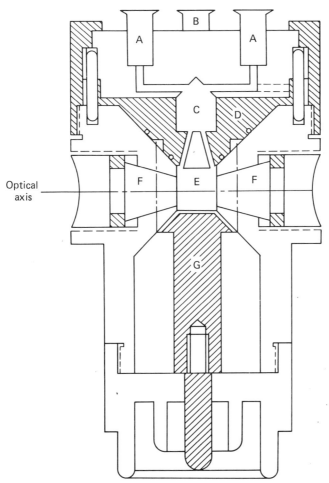

FIG. 4-27. Combined stopped-flow temperature jump cell. A, drive syringes; B, stopping syringe; C, mixing chamber; D, ground electrode; E, observation chamber; F, quartz lenses; G, high voltage electrode. (According to R. C. Patel, ref. 4-67.)

A block diagram of a combined stopped flow temperature jump apparatus is shown in Fig. 4.28 [4-67]. This arrangement allows a reaction to be initiated rapidly by a stopped flow method (Section 4.2) and a perturbation of stationary states two milliseconds after mixing of reagents, so that kinetic parameters of essentially irreversible reactions can be investigated. For example, if in the reaction (4.3-102) the first step (I $\rightleftharpoons$ II) reaches equilibrium much more rapidly than the second (II $\rightleftharpoons$ III), it would be possible to mix A and B rapidly and study equilibrium I $\rightleftharpoons$ II by a temperature jump using equation (4.3-100) before conversion to D by the second equilibrium (II $\rightleftharpoons$ III) can take place as discussed in Section 4.3.4.1. This method is also very useful for the generation of highly reactive or air sensitive intermediates in the stopped flow temperature jump cell (Fig. 4-27) and investigation of the kinetics of reactions of these species by means of a tem-

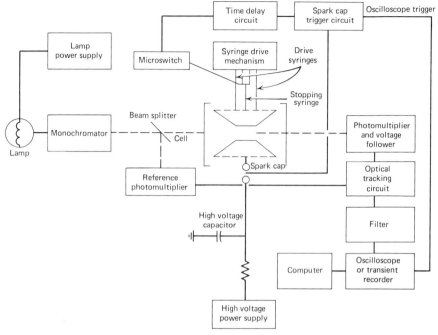

FIG. 4-28. Block diagram of combined stopped-flow temperature jump apparatus (According to R. C. Patel, ref. 4-67.)

perature perturbation. The sample cavity (E in Fig. 4-27) between the two electrodes is between 0.1 and 0.2 ml. In this cell a concentric flow arrangement is achieved. It is ideally suited for fluorescence measurements including polarized fluorescence. Evaluation of the relaxation time, τ, and the amplitude, $\bar{X}_{A,\infty}$, from the observed physical property can be carried out as shown in Fig. 4-24.

Although the capacitive discharge is the most commonly used method of heating, the use of a high-voltage cable discharge has been shown to decrease the heating time to approximately 10 nanoseconds. Other heating methods include the use of microwaves, lasers, and for very slow reactions, a dual thermostat with a rapid-switching arrangement. In these cases, it is not necessary for the solutions to be electrically conducting, so that nonaqueous systems can also be investigated. Although the presence of a conducting supporting electrolyte may appear to be a limitation for organic solvents when the capacitive discharge temperature jump is used, a number of reactions have been studied in solvents such as methanol, alcohol–water mixtures, dioxane, and dimethylsulfoxide [4-61].

The most popular method of following relative concentration changes as a function of time has been by absorption spectrophotometry. Since relaxation methods involve only small perturbations, very high sensitivity is required of the detection system. For example, total changes in absorbance of about 0.01 are relatively large for temperature jump experiments, and signal/noise (S/N) ratios $\geq 10^4$ have been achieved in good instruments. It is important to use as high an intensity of the light source (usually a mercury-xenon or xenon discharge lamp) as practical, and to use the least number of dynodes of the photomultipler. Further improvement in the accuracy of data can be achieved if the phototube signal is digitized by means of commercially available analog-to-digital, digital-to-analog converters (e.g., Biomation, Cupertino, California) so that over 2000 data points with a resolution of better than 0.1% can be processed in a matter of minutes in conjunction with a minicomputer such as an IBM instrumentation computer. Other detection methods include light scattering, polarimetry, and conductance, although the latter loses sensitivity when the capacitive discharge method is used, as a result of the high background conductance due to the supporting electrolyte.

Spectrophotometry and fluorescence are highly specific, since they enable individual components in a reaction mixture to be followed separately. Recent developments in resonance Raman spectroscopy are likely to provide a highly selective and sensitive (although expensive) detection capability. Optical detection can also be used for reactions involving optically inactive species by coupling the reaction with an appropriate, fast responding, indicator reaction (e.g., a colored pH indicator for an acid–base reaction involving no color change), as discussed in Section 4.3.4.3.

An example of a typical reaction which has been studied [4-68] is the oxidation reduction reaction of quinone with the hydroquinone dianion:

$$(4.3\text{-}123)$$

where both k_1 and k_{-1} were found to have a similar value of about 10^8 liter mol^{-1} s^{-1}. An overview of the classes of reactions which have been investigated by the temperature jump technique is given in Table 4-5.

TABLE 4-5
Some Systems Studied with the Temperature-Jump Method [4-1, 4-61]

Acid–base equilibria	Micelle reactions
Intermolecular proton transfer	Phospholipid dispersions
Hydrolysis of halogens	Protein and enzyme small-molecule interactions
Hydration	Enzyme catalysis
Metal complex formation	Enzyme regulation
Electron transfer reactions	Polynucleotide small-molecule interactions
Organic substitutions	Polynucleotide interactions

4.3.4.5. The Pressure Jump Method [4-1], [4-57], [4-69]

As in the temperature jump method, reactions can also be displaced from the equilibrium state by a sudden change in pressure,

$$\left(\frac{\partial \ln K}{\partial P}\right)_T = \frac{-\Delta V^\circ}{RT} \tag{4.3-124}$$

where ΔV° is the standard molar volume change of the reaction; P, the pressure in atmospheres; and R, the gas constant (82 ml atm mol^{-1} deg^{-1}).

In aqueous solutions, if $\Delta V^\circ = 0$ or is very small this method is not practical. In general, ion forming or ion consuming reactions are most likely to be accompanied by relatively large volume changes as a result of solution and electrostriction effects. For example, for the acetic acid dissociation reaction $CH_3COOH \rightleftharpoons CH_3COO^- + H^+$, the value of ΔV° is -10.9 ml. In a typical experiment, with a ΔP of 60 atm, and a $\Delta V = 10$ ml, the relative change in the equilibrium constant ($\Delta K/K$) can be calculated from equation (4.3-124) to be 2.45%. The highly sensitive conductometric detection has been used mostly to monitor the comparatively smaller concentration changes in pressure jump experiments. In a number of nonaqueous solvents the equilibrium displacements may be larger, because the pressure change takes place adiabatically, and the exact equation for the pressure dependence of the equilibrium constant is given by:

$$\left(\frac{\partial \ln K}{\partial P}\right)_S = \frac{-\Delta V^\circ}{RT} + \frac{\alpha}{\delta C_p}\frac{\Delta H^\circ}{RT} \tag{4.3-125}$$

where δ is the density, C_p is the specific heat at constant pressure, and α is the thermal expansion coefficient.

An additional correction term resulting from the compressibility and corresponding concentration changes of the solution has to be added to equation (4.3-125). For aqueous solutions this correction is usually quite small ($\approx 5\%$ of the total effect). Especially for nonaqueous solvents, the second term in equation (4.3-125) may be quite large, and the adiabatic change in temperature can be calculated from:

$$\left(\frac{\partial T}{\partial P}\right)_S = \left(\frac{T}{C_p}\right)\left(\frac{\partial V}{\partial T}\right)_P \tag{4.3-126}$$

Some representative values of $(\partial T/\partial P)_S$ are shown in Table 4-6.

A convenient, rapid change in temperature can be achieved in favorable cases. This can be particularly useful for studies in some of the nonpolar solvents shown in Table 4-6, in which a conventional temperature jump study may be difficult due to the requirements for an electrically conducting solution.

A typical, modern pressure jump apparatus is shown in Fig. 4-29 [4-70]. The sample and reference cells are both inside the pressure autoclave and form two

TABLE 4-6
Adiabatic Temperature Change
$(\partial T/\partial P)_S$ for Some Solvents

Substance	Temperature Rise (°C/100 atm)	
Pentane	3.5	
Ethyl ether	3.2	
Benzene	2.5	
Paraffin	0.8	
Glycerol	0.5	
Water	0.07	(20°C)
Water	0	(4°C)

arms of an alternating current (40 kHz or higher) Wheatstone bridge whose off-balance voltage is read with an oscilloscope. The highly sensitive measurement of conductance changes has been generally used to monitor the comparatively small concentration changes resulting from the pressure perturbation, which is achieved by the sudden bursting of a thin metallic membrane (10) located above the sample and reference cells (1). It is convenient to increase the pressure in the experimental chamber by pumping in a pressurizing liquid such as kerosene. A pressure transducer (5) triggers the oscilloscope to start the recording when the diaphragm bursts. The use of shock waves in a shock tube has enabled shorter relaxation times ($\approx 1\mu s$) to be measured compared with the $\approx 30\ \mu s$ for the conventional apparatus.

Although the pressure jump method is relatively simple and inexpensive, it has not found widespread application, partly because of the nonavailability of a commercial apparatus, and partly due to the requirements for a relatively large $\Delta V°$. The determination of thermodynamic parameters ($\Delta V°$, K) from pressure jump amplitudes can be undertaken as described in Section 4.3.4.2. The reactions which have been studied by this method include metal complexation reactions, and hydration of carbonyl compounds, for example of pyruvic acid [4-71].

4.3.4.6. Method of Electrical Impulses [4-72]

As a result of the so-called second Wien effect [4-73], the dissociation of weak electrolytes increases in a strong electric field. This effect can be made use of in a way similar to the two preceding methods if a solution containing a weak electrolyte is suddenly exposed to a strong electric field.

For a univalent electrolyte (e.g., the dissociation of acetic acid), it can be shown that for small perturbations the relative change in the equilibrium constant is given by:

$$\frac{\Delta K}{K} = \frac{9.64}{E\ T^2}\ |\Delta E| \qquad (4.3\text{-}127)$$

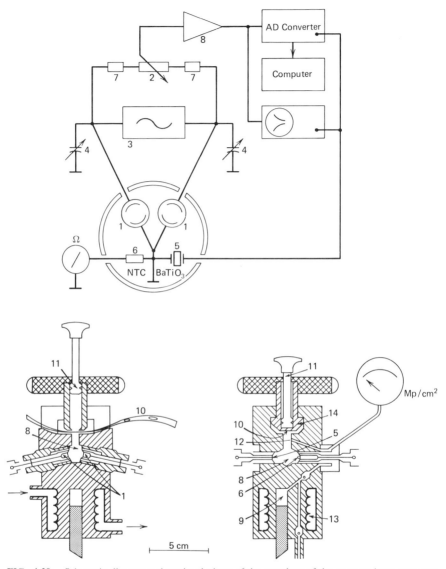

FIG. 4-29. Schematic diagram and sectional views of the autoclave of the pressure-jump apparatus by Knoche and Wiese [4-70]: 1, conductivity cells; 2, potentiometer; 3, 40 kHz generator for Wheatstone bride; 4, tunable capacitors; 5, piezoelectric capacitor; 6, termistor; 7, 10-turn helipot for tuning bridge; 8, experimental chamber; 9, pressure pump; 10, rupture diaphragm; 11, vacuum pump; 12, pressure inlet; 13, heat exchanger; 14, bayonet socket. (Reproduced by permission of Marcel Dekker, Inc.)

where $|\Delta E|$ = electric field strength in V/cm. Equation (4.3-127) has been derived using the treatment of Onsager [4-74] under appropriate conditions. For water, with a dielectric constant, ϵ, of 78, at 25°C, and $|\Delta E|$ = 100 kV/cm, the value of K for acetic acid changes by about 14%.

To obtain measurable displacements from the initial equilibrium state it is necessary to apply fields of considerable intensity. If such fields are applied over a prolonged period of time, a temperature increase necessarily results. It is thus necessary to apply very strong fields over a short time period.

In practice a condenser or coaxial cable is charged to 20 kV and discharged over two electrodes in the reaction cell. The product of resistance and capacitance should be large in order to achieve a practically constant intensity of the field after it is applied and the electrical resistances of the cells should be large to avoid Joule heating. Two vessels are used—one for the reaction mixture and the other for the reference solution—and the difference of a characteristic property, conductivity or absorbance for example, is recorded.

For dilute solutions and pure solvents with high resistance, the field is applied for up to 10^{-4} s and relaxation times of the order of 10^{-9} to 10^{-4} s can be measured. For other solutions, whose conductivity cannot be neglected, it is necessary to limit the application of the electric field to 10^{-7} to 10^{-5} s to prevent heating of the solution. Such fields are produced with damped harmonic impulses and are used for the study of acid–base reactions. It should be pointed out that the use of intense laser beams, which can be strongly focused for spectrophotometric detection, enables the construction of cells [4-75] where the electrodes are separated by very small distances, thus reducing the requirements for high voltages to produce high electric field strengths. Results obtained with these methods show that the structure of the acid has little effect on the rate of the proton recombination reaction [4-76]. Some of the observed changes can be ascribed to steric effects.

4.3.4.7. Comparison of Relaxation Methods Based on One Displacement from Equilibrium

Of the three methods described, temperature jump can be used for the widest range of reaction rates—from $\tau \approx 5 \times 10^{-9}$ (cable discharge) to $\tau = 1$ s or longer [4-77]. Methods using electrical impulses are preferably used for fastest reactions. Because of the heating of the solution they are not recommended for $\tau > 2 \times 10^{-4}$ s (for rectangular pulse) or $\tau > 10^{-5}$ s (for damped oscillations). Pressure jump methods are suitable for application only for slower reactions with τ between 3×10^{-5} s (1 μs using pressure shock tubes [4-78]) and 50 s, provided that the value of ΔV is sufficiently high. In general, however, this difficulty can be overcome by coupling with a reaction with a large $\Delta V°$, as described in detail for the temperature jump method.

Relaxation techniques do not pose any limitation on the nature of the reacting species when the pressure jump method is involved and very little when the temperature jump method is used in connection with spectrophotometry.

If electrometric methods are used for analysis, ionizing substances must be involved. In the temperature jump methods, supporting electrolyte must be added to maintain high electrical conductivity. Methods based on electric impulses are generally restricted to studies of equilibria of electrolytes. Nevertheless, the study of buffered systems is often excluded.

Relatively small volumes are required for all three methods (1 ml or less), and in the temperature jump method using a microcell about 40 μl are sufficient.

When selectivity is compared, the temperature jump method with spectrophotometric detection of concentration changes of individual species is advantageous. In other instances, quantities (such as conductance) are measured which are not specific, but rather general properties of the change in the studied solutions. Finally, where instrumentation is concerned, all of the above methods require special equipment. Some of the apparatus—for example, for temperature jump—is commercially available, although rather expensive [4-61].

4.3.4.8. Periodical Displacement from Equilibrium State [4-57, 4-79–4-83]

If the quantity causing the displacement of the studied system from equilibrium is not applied in a sudden jump, but if the system is displaced from equilibrium

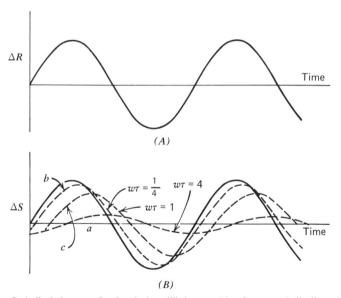

FIG. 4-30. Periodical changes of a chemical equilibrium resulting from a periodically varied external parameter. (A) Dependence of the variation in the external parameter ΔR (such as electrical field, temperature or pressure) on time; (B) dependence of variation in reaction parameter ΔS (such as degree of dissociation) on time: (a) slowly established equilibrium; (b) rapidly established equilibrium; (c) equilibrium established within a relaxation time comparable with that of the variation in the external parameter. (According to E. F. Caldin, *Fast Reactions in Solution*, Wiley, New York, 1964, p. 82.)

periodically, then the effect of such perturbation depends on the ratio of relaxation time to the frequency of the displacing quantity [4-57] (Fig. 4-30).

If the equilibrium system is influenced by a periodically changing external parameter (pressure, electrical field, etc.), it can be shown that the time change of the degree of conversion (ΔX, e.g., degree of dissociation) depends partly on the relaxation time τ and partly on the frequency (ω) of the periodic perturbation ($\omega = 2\pi f$, where f is the angular frequency).

At constant frequency of the periodically changing quantity E for rapidly established equilibria (where $\tau \rightarrow 0$) the change ΔX follows in magnitude and direction the changes in E and remains in phase with it. For slower reactions ($\omega\tau = 1/4$ to $\omega\tau = 4$) when the relaxation time of the reaction is comparable to the value of ω, the establishment of the equilibrium follows the change in quantity E but is delayed. The observed plot $\Delta X = f(t)$ is shifted in phase when compared with $E = f(t)$. When the establishment of the equilibrium is very slow ($\tau \rightarrow \infty$), the position of the equilibrium in the course of the individual changes of the field is so little affected that the value of ΔX approaches zero. Then the slowly established system cannot respond to a perturbation that changes direction at a much higher frequency.

In the region of comparable relaxation time τ and frequency ω, that is, in the region $1/4 < \omega\tau < 4$, the gradual increase in frequency results in the absorption of energy by the equilibrium system, which first increases and then again decreases. This energy absorption is a consequence of the irreversible work that has to be done to shift the system to new equilibrium conditions (i.e., corresponding to new T, P, E, etc.). Maximum absorption in one cycle is achieved at a frequency when $\omega\tau = 1$. Determination of the angular frequency ω_{max} at which maximum absorption occurs enables a calculation of the value of τ.

In practice the periodically changing quantity is either temperature and pressure in the case of ultrasonic waves or electric field when a low amplitude alternating electric field superimposed on a high dc field is applied to the system under investigation.

Ultrasonic waves are propagated adiabatically and result in small periodic changes of temperature (of the order of $\pm 0.001°C$) and pressure (≈ 0.03 atm) in the medium through which they pass. For reactions with equilibrium constants which depend sufficiently on temperature and pressure (i.e., have sufficiently large values of $\Delta H°$ and $\Delta V°$) the system will be periodically displaced from equilibrium, if the relaxation time is comparable to the periodicity of the temperature and pressure changes. In aqueous solutions the contribution from the $\Delta V°$ term is generally predominant.

When the frequency of the ultrasonic waves (f) is varied, the absorption of the sound (which is measured as absorption coefficient per 1 cm and denoted as α) reaches its maximum at a certain frequency (f_{max}). Experimentally, the attenuation of sound intensity follows a law which is similar to Beer's law. At a given frequency, α is obtained by measuring the change in intensity of the ultrasonic waves as a function of the distance between a source and a detector, which is placed in the solution. Piezoelectric transducers (quartz crystals) are used to generate and detect the sound waves. The dependence of the sound absorption coefficient on frequency can be written in the form:

$$\frac{\alpha}{f^2} = \frac{A}{1 + \left(\dfrac{f}{f_{max}}\right)^2} \qquad (4.3\text{-}128)$$

where A is a constant which is practically frequency independent.

The absorption of ultrasonic waves is caused—in addition to the relaxation process—also by the viscosity of the medium, heat convection, and possibly by relaxation processes with shorter relaxation times, such as orientation of dipoles and translational-vibrational energy exchange. This contribution of residual absorption is expressed as a constant B:

$$\frac{\alpha}{f^2} = \frac{A}{1 + (f/f_{max})^2} + B \qquad (4.3\text{-}129)$$

From the experimentally obtained plot of α/f^2 versus log f (Fig. 4-31) it is possible to find f_{max} as the inflection point of the S-shaped portion of the curve. From f_{max} it is possible to calculate the relaxation time τ by means of equation

$$\tau = \frac{1}{2\pi f_{max}} \qquad (4.3\text{-}130)$$

The value of B in the reaction mixture is usually sufficiently similar to the value of B for the pure solvent, and is small for aqueous solutions. The value of the constant A depends on the concentration of the reacting components, on the relaxation time τ, and on the thermodynamic properties of the studied equilibrium. Concentration ranges in which it is possible to determine relaxation times with sufficient accuracy depend strongly on the studied system [4-82, 4-83]: for aqueous solutions of strong electrolytes 0.01 M solutions are usually sufficient; for acetic acid in toluene it is necessary to use 0.1 M solutions; and for the reaction of trinitrobenzene with diethylamine in acetone, 1 M solutions.

Agreement of experimental points with the theoretical curve (Fig. 4-31) is considered to be good evidence that a simple relaxation process is involved. In such systems it is possible to determine the value of τ at varying concentrations of reacting substances and calculate the values of rate constants.

Deviations of experimental points from the theoretical curve (Fig. 4-31), frequently observed in solutions of ionized species, are considered to indicate that several relaxation processes are operating and the following expression applies:

$$\frac{\alpha}{f^2} = \sum_j \left[A_j \frac{1}{1 + (f/f_{max})_j^2} \right] + B \qquad (4.3\text{-}131)$$

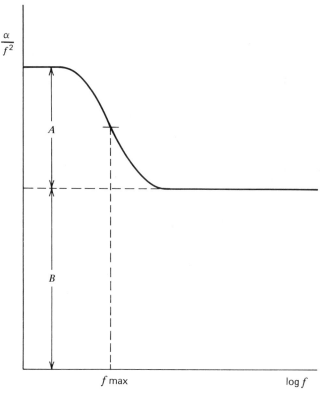

FIG. 4-31. Dependence of the sound absorption (α/f^2) on the logarithm of the frequency (f). *A* and *B* are constants; f_{max} is the frequency at which the sound absorption is maximum. (According to E. F. Caldin, *Fast Reactions in Solution*, Wiley, New York, 1964, p. 88.)

By measurement of absorption of ultrasonic waves for individual components it is frequently possible to distinguish relaxation resulting from a chemical reaction from effects of internal rotation, transfer of energy by the solvent, or interactions between solvent molecules.

Even for the simple system where agreement of experimental points with the theoretical curve indicates that only one relaxation process is involved, it is necessary to determine the value of f_{max} to study absorption over at least a tenfold change in frequency. When more than one relaxation process is involved it is necessary to cover a still larger range of frequencies.

Several different methods have been developed for measuring ultrasonic absorption over the range from 10^4 to 10^9 c/s, but each of these methods can be used only over a certain range of frequencies. Individual techniques will be only enumerated in Table 4-7.

An example of successful application is the dimerization of carboxylic acids [4-84–4-87]:

$$2R\ COOH \underset{k_d}{\overset{k_a}{\rightleftharpoons}} (R\ COOH)_2$$

For example, for benzoic acid in two solvents, the following results were obtained:

	$k_a[\text{liter mol}^{-1} \text{ s}^{-1}]$	$k_d[\text{s}^{-1}]$
CCl_4	4.7×10^9	7.4×10^5
$C_6H_5CH_3$	1.6×10^9	3.7×10^6

Ultrasonic methods can be used independently of the nature of compounds participating in the equilibrium. This can be compared with the requirements for ϕ_i in the single-step perturbation methods, (cf. Section 4.3.4.2.). This advantage is also a limitation as the nature of the reaction studied must be well understood beforehand.

The individual techniques can be used for relaxation times τ ranging from 10^{-4} s to 10^{-9} s and can be adapted to low temperatures (e.g., the pulse technique up to $-80°C$). The equilibrium studied must be sufficiently sensitive to changes in pressure and temperature. According to the nature of the substance, $0.01\ M$ to 1 M solutions must be used so that solubility can be a limiting factor. Relatively large solution volumes are required (Table 4-7). All individual techniques based on ultrasonic processes need special equipment, although assembly of ultrasonic instrumentation from commercially available components is comparatively simple.

TABLE 4-7
Comparison of Ultrasonic Techniques [4-83]

	Frequency (f) [c/s]	Volume (in liters)
Pulse technique	$10^6 - 2.5 \times 10^8$	$0.01 - 0.5$
Debye–Sears effect (interference of light)	$1.5 \times 10^6 - 1 \times 10^8$	
Streaming method	$1.3 \times 10^5 - 2 \times 10^6$	< 1
Resonant sphere (damped oscillations)	$5 \times 10^3 - 1.5 \times 10^6$	$1 - 50$

4.3.4.9. Evaluation of Relaxation Methods Proper

Relaxation methods are suitable for measurement of reactions with half-times ranging from 1 s to 10^{-9}s and are restricted generally to reversible reactions (compare also the combined stopped flow temperature jump method described earlier). A comparison of relaxation methods and the accessible time ranges is given in Fig. 4-32. Unless spectrophotometric or other specific techniques are used, doubts can arise in the interpretation of the relaxation time and the associated reaction. In particular for more complex reactions with more than one relaxation time (e.g., in enzymatic reactions) more than one interpretation is possible.

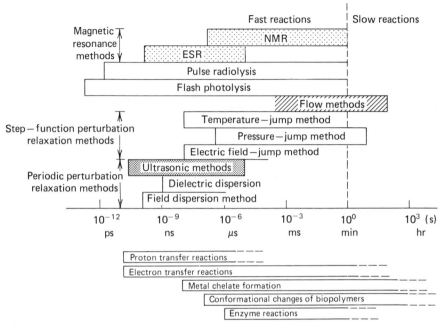

FIG. 4-32.   Time ranges for fast reactions in solution and methods for studying them. (According to Hiromi, *Kinetics of Fast Enzyme Reactions*, Halsted Press, Division of Wiley, New York, 1979, p. 66.)

4.3.4.10.   Methods of Flash Photolysis [4-88–4-90]

The method of flash photolysis, in which the reaction mixture is illuminated by a brief (10^{-6}–10^{-4} s) but very intense flash of light and in which the changes resulting from rapid photochemical processes are followed, was originally developed for reactions in the gas phase, but proved useful even for reactions in solutions with half-times up to 10^{-5} s. The perturbing parameter is, in this case, the intense light pulse, as compared with temperature in the temperature jump method. By using mode locked lasers it has become possible to extend measurements to picoseconds.

In the experiment the reaction mixture is placed in quartz vessels 10–20 cm long, 2–4 cm in diameter, parallel to which is placed a powerful flashbulb as shown in Fig. 4-33. Absorption of light of a given wavelength from the monochromator or emission in the reaction mixture is measured by photomultipliers and recorded by a transient recorder, oscilloscope, or photographic plate.

Recorded dependence of optical density as a function of time or of whole spectra ($A = f(\lambda)$) can offer information on concentration changes of starting material and final products, as well as on evidence for existence, nature, and reactivity of some intermediates.

The highest second-order rate constant obtained by conventional flash photolysis is $10^7 \, \epsilon$ liter mol^{-1} s^{-1}. If the extinction coefficient is typically 10^4 liter mol^{-1}

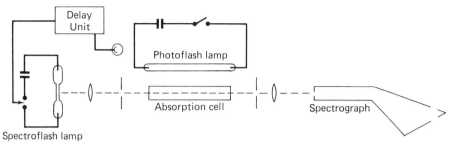

FIG. 4-33. Block diagram illustrating the principles of flash photolysis and flash spectroscopy.

s^{-1}, it follows that rate constants up to 10^{11} liter mol^{-1} s^{-1} can be determined. In the other extreme flash photolysis enables us to follow reactions taking place in several milliseconds.

Flash photolysis has been also applied to equilibrium reactions, for which the return to equilibrium after a considerable shift from the original equilibrium has been followed in addition to some irreversible processes.

By flash photolysis either a new compound or a new electronically excited state can be formed.

Because numerous organic acids are stronger acids in the excited state (HA*) than in the ground state, flash photolysis can result in an increase in the degree of dissociation. Excited anions A*$^-$ are formed which are deactivated by fluorescence (A*$^-$ $\longrightarrow$ A$^-$ + $h\nu$). The recombination of anions A$^-$ which are present in excess of equilibrium conditions is followed. This method was used, for example, for a study of substituent effects on proton transfer for substituted sulfonated phenols [4-91].

Another example studied by this method is the reaction of radical anions of aromatic hydrocarbons. The blue solutions of alkali metals in tetrahydrofuran react with aromatic hydrocarbons RH$_2$ (such as naphthacene) and form radical anions and dianions [4-92]:

$$RH_2 + e \longrightarrow RH_2^{\bar{\cdot}} \tag{4.3-132}$$

$$RH_2^{\bar{\cdot}} + e \longrightarrow RH_2^{-} \tag{4.3-133}$$

These negatively charged species show characteristic absorption bands. After irradiation of a solution containing anions RH$_2^-$, it is possible to observe a decrease in the absorbance of RH$_2^-$, and an increase in that of RH$_2^{\bar{\cdot}}$, attributed to the reaction (4.3-134):

$$RH_2^- + h\nu \longrightarrow RH_2^{\bar{\cdot}} + e + \text{Energy} \tag{4.3-134}$$

After interruption of irradiation an increase of the absorption band of RH$_2^-$ with time is observed, attributed to the reaction

$$RH_2^{\bar{\cdot}} + e \xrightarrow{k} RH_2^{-} \tag{4.3-135}$$

The time dependence of the absorbance indicated that this increase corresponds to a second-order reaction with constant $k = 4 \times 10^9$ liter mol^{-1} s^{-1}.

Other examples of this type of study include deactivation of triplet states (e.g., deactivation of the triplet state of benzophenone by interaction with naphthalene has a $k > 10^8$ liter mol^{-1} s^{-1}) [4-93], reactions of semiquinones [4-94] and transformation of benzyl radicals generated by irradiation of benzyl chloride ($k = 10^9$ liter mol^{-1} s^{-1}) [4-95].

A principle, similar to one applied to flash photolysis is applied in pulse radiolysis where the intense flash of light is replaced by a pulse of high energy radiation (such as X-rays or an electron beam) [4-96]. The pulse must deliver 100 Joules during a maximum of 50 μs. By this method, for reactions involving cleavage of radicals of aromatic hydrocarbons rate constants of the order of 10^8 to 10^9 liter mol^{-1} s^{-1} have been found.

Flash photolysis and related techniques enable measurements of rates of even the fastest reactions. In combination with recording of the electronic spectra this technique enables identification of unstable intermediates, even when the attribution of structure is not always equivocal due to the inaccessibility of model substances. Measurements can be adapted to low temperatures (up to $-196°C$). Because of the sensitivity of optical methods, especially for optical paths of 10–20 cm, it is possible to study very dilute solutions. The method is restricted to the study of reactions involving photochemical change. The possibility of changes of the studied system by the analyzing light beam in the evaluation of concentration changes in the reaction mixture should be kept in mind.

4.3.4.11. Evaluation

Comparison of the range of applications of individual techniques for studies of fast reactions is given in Fig. 4-32.

REFERENCES

[4-1]   G. G. Hammes, Ed., *Investigation of Rates and Mechanisms of Reactions* in *Techniques of Chemistry* (A. Weissberger, Ed.), Vol. 6, Part 2, Wiley-Interscience, New York, 1974.

[4-2]   R. P. Bell and E. N. Ramsden, *J. Chem. Soc.* 1958, 161; R. P. Bell and T. Spencer, *Proc. Roy. Soc.* A 251, 41 (1959), *J. Chem. Soc.* 1959, 1156.

[4-3]   L. Onsager, *J. Chem. Phys.* **2**, 599 (1934); P. Debye, *Trans. Electrochem. Soc.* **82**, 265 (1942).

[4-4]   E. H. Cordes and W. P. Jencks, *J. Am. Chem. Soc.* **84**, 4319 (1962).

[4-5]   K. Veselý and R. Brdička, *Collect. Czechoslov. Chem. Commun.* **12**, 313 (1947); R. Bieber and G. Trümpler, *Helv. Chim. Acta* **30**, 706, 971, 1109, 1286, 1534, 2000 (1947).

[4-6]   D. Barnes and P. Zuman, *J. Electroanal. Chem.* **46**, 323 (1973); P. Zuman, *J. Electroanal. Chem.* **75**, 523 (1977).

[4-7]   R. G. Pearson and R. L. Dillon, *J. Am. Chem. Soc.* **75**, 2439 (1953); see also J. R. Jones, *The Ionisation of Carbon Acids,* Academic, London 1973, p. 28ff.

[4-8]   R. W. Taft, Jr. and E. H. Cook, *J. Am. Chem. Soc.* **81,** 46 (1959); F. Basolo and R. G. Pearson, *Mechanisms of Inorganic Reactions,* 1st ed., Wiley, New York, 1958, p. 112.

[4-9]   R. P. Bell and P. J. Evans, *Proc. Roy. Soc.* (London) **A 291,** 297 (1966).

[4-10] H. W. Brown and G. C. Pimentel, *J. Chem. Phys.* **29,** 883 (1958).

[4-11] H. Hartridge and F. J. W. Roughton, *Proc. Roy. Soc.* (London) **A 104,** 376 (1923); **B 94,** 336 (1923).

[4-12] B. Chance, ref. 4-1, pp. 5–62; K. Hiromi, *Kinetics of Fast Enzyme Reactions,* Halsted Press, Division of Wiley, New York, 1979.

[4-13] R. L. Berger, B. Balko, and H. F. Chapman, *Rev. Sci. Instrum.* **39,** 493 (1968).

[4-14] G. D. Owens, R. W. Taylor, T. Y. Ridley, and D. W. Margerum, *Anal. Chem.* **52,** 130 (1980).

[4-15] Q. H. Gibson, *Disc. Faraday Soc.* **17,** 133 (1954).

[4-16] H. Shimada, R. Ueno, T. Iizuka, and Y. Ishimura, Abstracts for the 6th International Biophysics Congress, p. 403 (1978); Y. Talmi, ACS Symposium Series, No. 102, Y. Talmi, Ed., 1979, p. 1.

[4-17] B. Balko, P. Bowen, R. L. Berger, and K. Anderson, *J. Biochem. Biophys. Methods,* **4(1),** 1 (1981).

[4-18] E. G. Ball and T. T. Chen, *J. Biol. Chem.* **102,** 691 (1931); E. G. Ball and W. M. Clark, *Proc. Natl. Acad. Sci. U.S.* **17,** 347 (1931); A. Tockstein and F. Skopal, *Collect. Czechoslov. Chem. Commun.* **39,** 3430 (1974).

[4-19] B. Chance, *Biochem. J.* **46,** 387 (1950); N. C. Staub, J. M. Bishop, and R. E. Foster, *J. Appl. Physiol.* **16,** 511 (1961); **17,** 21 (1962).

[4-20] B. Chance, *J. Franklin Inst.* **229,** 455, 613, 737 (1940); *Rev. Sci. Instr.* **22,** 619 (1951).

[4-21] E. Bauer and H. Berg, *Naturwiss.* **51,** 460 (1964).

[4-22] L. Meites, *Polarographic Techniques,* 2nd ed., Wiley, New York, 1965.

[4-23] J. Heyrovský and J. Kůta, *Principles of Polarography,* Academic, New York, 1965.

[4-24] S. G. Mairanovskii, *Catalytic and Kinetic Waves in Polarography,* Plenum, New York, 1968.

[4-25] P. Zuman, *The Elucidation of Organic Electrode Processes,* Academic, New York, 1969.

[4-26] Z. Galus, *Fundamentals of Electrochemical Analysis,* E. Harwood, Chichester, U.K., 1976.

[4-27] A. M. Bond, *Modern Polarographic Methods in Analytical Chemistry,* Dekker, New York, 1980.

[4-28] A. J. Bard and L. R. Faulkner, *Electrochemical Methods,* Wiley, New York, 1980.

[4-29] R. Brdička, V. Hanuš, and J. Koutecký, *Progr. in Polarography* (P. Zuman and I. M. Kolthoff, Eds.), **1,** 145 (1962).

[4-30] P. Zuman, *Adv. Phys. Org. Chem.* (V. Gould, Ed.), **5,** 1 (1967).

[4-31] P. Zuman, *Fast Reactions* (K. Kustin, Ed.), *Methods in Enzymology* (S. P. Colowick and N. O. Kaplan, Eds.), Vol. 16 Academic, New York, 1969, p. 121ff.

[4-32] Ref. 4-25, p. 47.

[4-33] S. G. Mairanovskii and L. I. Lishcheta, *Collect. Czechoslov. Chem. Commun.* **25,** 3025 (1960).

[4-34] P. Zuman, J. Chodkowski, and F. Šantavý, *Collect. Czechoslov. Chem. Commun.* **26,** 380 (1961).

[4-35] C. D. Ritchie and H. Fleischhauer, *J. Am. Chem. Soc.* **94,** 3481 (1972).

[4-36] G. Nişli, D. Barnes, and P. Zuman, *J. Chem. Soc.* **B 1970,** 764, 771, 778.

[4-37] W. A. Szafranski and P. Zuman, *J. Electroanal. Chem.* **64,** 255 (1975); W. Szafranski, M. Sc. Thesis, Clarkson College of Technology, Potsdam, NY, 1976.

[4-38] W. J. Scott and P. Zuman, *Anal. Chim. Acta* **126,** 71 (1981).

[4-39] A. C. Riddiford, *Adv. Electrochem. and Electrochem. Eng.* (P. Delahay, Ed.), **4,** 47 (1966).

[4-40] R. N. Adams, *Electrochemistry at Solid Electrodes*, Dekker, New York, 1969, p. 67ff.

[4-41] W. J. Albery and M. L. Hitchman, *Ring-Disc Electrodes*, Clarendon Press, Oxford, 1971.

[4-42] Yu. V. Pleskov and V. Yu. Filinovskii, *The Rotating Disc Electrode*, Consultants Bureau, New York, 1976.

[4-43] V. G. Levich, *Acta Physichim.* USSR, **17**, 257 (1942); **19**, 117, 133 (1944).

[4-44] W. Vielstich and D. Jahn, *Z. Elektrochem.* **64**, 43 (1960); J. Albery and R. P. Bell, *Proc. Chem. Soc.* **1963**, 169.

[4-45] M. Eigen and E. M. Eyring, *J. Am. Chem. Soc.* **84**, 3254 (1962).

[4-46] V. Plichon and G. Faure, *J. Electroanal. Chem.* **44**, 275 (1973).

[4-47] P. Delahay and S. Oka, *J. Am. Chem. Soc.* **82**, 329 (1960).

[4-47a] L. Meites in *Physical Methods of Chemistry*, 5th ed. (B. W. Rossiter and J. F. Hamilton, Eds.), Wiley, New York, in press.

[4-48] M. Eigen and L. De Mayer, *Z. Elektrochem.* **60**, 1037 (1956).

[4-49] H. J. S. Sand, *Phil. Mag.* **1**, 45 (1901).

[4-50] P. Delahay and T. Berzins, *J. Am. Chem. Soc.* **75**, 2486 (1953).

[4-51] H. B. Mark and F. C. Anson, *Anal. Chem.* **35**, 722 (1963).

[4-52] J. G. Calvert and J. N. Pitts, *Photochemistry*, Wiley, New York, 1966.

[4-53] H. W. Melville, *J. Chem. Soc.* **1947**, 274; G. M. Burnett and H. W. Melville, *Nature*, **156**, 661 (1945); H. W. Melville, *Proc. Roy. Soc.* (*London*) A **237**, 149 (1956); A. Shepp, *J. Chem. Phys.* **24**, 939 (1956).

[4-54] R. C. Chen and H. Edelhoch, Eds., *Biochemical Fluorescence Concepts*, Vols. 1 and 2, Dekker, New York, 1975 and 1976; E. L. Wehry, *Modern Fluorescence Spectroscopy*, Plenum, New York, 1976.

[4-55] C. A. Parker, *Photoluminescence of Solutions, With Applications to Photochemistry and Analytical Chemistry*, Elsevier, Amsterdam, 1968.

[4-56] R. L. Ward and S. I. Weissman, *J. Am. Chem. Soc.* **79**, 2086 (1957).

[4-57] M. Eigen and L. De Maeyer, in *Investigation of Rates and Mechanisms of Reactions* (S. L. Friess, E. S. Lewis, and A. Weissberger, Eds.) *Technique of Organic Chemistry* (A. Weissberger, Ed.), Vol. 8, Part 2, Wiley, New York, 1963, p. 895.

[4-58] R. Brouillard, *J. Chem. Soc. Faraday I*, **76**, 583 (1980).

[4-59] J. Meixner, *Kolloid-Z.*, **134**, 3 (1953).

[4-60] W. Müller and D. M. Crothers, *J. Mol. Biol.* **35**, 251 (1968).

[4-61] C. F. Bernasconi, *Relaxation Kinetics*, Academic, New York, 1976.

[4-62] H. H. Trimm, R. C. Patel, and H. Ushio, *J. Chem. Ed.* **56**, 762 (1979).

[4-63] R. C. Patel, H. H. Trimm, H. Ushio, and M. Zemany, *J. Sol. Chem.* **10**, 1 (1981).

[4-64] D. Thusius, *J. Am. Chem. Soc.* **94**, 356 (1972),.

[4-65] U. Strahm, R. C. Patel, and E. Matijević, *J. Phys. Chem.* **83**, 1689 (1979).

[4-66] R. Winkler-Oswatitsch, and M. Eigen, *Angew. Chem. Int. Ed.* **18**, 20 (1979).

[4-67] R. C. Patel, *J. Chem. Instr.* **7**, 83 (1976).

[4-68] H. Diebler, M. Eigen, and P. Matthies, *Z. Elektrochem.* **65**, 634 (1961).

[4-69] H. Strehlow and W. Knoche, *Fundamentals of Chemical Relaxation*, Verlag Chemie, Weinheim, DBR, 1977.

[4-70] W. Knoche and G. Wiese, *Chem. Instrum.* **5**, 91 (1973–74).

[4-71] H. Strehlow, *Z. Elektrochem.* **66**, 392 (1962).

[4-72] A. Persoons, L. Hellemans, *Biophys. J.*, **24(1)**, 119 (1978); A. Persoons, *J. Phys. Chem.* **78**, 1210 (1974); L. De Maeyer and A. Persoons, in ref. 4-1, p. 211.

[4-73] M. Wien, *Phys. Z.* **32**, 545 (1931).

[4-74] L. Onsager, *J. Chem. Phys.* **2**, 599 (1934).

[4-75] G. Ilgenfritz, in *Probes of Structure and Function of Macromolecules and Membranes*, (B. Chance, Ed.), Vol. 1, Academic, New York, 1971, p. 505.

[4-76] (a) M. Eigen and K. Kustin, *J. Am. Chem. Soc.* **82**, 5952 (1960); (b) M. Eigen and E. M. Eyring, *J. Am. Chem. Soc.*, **84**, 3254 (1962); (c) M. Eigen, G. G. Hammes, and K. Kustin, *J. Am. Chem. Soc.* **82**, 3482 (1960).

[4-77] F. M. Pohl, *Europ. J. Biochem.* **4**, 373 (1968).

[4-78] A. Jost, *Ber. Bunsenges. Phys. Chem.* **70**, 1057 (1966).

[4-79] K. F. Herzfeld and T. A. Litovitz, *Absorption and Dispersion of Ultrasonic Waves*. Academic, New York, 1959.

[4-80] J. Stuehr and E. Yeager, *Phys. Acoust.* **11A**, 351 (1965).

[4-81] M. J. Blandamer, *Introduction of Chemical Ultrasonics*, Academic, New York, 1973.

[4-82] J. Rassing, in *Chemical and Biological Applications of Relaxation Spectrometry* (E. Wyn-Jones, Ed.), Reidel, Dordrecht-Holland, 1975, p. 1.

[4-83] J. Stuehr, ref. 4–1, p. 237.

[4-84] J. E. Piercy and J. Lamb, *Trans. Faraday Soc.* **52**, 930 (1956).

[4-85] W. Maier, *Z. Elektrochem.* **64**, 132 (1960).

[4-86] J. Rassing, O. Osterberg, and T. A. Bar, *Acta Chem. Scand.* **21**, 1443 (1967).

[4-87] J. Rassing, *Adv. Mol. Relax. Proc.* **4**, 55 (1972).

[4-88] G. Porter and M. A. West, ref. 4–1, p. 367; W. J. Chase and J. W. Hunt, *J. Phys. Chem.* **79**, 2835 (1975); H. Ruppel and H. T. Witt, in *Fast Reactions* (K. Kustin, Ed.), *Methods in Enzymology* (S. P. Colowick and N. O. Kaplan, Eds.), Vol. 16, Academic, New York, 1969, p. 316.

[4-89] T. L. Netzel, W. S. Struve, and P. M. Rentzepis, *Ann. Rev. Phys. Chem.* **24**, 473 (1973); A. Laubereau and W. Kaiser, *Ann. Rev. Phys. Chem.* **26**, 83 (1975).

[4-90] N. J. Turro, *Molecular Photochemistry*, Benjamin-Cummings Pub. Co. Menlo Park, California, 1965.

[4-91] K. Breitschwerdt and A. Weller, *Z. Elektrochem.* **64**, 395 (1960).

[4-92] H. Linschitz and J. Eloranta, *Z. Elektrochem.* **64**, 169 (1960); J. Eloranta and H. Linschitz, *J. Phys. Chem.* **38**, 2214 (1963).

[4-93] G. Porter, *Proc. Chem. Soc.* **1959**, 291; G. Porter and F. Wilkinson, *Trans. Faraday Soc.* **57**, 1686 (1961).

[4-94] N. K. Bridge and G. Porter, *Proc. Roy. Soc.* **A. 244**, 259, 276 (1958).

[4-95] G. Porter and M. W. Windsor, *Nature* **180**, 187 (1957).

[4-96] M. Ebert, J. P. Keene, A. J. Swallow, and J. H. Baxendale (Eds.), *Pulse Radiolysis*, Academic, London, 1965.

BIBLIOGRAPHY

(a) General on Fast Reaction Techniques

S. C. Friess, E. S. Lewis, and A. Weissberger, Eds., *Investigation of Rates and Mechanisms of Reactions*, in *Technique of Organic Chemistry* (A. Weissberger, Ed.), 2nd ed., Vol. VIII, Part 2, Wiley, New York, 1963.

E. F. Caldin, *Fast Reactions in Solution*, Wiley, New York, 1964.

G. G. Hammes, (Ed.), *Investigation of Rates and Mechanisms of Reactions*, in *Techniques of Chemistry* (A. Weissberger, Ed.), Vol. 6, Part 2, Wiley-Interscience, New York, 1974.

K. Hiromi, *Kinetics of Fast Enzyme Reactions*, Halsted Press, Division of Wiley, New York, 1979.

(b) Electrochemical Methods

R. Brdička, V. Hanuš, and J. Koutecký, *Progr. in Polarography* (P. Zuman and I. M. Kolthoff, Eds.), **1**, 145 (1962).

L. Meites, *Polarographic Techniques,* 2nd ed., Wiley, New York, 1965.

J. Heyrovský and J. Kůta, *Principles of Polarography,* Academic, New York, 1965.

P. Zuman, *Adv. Phys. Org. Chem.* (V. Gould, Ed.) **5**, 1 (1967).

S. G. Mairanovskii, *Catalytic and Kinetic Waves in Polarography,* Plenum, New York, 1968.

P. Zuman, *The Elucidation of Organic Electrode Processes,* Academic, New York, 1969.

P. Zuman, in *Fast Reaction in Enzymology* (K. Kustin, Ed.), *Methods in Enzymology* (S. P. Colowick and N. O. Kaplan, Eds.), Vol. 16 Academic, New York, 1969.

W. J. Albery and M. L. Hitchman, *Ring-Disc Electrodes,* Clarendon Press, Oxford, 1971.

Yu. V. Pleskov and V. Yu. Filinovskii, *The Rotating Disc Electrode,* Consultants Bureau, New York, 1976.

Z. Galus, *Fundamentals of Electrochemical Analysis,* E. Harwood, Chichester, U. K., 1976.

A. M. Bond, *Modern Polarographic Methods in Analytical Chemistry,* Dekker, New York, 1980.

A. J. Bard and L. R. Faulkner, *Electrochemical Methods,* Wiley, New York, 1980.

(c) Photochemistry

J. G. Calvert and J. N. Pitts, *Photochemistry,* Wiley, New York, 1965.

C. A. Parker, *Photoluminescence of Solutions, With Applications to Photochemistry and Analytical Chemistry,* Elsevier, 1968.

H. Rüppel and H. T. Witt, in *Fast Reactions* (K. Kustin, Ed.), *Methods in Enzymology* (S. P. Colowick and N. O. Kaplan, Eds.), Vol. 16, Academic, New York, 1969.

G. Porter and M. A. West, in *Investigation of Rates and Mechanisms of Reactions*, in *Techniques of Chemistry* (A. Weissberger, Ed.), Vol. 6, Part 2, Wiley-Interscience, New York, 1974.

R. C. Chen and H. Edelhoch, Eds., *Biochemical Fluorescence Concepts,* Vols. 1 and 2, Dekker, New York, 1975 and 1976; E. L. Wehry, *Modern Fluorescence Spectroscopy,* Plenum, New York, 1976.

N. J. Turro, *Molecular Photochemistry,* Benjamin-Cummings Pub. Co., Menlo Park, California, 1965.

(d) Pulse Radiolysis

M. Ebert, J. P. Keene, A. J. Swallow, and J. H. Baxendale (Eds.), *Pulse Radiolysis,* Academic, London, 1965.

(e) Relaxation Methods

K. F. Herzfeld and T. A. Litovitz, *Absorption and Dispersion of Ultrasonic Waves,* Academic, New York, 1959.

M. Eigen and L. De Maeyer, in *Investigation of Rates and Mechanisms of Reactions* (S. L. Friess, E. S. Lewis, and A. Weissberger, Eds.) in *Technique of Organic Chemistry* (A. Weissberger, Ed.), Vol. 8, Part 2, Wiley, New York, (1963).

M. J. Blandamer, *Introduction of Chemical Ultrasonics,* Academic, New York, 1973.

C. F. Bernasconi, *Relaxation Kinetics,* Academic, New York, (1976).

H. Strehlow and W. Knoche, *Fundamentals of Chemical Relaxation,* Verlag Chemie, Weinheim, DBR, 1977.

INDEX

R